DIGITAL IMAGE PROCESSING
Practical Applications of
Parallel Processing Techniques

ELLIS HORWOOD SERIES IN DIGITAL AND SIGNAL PROCESSING

Series Editor: D. R. SLOGGETT, Technical Director, Marcol Group Ltd, Woking, Surrey

G.D. Bergman	ELECTRONIC ARCHITECTURES FOR DIGITAL PROCESSING: Software/Hardware Balance in Real-time Systems
Z. Hussain	DIGITAL IMAGE PROCESSING Practical Applications of Parallel Processing Techniques
R. Lewis	PRACTICAL DIGITAL IMAGE PROCESSING
S. Lawson & A. Mirzai	WAVE DIGITAL FILTERS
J.D. McCafferty	HUMAN AND MACHINE VISION: Computing Perceptual Organisation

DIGITAL IMAGE PROCESSING
Practical Applications of Parallel Processing Techniques

ZAHID HUSSAIN
Senior Research Scientist
GEC-Marconi Hirst Research Centre, England

ELLIS HORWOOD
NEW YORK   LONDON   TORONTO   SYDNEY   TOKYO   SINGAPORE

First published in 1991 by
ELLIS HORWOOD LIMITED
Market Cross House, Cooper Street,
Chichester, West Sussex, PO19 1EB, England

A division of
Simon & Schuster International Group
A Paramount Communications Company

Printed and bound in Great Britain
by Redwood Press Ltd, Melksham, Wiltshire

British Library Cataloguing-in-Publication Data

Hussain, Zahid
Digital image processing: practical applications of parallel processing techniques. —
(Ellis Horwood series in digital and signal processing)
I. Title. II. Series.
004
ISBN 0–13–213273–7
ISBN 0 13 213281 8 pbk

Library of Congress Cataloging-in-Publication Data available

Contents

Preface

Image processing (IP) and image understanding (IU) have been fast-growing fields now for the last thirty years. Influence for its growth and advancement has arisen from studies in artificial intelligence, psychology, psychophysics, computer architecture and computer graphics. Application areas for IP/IU includes document processing, medicine and physiology, remote sensing, industrial automation and surveillance amongst many others.

Image processing (or picture processing) involves the various operations which can be carried out on the image data. These operations include preprocessing, spatial filtering, image enhancement, feature detection, image compression and image restoration. However, this list is not exclusive. Image compression [Gonzalez 84] is mainly used for image transmission and storage. Image restoration [Andrews 74, Andrews 77] involves processes which restore a degraded image to something close to the "ideal".

Computer vision (image understanding or scene analysis) involves techniques from image processing, pattern recognition and artificial intelligence. The process attempts to recognise and locate objects in the scene. This book concentrates on the parallel computer vision algorithms, parallel hardware and the image processing techniques required for image understanding.

There are now a number of good textbooks on image processing and computer vision. However, they are also quite restricted in the materials they cover. While some are very strong in their coverage of image processing, they neglect computer vision; and vice versa. While image processing is very computationally intensive and lends itself to massive parallelism, this aspect is rarely covered in many of these textbooks. With a few exception, very few existing books explore the application of image processing and scene understanding. While many aspects of image processing are not covered in this book and many details of computer vision have been left out, I hope it does succeed in filling the missing gaps by covering the inter-relationship between algorithm, parallel hardware and software and application.

Much of the material presented here has arisen from the author's PhD thesis involving biomedical image-processing and development of algorithms for a SIMD processor array. Most of the techniques described in Chapters 2, 3, 4, 5 and 7 have arisen from this work. Later, the author worked on use of robust image processing techniques for industrial inspection tasks and some of the techniques used and developed by the author are described in Chapters 5 and 7.

This book consists of eight chapters but it can essentially be divided into

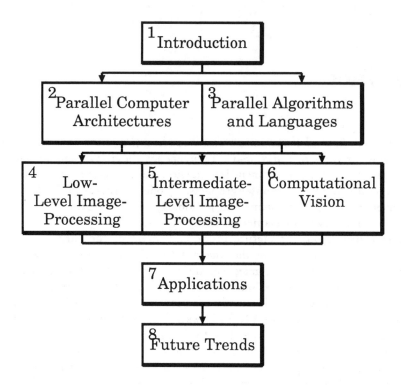

Figure 1: The organisation of the chapters for this book.

three parts: (i) parallel hardware and software techniques; (ii) image processing and image understanding algorithms; and (iii) applications of computer vision (see Figure 1).

In Chapter 1 we begin with a discussion of what types of processes and computations are involved in image processing and image understanding. We follow this up with a look at two vision theories, the computational theory of vision by Marr [Marr 82] and the ecological theory of vision due to Gibson [Gibson 79].

In Chapter 2 we describe the various parallel computer architectures which have been developed for image processing. We describe the various architecture taxonomies and follow this with description of SIMD, MIMD and pipelined (systolic) computers. However, IP/IU and scene interpretation involve different types of operations. While IP/IU involve operations which are massively data parallel, scene interpretation involves more task parallel methods. This leads to problems of *iconic-symbolic* transformation of data. This is considered by a description of some of the various hybrid architectures available.

In Chapter 3 we describe the parallel paradigm for image processing with an example and how an algorithm developed for a von Neumann machine can be mapped for SIMD and MIMD processors. It should be noted that a radically different approach to describing the problem is necessary to make an efficient use of parallelism. We next consider some general-purpose algorithm such as matrix multiplication, sorting and searching and their parallel implementations. Then we give an exposition of some general SIMD operations for image processing. Finally, Chapter 3 is concluded with a consideration of what is required in a computing language to exploit parallelism in the algorithms and the underlying hardware.

In Chapters 4, 5 and 6 we give descriptions of operations and algorithms for image processing and computer vision and their parallel implementation. In Chapter 4 we consider operations for preprocessing and filtering of images. In Chapter 5 we describe techniques for image segmentation, feature detection and object description.* In Chapter 6 we describe parallel algorithms which are more associated with computer vision such as stereo, optical flow, and shape from X methods. The general theme of Chapter 6 is that these techniques require either the computation of disparity maps using parallel searching and matching techniques or use the calculus of variation techniques leads to differential equations which can be solved using parallel iterative techniques.

In Chapter 7 we give descriptions of various applications of image-processing and image-understanding techniques in the domain of biomedical image processing, industrial automation and remote sensing. We also describe some general-purpose vision systems. Finally in Chapter 8 we consider what hurdles are still to be overcome in both theories, algorithms and hardwares which will lead to more use of computer vision in wider application domains and more robust vision systems. We conclude by looking at artificial neural networks as solving some of the problems in vision for which procedural algorithms are difficult to generate; and we consider optical computing as a means to providing the greater computing power required for real-time and more general-purpose vision systems.

---

* Part of Chapter 5 is reprinted from [Hussain 1988b] with acknowledgement to the copyright holders, Elsevier Science Publishers BV.

I would like to express my thanks to my colleagues and friends at the Image Processing Group (University College London) and Machine Vision Group (Royal Holloway and Bedford New College, University of London) for many fruitful discussions and assistance which planted the seed for the interest in the subject. Thanks are also due to Dr. E. R. Davies, RHBNC (University of London) for his support and making available computing facilities in his laboratory without which writing this book would not have been possible. Thanks to David Rooks (UCL) for printing the photographs.

Thanks also to Mike Still and Steve Maybank, Hirst Research Centre, for reading draft copies of the manuscript, for their helpful suggestions and for their correction of my English. Any grammatical errors remaining is entirely the fault of the author.

Finally, my thanks to my family: to my parents for their love, understanding and support, and to my wife Masmika for forbearance, support and sacrifices while I spent most of my evening and weekends for the past year and a half preparing the manuscript. This book is especially written for my daughter Nasreen for providing the music, the illumination and the the occasional distractions.

Z. H.

Ascot, England
August 1991

Chapter 1

Introduction

1.1 What is Image Processing?

What is the purpose of vision? Vision allows us to carry out three tasks: (1) perception of our world; (2) conception of a strategy to carry out certain actions; and (3) carrying out that action. People involved in both biological and machine vision have been aware of the parallelism involved in the task and have been designing hardware for some time. Early (or low-level) vision involves such tasks as filtering, segmentation, feature detection, stereopsis and optic flow, all involve local computation on the image array and these operations are highly parallel. The information extracted from these processes, which can be carried out almost context-free, or data-driven, can be passed on to a smaller number of specialised processors to carry out scene interpretation. However, the question remains to what extent we require both bottom-up (data-driven) and top-down (context-driven) processes. Hubel and Weisel from their study of cat and monkey cortices reported that there was a hierarchy to the processing [Hubel 62, Hubel 77]; although later works by Stone *et al.* [Stone 79] indicate that there are some direct connection for dataflow between the different parts of the visual cortex rather that data streaming through in a single path.

There is a belief shared between machine vision practitioners and those studying psychophysics about what is generally involved in low-level (early) vision and that the processing which takes place is massively parallel. Feldman has also proposed a model for parallelism in high-level vision [Feldman 85]. However, the fundamental problem in computer vision remains the identification and location of objects in a 2D image of a 3D scene.

Many algorithms have been devised for machine vision tasks but they have been heavily influenced by the von Neumann architecture on which they have been developed. Some influences are: (1) the way the problem and data can be represented; (2) the description of the problem to be solved; (3) the data-structures which can be used; (4) the way the problem can be mapped to the architecture; (5) the data and control flow allowed by the system; and (6) that the algorithm is optimised to run on the available architecture. One of the expositions of this book will be to explore the various methods for implementing algorithms on various types of parallel architecture. The reason for developing a new architecture should be: (1) it will perform existing algorithms better; (2) it will allow execution of algorithms previously not possible; and (3) it will stimulate the development of new (exotic) algorithms. There is close inter-dependency between algorithms, architectures and parallelism; study of these will be the main theme of this book.

The computational model of the visual process has grown from the study of physiology and psychology and the construction of parallel computers dedi-cated to image processing. In designing a computer vision system, we need to consider:

- What types of information are we attempting to extract from the image?

- What is the form of this information in the image?

- What *a priori* knowledge do we require to extract this information?

- What types of computational processes do we require?

- What are the forms for the knowledge and data representation required?

There are various ways in which images and data can be represented. Images are typically represented using a 2D array of picture elements (*pixels*) which represent the intensity values (*grey-levels*). The image can itself be of various types: it may be the result of filtering, or it may be vector fields representing surface orientation or velocity, or it could be the result of transformations (such as FFT), or it could be an intrinsic image containing such information as shadows, reflectance and depth. The images could also be represented using a set of coordinates, chaincodes, quadtrees or medial axis transforms. To extract information from these images different types of operation are required. These could involve local operations such as convolution or morphological operations. The computation of statistics such as histogramming and texture measurement could involve both local and global operations. Many geometric operations, such as projections, Fourier and Hough transforms, are required for feature detection. Segmentation can be achieved by boundary following (chaincode) or region growing (quadtree). Many operations are boolean operations such as determining the union, difference, and intersection of regions. There are more complex tasks involved such as determining the convex hull, Voronoi diagrams and medial axis transformation, or determining geometric properties of shapes using computation of moments, or determining the connectivity of objects.

What architectures do we require for carrying out these varied operations? Many have developed pipelined computers (such as "systolic array") which gives a limited concurrency of operations but a good rate of data throughput; others have developed mesh connected arrays because of their efficient map-ping of images. Many have augmented these mesh architectures by building tree or pyramid architectures because they give the hierarchy of operations be-lieved to be involved in biological vision systems. Others have developed more general parallel machines based on shared memory or hypercube connectivity. This book explores the types of algorithms required for image processing and

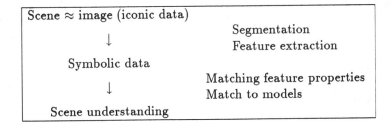

Figure 1.1: Two-dimensional paradigm for computer vision.

scene understanding and the computer architectures required for the efficient computation of vision algorithms.

In many *image-understanding* or computer vision applications, the scene to be processed is essentially two-dimensional, such as document processing, character recognition, surface inspection, and many instances of remote sensing. An object consists of features (shapes, texture, colour, etc.) and relationships (relative position and size) between them.

The process for detecting constituent parts of an image is called *segmentation* or *feature detection*. Feature detection may use template matching, correlation or convolution techniques to detect such item as edges, corners, holes and textures. The segmentation process attempts to distinguish between background and objects based upon detection of "homogeneous" regions of the scene. Thus the segmentation process takes iconic data and outputs symbolic data (position of features or regions of objects). The recognition process is to "find" the relationship between these symbolic data. This two-dimensional image analysis paradigm can be represented as in Figure 1.1.

Other instances where this two-dimensional scene-analysis paradigm applies is in recognition of a three-dimensional scene from a constant or known viewpoint. This technique is applicable in many industrial inspection tasks or in bio-medical image processing involving various types of radiograph.

1.2   Image and Vision Theory

While the vision task is very easy for us humans, it has proven to be very difficult to teach a computer even the rudiments of a vision system. There are many reasons for this. While we can see, we still cannot explain this perception process. We cannot recover all the information in the scene from just the intensity. The intensity map in an image is the result of a combination of many factors, such as the surface property of objects, the illumination source and its direction, ambient light, the atmospheric condition, and the properties of the detector amongst many. Just by looking at the grey level value at a pixel, we cannot differentiate the contribution from each of these factors. An image is a projection of a 3D scene on to a 2D surface. The depth information has been collapsed and the image is underconstrained. We cannot recover the depth information without additional information, such as the size and position of some objects in the scene. Furthermore, image interpretation requires *a priori* domain knowledge.

Finally, *early* vision requires processing of vast amounts of data. As we have learned of the requirements, the vision systems have developed to meet these demands.

As behavioural experiments have shown, we humans have little difficulty in recognising complex structures in a scene (e.g. people, furniture, vehicles) in considerably less than a second. Experiments on macaque monkeys show their neurons respond to stimuli such as a face in 70–200 ms; this is despite the fact that neurons are slow devices. A signal can pass from one neuron to another across a synaptic gap in 1–2 ms.

In a human brain there are probably $10^{10} - 10^{11}$ neurons, each synapsing with possibly thousands of others. A neuron will fire if the sum of the input signal is above a threshold. In its rest state, a neuron may only fire at 2–20 times a second; however, this may rise to 1000 times a second if it receives an input signal. A modern microprocessor may have upwards of a million transistors on a chip and it can execute an instruction in better than 20 ns. Yet, these processors are not *fast* enough to carry out many of the rudiments of image processing in *real-time*. This is because while each individual neurons is *slow*, it acts with others in a massively parallel manner. Since an image frame sequence with some moving objects must be processed about every 40 ms, there can only be 15–30 steps involved in the serial processing involving the neurons. For more complex scene understanding, the processing time may be about 1 s. This implies that only 50–100 steps are involved in the serial depth of the human *visual processing algorithm*. This is regarded as strong evidence that such shallow serial depth is only possible if the simple neuron processors are acting in a massively parallel manner. Compare this with a micro processor in which most of its many thousands of devices are idle at any given time. These considerations have led many computer vision practitioners to design and build massively parallel array computers with many very simple computing engines.

In the eye an image is formed on the retina by the lens. The retina consists of about 6–10 million cones, which are sensitive to colour, and about 100 million rods which are sensitive to motion and intensity. The necessary information is extracted in the retina and, using the visual channels, the information is passed to the visual cortex for perception and cognition. Many of the biological vision processes can be modelled by mathematical functions such as convolutions and discriminant operators [Gupta 89]; however, many other processes, such as the receptive field in the retina and the visual cortex, cannot be modelled [Gupta 88].

By studying the rods and cones and neurons we cannot understand the vision process. Marr puts it succinctly:

    ... trying to understand perception by studying only neurons is like trying to understand bird flight by studying only feathers: it just cannot be done. In order to study bird flight we have to understand aerodynamics; only then do the structures of feathers and different shapes of bird wings make sense. ([Marr 82],p. 27).

To explain the visual process, Marr put forwards the *computational* theory of vision. This consists of three parts: (1) describe the tasks that are required by a vision system, (2) devise the algorithms for carrying out the tasks, and (3) discover how these algorithms are implemented in neurons. Marr's computational theory of vision attempts to explain how light falling upon the retinal cells is transformed to symbolic representation of position, orientation and movement of surfaces. The first stage of this transformation is referred to as the *raw primal sketch*, which extracts the edges, texture and surfaces from the light pattern. Edges have different gradient, therefore all edges cannot be detected by using a single mask size which measures intensity difference in its two equal symmetric regions [Marr 76]. To locate edges of different gradient, many masks of different size must be applied in parallel. The different *space scale* images are created by applying Gaussian filters of different width to the image. Edges in each of these *scale space* images is extracted by locating *zero-crossings* after application of the Laplacian operator. Finally, the raw primal sketch is created by combining the location of the edges in the different scale space called *channels* [Marr 80]. Marr and Hildreth argue that edges at natural surfaces will give rise to edges in many different channels. Matching zero-crossing segments in different channels gives rise to edge-segments. Closed edge segments are called *blobs*. Antiparallel zero-crossing segments in a narrow channel are called *bars*; the ends of bars are called *terminals*.

Marr and Hildreth further argue that the Laplacian of the Gaussian is computed in the X retinal ganglion cells since they respond to light intensity which are difference of Gaussian. The *edge-segments*, *blobs*, *bars* and *terminals* are perceptually grouped to form larger structures. This refined description of the scene is referred to as the *full primal sketch*. Movement of edges can be detected by computing the difference of the Laplacian of the Gaussian of the image with respect to time [Marr 81]. It is argued by Marr and Ullman that the time derivative of the Laplacian of Gaussian is computed in the retina and signalled by the **Y** retinal ganglion cells. Marr's paradigm for visual computation can be represented by Figure 1.2.

From Marr's theory we note that a very large amount of processing must continually be carried out to build complex symbolic descriptions of the scene. An alternative theory, referred to as the *ecological approach* was developed by J. J. Gibson [Gibson 50, Gibson 66, Gibson 79]. He proposes that an observer actively samples for the spatial and temporal information in the optic array to "pick up" regions and events of interest. However, the *ecological theory* can be regarded as being complementary to Marr's *computational theory* — operating on a more global level.

Gibson states that the visual process does not start with the retinal image, rather, it is the structure in the light in the form of texture and surface which

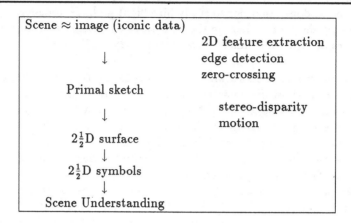

Figure 1.2: Marr's paradigm for visual computation.

provides information about surfaces and objects. In the ecological theory of vision, for perception to occur, the light must be structured and there must be motion. With Marr and other students of visual information processing, their starting position is with the retinal image. This contrasts markedly with Gibson whose starting point is the *optic array*. The optic array is a structured pattern of light reaching the observer and contains spatio-temporal information. The structuring of the light is a result of its *interaction* with surfaces and objects, and of their and the observer's movement in the world. The optic array contains *invariant* information. The texture density (or gradient) information can be obtained directly from the optic array and this provides surface orientation information which can be useful in scene interpretation.

Surface information does not only arise from texture gradient but the optic array also contains information from gradients of colour, intensity and disparity. The motion of the objects or the observer is also important for scene analysis; the motion causes changes in the optic array, this Gibson called *optic flow*. The variant information about the world is obtained through motion. While in computational vision, static sequences of frames are taken, features computed and matched to determine motion, one can start with the optic flow and recover features [Clocksin 80].

Traditionally, size invariance is viewed as being imposed by computing distances of objects and then interpreting the scene by scaling objects appropriately. However, according to Gibson, the scaling is not a problem because the relative size of an object can be inferred from the texture gradient of the background to which the object is *attached* [Gibson 50]. Also, traditionally, stereopsis is made by detecting and matching local features to compute the disparity map. Proponents of ecological vision would argue that the retinal image comparison and fusion is unnecessary [Michaels 81]. Again, the invariant properties of optic array would specify the distance to objects.

1.3   Concluding Remarks

There has been, until recently, a wide divergence in the views (often in the absence of facts) of what are the important processes for perception [Marr 82, Gibson 50], although, many now accept the complementarity in the perception methods proposed by Marr and Gibson. However, a fact that has never been in dispute is that the computation involved in processing retinal images at the low level is massively parallel. This has been a strong motivational point in the development of cellular array parallel computers (such as CLIP4 and MPP) dedicated to image processing and computer vision.

There are four main aspects to the type of processing required for image processing and computer vision: (a) pre-processing of the image data; (b) object feature detection; (c) iconic to symbolic data transformation; and (d) scene interpretation. Each of these tasks requires different data representation and has different computational requirements. The pre-processing of images mainly consists of filtering, enhancement and restoration if the image is degraded. These operations can be often carried out by use of convolution techniques which are *local* operations. A local operation is one in which the result computed at a pixel point is dependent only on the value of its neighbouring points. These local operations involve *image-to-image* transformations, and hence are efficiently mapped to massively parallel array computers. Feature detection involves operations to detect edges, corners, and texture or carry out medial axis transformations. These operations are also *local*. There are also some requirements for *global operations* (those involving whole or large parts of the image) such as histogramming, statistical measurements or computation of moments. Image analysis of a three-dimensional scene is more complex where there may be varying illumination and shadows. Various parts of objects may be in occlusion because of the orientation and the presence of other objects. A typical technique for processing such images is to infer surfaces; there are several techniques for recovering surface orientation called "shapes from X".

Chapter 2

Parallel Computer Architectures

2.1   Introduction

Apart from the large data throughput required for image processing, computer vision tasks also require many types of processing: from filtering and registration using 2D correlation and convolutions in the iconic domain, and feature extraction, and symbolic representation, through to geometric and graph analysis. These processes need different data types and representations and hence place different requirements on the computer hardware.

Most computers today still use the original von Neumann architecture. It consists of a single processing unit and a single memory unit for both the instruction and data. This implies that only a single memory location can be accessed at any one time. Therefore, for execution of a sequence of operations, much of the processor is idle. Modern chip architectures have advanced to improve upon this model. One method is to separate the instruction and data spaces; allowing instruction and data to be fetched simultaneously. This is called the Harvard architecture, developed by Howard Aiken [Chen 75, Willis 86]

It has also been noted that a few instructions are executed very often by a microprocessor. Therefore, a method for optimising is to ensure that these few instructions are executed more efficiently. RISC (*reduced instruction set computer*) Architecture started back in the mid 1970s with research carried out by IBM [Radin 83] and Inmos. The term "RISC" was coined by Patterson and Ditzel [Patterson 80]. The reasons for and the implementation of RISC are discussed in [Patterson 82, Hennessy 84]. Some of the important properties are listed below:

- small number of instructions (<100) with single cycle execution,

- small number of addressing modes (typically 1 or 2),

- small number of instruction formats, all of same length, and

- large number of registers.

As a result of all this, the control unit ("control area") is considerably reduced; in the Motorola MC68020 the control area forms 68% of the chip area; this is typically reduced to under 10% in a RISC processor.

Yet another technology which has arrived is the VLIW (*very long instruction word*) machine [Fisher 84]. These machines are highly parallel but differ from multiprocessors because they are tightly coupled with very parallel horizontal microcode single flow control mechanisms. Because a large number of processors are controlled by a single microcode, the microcode needs to be constructed using a technique called *trace scheduling* [Fisher 81]. Each microcode contains a field to control each individual processor but each processor may compute different functions.

The von Neumann model of computation has no concept of parallel processing beyond what has already been discussed. However, apart from building faster and faster processors (and there must be physical and financial limits to this), there is another mode for extracting a speed increase: distributing a problem to be solved on multi-computers. Through the 50s, 60s and 70s, Grosch's law [Grosch 85] showed empirically that uniprocessor performance increased more rapidly than its cost; the processor performance was proportional to the square of its cost. Thus it appeared uneconomical to build multiprocessor systems. With the advent of VLSI technology, this situation has changed. Ein-Dor's study [Ein-Dor 85] showed that Grosch's law still applied within different categories of computers. However, some recent studies have shown [Mendelson 87] that there is a constant average cost per MIPS (*million instructions per second*) for uniprocessors.

It is undeniable that many low-level image-processing operations are highly parallel; because of their spatial nature, they can be implemented on massively parallel spatially orientated computer architectures [Unger 58]. However, many intermediate and high-level image processing and computer vision algorithms are not wholly spatially orientated, but rather are serial searches and data-dependent execution are the normal. Thus there has been a certain scepticism centering on Amdahl's law [Amdahl 67] regarding the use of massively parallel computers for solving these problems.

Gene Amdahl argued that an algorithm has two components, a serial part s and a parallel part p. The parallel part p can be distributed over N processors; however, the serial part must be on the control processor. The *speedup* by using N processors according to Amdahl's law is

$$
\begin{aligned}
\text{speedup} \quad &= \quad (s+p)/(s+p/N) \\
&= \quad 1/(s+p/N) \quad\quad\quad\quad\quad (2.1)
\end{aligned}
$$

where $s+p=1$ is the normalised time. This indicates that if $s=0$, then the *speedup* is by a factor N. However, the speedup falls rapidly as the serial component increases. Consider, when s is 10% of the total time, then using $N=1024$ processors, the speedup is $\approx 9.9\times$, and for $N=128$ processors the speedup is $\approx 9.1\times$. For $s=0.1$, the cost of using 1024 processors may not be justifiable against using 128 processors.

This proposition has assumed that p is independent of N, and the problem size is fixed. However, experience shows us, as available processing power increases, the problem gets scaled to fill the available resources; we rarely have a fixed size problem. The serial part of many algorithms (such as control and I/O) do not increase proportionately with the problem size. Rather, Gustafson *et al.* note, it is the "run time, not problem size" which is constant [Gustafson 88b].

We need to ask "how long would a given parallel program take to run on a serial processor"? If on a parallel system with N processors, s' and p' are the serial and parallel components respectively, then an uniprocessor would require $s' + p'N$ time to perform the same task. According to Gustafson's law [Gustafson 88a].

$$
\begin{aligned}
\text{Scaled speedup} \quad &= \quad (s' + p'N)/(s' + p') \\
&= \quad N + (1 - N)s' \quad\quad\quad (2.2)
\end{aligned}
$$

However, in many problem domains, increasing the number of processors does not increase the 'scaled speedup'. For an increasing number of processors, the performance may level off and eventually decrease. This is because, as the number of processors increases, so does the memory contention if memory is shared, or the bus saturation due to interprocessor communication. This is referred to as the *saturation effect* [Jenny 77]. There are opportunities for identifying parallelism both in the control sequence and in the data. There are parts of the control sequence which could operate independently. This is called *control parallelism* and forms the basis of programming many multicomputer architectures. In many other problems, such as low-level image processing, there is a large number of independent data elements which could be processed by assigning them each to their own processor and processing them in parallel; this is referred to as *data parallelism* [Hillis 85].

Massively parallel computers with many thousands of processors exploit the *data parallel* model of computing. Examples of this class of computers include University College London's Cellular Logic Image Processors [Duff 78, Fountain 88] (CLIP), AMT's Distributed Array Processors (DAP) [Reddaway 79], Goodyear's Massively Parallel Processor (MPP) [Batcher 80, Potter 85], Thinking Machine's Connection Machine (CM) [Hillis 85], and Columbia University's Non-Von [Shaw 84]. We will review some of these architectures in §2.3.

Duff states [Duff 89], there are five factors which determine the design of a computer architecture:

(a) To obtain greater data processing at a lower cost; it may not be sufficient to just maximise the speed/cost ratio. There may be a minimum performance requirement or there may be a maximum acceptable cost.

(b) To obtain better performance within application system constraints, such as its size, weight and power consumption.

(c) To offer hardware support which would allow for a simpler implementation of algorithms and higher performance.

(d) To develop new architecture which allows not only a higher performance
   but also the exploration of novel algorithms.

(e) To provide a system which is adaptive and trainable (such as associative
   memory machines and neural networks).

None of these implies that every processor should be usefully employed one
hundred per cent of the time, nor do any of them require that the architecture
should perform equally well for all data types, or that a few higher-performance
operations should compensate for inadequate performance for others.

2.2   Architecture Taxonomies

Recently, there has been a proliferation in parallel processing technology. With
this diversity, there has arisen a need for classification, or a *taxonomic* system.
With computer architecture, it is important to distinguish between the logical
and the functional structure observed by a programmer and the logical and
functional structure and their interconnection at the circuit level. These have
been referred to as *exoarchitecture* and *endoarchitecture* [Dasgupta 84, Dasgupta 90].

Why is classification required? Classification allows the grouping of existing architectures and indicates the possibility (not all may be useful) of other
architectures. It also allows us to create models for performance analysis. In
this section we will look at a number of architecture taxonomies.

Flynn categorizes computers into four major types of organisation [Flynn
72, Flynn 66]:

SISD: Single-instruction stream — Single-data stream, this represents traditional von Neumann computers.

SIMD: Single-instruction stream — Multiple-data stream, processors acting
   synchronously to execute an instruction on their local data.

MISD: Multiple-instruction stream — Single-data stream, multiple processors
   applying different instructions to the same datum. This represents a
   pipeline computer [Maresca 88] (however, others [Hwang 85] state there
   is no realisation of this architecture), where different processors execute
   different instructions on the input data stream, and output a data stream
   to the next processor in the pipeline. The data stream is often in a raster-
   scan mode.

MIMD: Multiple-instruction stream — Multiple-data stream, multiple processors acting autonomously, each executing different instructions on their
   local data.

However, there is no dichotomy between SIMD and MIMD processors; between these two ends, there is almost a continuum of architectures which do not fit well within Flynn's taxonomy.

While Flynn bases his classification upon instruction and data streams, Shore [Shore 73] bases his classification on the organisation of the constituent parts of the computer. A computer consists of four constituent parts, a control unit (CU), a number of processing units (PU), instruction memory (IM) and data memory (DM).

Skillicorn's taxonomy of computer architecture [Skillicorn 88] is based upon the functional structure of the architecture and the data flow between its component parts. In his model, a computer has four functional units:

- An instruction processor (IP).

- A data processor (DP).

- A memory hierarchy (for both data (DM) and instructions (IM)).

- A switching network (SW).

The instruction processor is Shore's control unit and the data processors are the processing units. The functions of an IP are to:

- compute the address of the next instruction to be performed,

- fetch the instruction and decode it,

- broadcast the instruction to the data processors,

- broadcast addresses of operands to the data processors,

- receive acknowledgement of process completion from data processors.

The functions of DP are to:

- receive instructions from IP,

- receive operand addresses from IP,

- fetch operands from memory (DM),

- execute instructions,

- write results back to DM,

- return status value to IP.

Skillicorn's taxonomy system has two levels which form a super set of Flynn's classification system. The first-level is determined by

- the number of instruction processors (nIP),

- the number of data processors (nDP),

Table 2.1: Computer taxonomy due to Skillicorn.

Class	*nIP*	*nDP*	*IP-DP*	*IP-IM*	*DP-DM*	*DP-DP*
6	1	1	1-1	1-1	1-1	none
8	1	n	1-n	1-1	n-n	$n \times n$
9	1	n	1-n	1-1	$n \times n$	none
13	n	n	n-n	n-n	n-n	none
14	n	n	n-n	n-n	n-n	$n \times n$
15	n	n	n-n	n-n	$n \times n$	none

Class 6: von Neumann uniprocessor
Class 8: Type 1 array processor
Class 9: Type 2 array processor
Class 13: separate von Neumann uniprocessors
Class 14: loosely coupled von Neumann
Class 15: tightly coupled von Neumann

- the number of memory units,

- the connectivity between IPs and DPs (IP–DP),

- the connectivity between DPs (DP–DP),

- the connectivity between IPs and instruction memories (IP–IM),

- the connectivity between DPs and data memories (DP–DM).

On this basis, Skillicorn [Skillicorn 88] determines that there are a possible 28 classes of architectures. A few of these are given in Table 2.1.

   While Skillicorn's taxonomy shows the coupling between processors, it does not explicitly indicate the autonomy present within various parts of the architecture (especially prevalent in many new generation SIMD computers). Yet another taxonomy based on the degree of autonomy of each processor has been proposed [Fountain 87a] with refinements made by Maresca *et al.* [Maresca 88]; these include the use of network topology and data width. Within Flynn's SIMD model, three types of processor autonomy can be introduced: (1) Operation autonomy (allowing local branching on a result). (2) Addressing autonomy (allowing local generation of addresses for fetching operands). (3) Connectivity autonomy (allowing the network topology to be changed dynamically).

Operation Autonomy

Operation autonomy does not imply that SIMD processors are able to execute separate programs as in a MIMD paradigm; however, it does allow the logical division of processors under software control using an "activity" (or "masking") register. This allows conditional operation of the type:

if <condition> then <action>

This could prevent such mishaps as "division by zero". On massively parallel computers, this appears to be the most common and cost-effective method for allowing a degree of local autonomy. Using this "activity-bit", Maresca *et al.* [Maresca 88] show how a linked-list structure can be implemented on a SIMD computer. On CLIP7 [Fountain 87a], the concept of operation autonomy is further enhanced: there is a multi-bit "activity" register which allows each processor to interpret the broadcast signal differently.

Addressing Autonomy

A few SIMD, MSIMD (Multiple-SIMD) and MISD processors such as CLIP7 [Fountain 87a, Fountain 83a] and WARP [Annaratone 87, Kung 84] support addressing autonomy. These processors typically have word-parallel [Hwang 85, Feng 74] arithmetic logic units. One of the uses of addressing autonomy is in parallel table lookup; an other use is in accelerating floating-point arithmetic [Batcher 80]. Also Cypher *et al.* have indicated its use in Hough transforms on massively parallel array processors [Cypher 87].

Connection Autonomy

This is important in efficiently mapping algorithms to the hardware by dynamically mapping the network topology. However, this is different to reconfigurable computers because of the granularity, network topology, and control mechanism between MIMD and massively parallel SIMD computers. Computers such as the MPP [Batcher 80, Potter 83] and CLIP7 [Fountain 87a] support a limited amount of connection autonomy. Greater connection autonomy is exhibited in the Connection Machine [Hillis 85, Hillis 87a] and Polymorphic-Torus [Li 87]. These we will consider in a greater detail later in this chapter.

Danielsson and Levialdi [Danielsson 81] distinguishes that a computer may exhibit four types of parallelism:

(a) *Operator parallelism:* This is pipeline processing. Thus, concurrently, different operations may be executed in different processors; a processor receives data from a preceding processor or memory, operates upon it and passes the result to another processor.

(b) *Image parallelism:* A number of processors operate upon different parts of the same image concurrently.

(c) *Neighbourhood parallelism:* A processor can fetch simultaneously the data from its neighbouring pixels and operates upon them.

(d) *Pixel-bit parallelism:* This is conventional word-parallel computing as opposed to bit-serial processing.

The "total parallelism" is defined to be the product of the degrees of parallelism, k_o, k_i, k_n, k_p respectively for the four types of parallelism described above.

$$K = k_o \times k_i \times k_n \times k_p$$

2.3   SIMD Architecture

An SIMD architecture typically has a single control unit which broadcasts the same instruction to all the processor elements (PEs). There is an interconnection network (typically nearest neighbourhood connectivity) for processor-to-processor or processor-to-memory connections. The instruction broadcast by the control unit is executed in a lockstep fashion by each PE on its local memory. The interconnection network allows a result calculated at one pixel to be passed to another processor to be used as an operand for subsequent processing.

The two-dimensional aspect of a digitised image makes it natural to be processed on such architectures [Unger 58]. In such cases, each processor is responsible for each pixel of the image with the neighbouring processors responsible for the neighbouring pixels, with hardwired interconnections between neighbouring processors for local operations. Each of these processors can be loaded up with the same or different data but they all perform the same instruction simultaneously.

In this section, a brief discussion of such an architecture is given with respect to three large SIMD structures built, the CLIP4 [Duff 78], DAP [Hunt 81] and MPP [Batcher 80]. Some comparison between these three computers will be made. Also a brief discussion of the advantages and disadvantages of using current SIMD array computers is given.

Although "cellular automata" had been discussed by Von Neumann [Neumann 66] and Unger [Unger 58], it was not until the late 1970s that we saw the development of large SIMD processor arrays. The three computers under discussion all have large numbers of relatively simple, bit-serial processing elements (PEs).

The general architecture of all these computers is shown in Figure 2.1; it consists of a serial controller, parallel array and local array memory. The single serial control unit supplies sequences of PE-instructions and PE-memory addresses to all the PEs in the array; the instructions are then executed simultaneously by all the PEs using the same (local) PE-memory address.

Before looking at SIMD cellular arrays in detail, some advantages and disadvantages of current processor-array implementations will be noted.

Advantages:

- Data access from the nearest neighbours is implicit in the architecture; this mitigates the need for indexing or special addressing to operate on the values of the nearest neighbours.

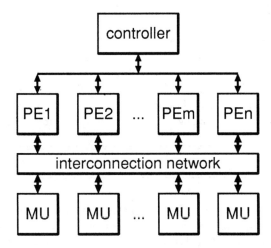

Figure 2.1: The general architecture of a SIMD machine.

- Local memory speeds up the addressing involved during fetching and storing of operands and results.

- Inherent parallelism is visible and not obscured by scan simulation.

- Bit-oriented PEs allow flexible precision and efficient storage.

Disadvantages:

- Because of the large number of PEs involved, the PEs need to be relatively simple and the PE-memory often is very limited.

- Although operations involving (3×3) neighbourhoods are efficient, communication over larger distances is not usually provided; this means that to implement operations requiring data from further away processors, the data needs to be shifted in separately and not via some local neighbourhood gating logic.

In Table 2.2 the major differences between the implementations of CLIP4, DAP and MPP are given. For a more detailed comparison between these computers, the reader is referred to [Gerritsen 83].

In a serial computer, pointers could be used to implement large complex data structures. On data-parallel architectures this is not practicable. Rather, each element of the data structure needs to be placed on an individual processor and the relationship between them provided by the connectivity between the processors.

A special case of the SIMD array processor is the associative array processor. The difference between them is the method used for accessing memory

Table 2.2: Difference between CLIP4, DAP and MPP.

Feature	CLIP4	DAP	MPP
size	(96×96)	(64×64)	(128×128)
PE-intercon	4,6,8 (selectable)	4	4
memory - bits	32	4096	1024
Additional	Neighbourhood	PE-memory	shift register
feature	gating logic	mapped to	for multiplic-
		host	ation & division

in *random access memory* (RAM) and associative memory (AM). While RAM is accessed by specifying word addresses, AM is content addressed (*content-addressable memory* (CAM)); allowing parallel access to many memory locations. This allows parallel search and parallel comparisons. However, currently CAM is considerably more expensive than RAM.

2.3.1   Array Processors

CLIP Systems

CLIP4 (Cellular Logic Image Processor) [Duff 78, Duff 76, Duff 73, Fountain 81] is a 96×96 array of two boolean processors with 35 bits of local memory each (32 bits are addressable RAM). When built in 1980, CLIP4 was the first large array assembled using custom-designed integrated circuits. Each of the processors can be loaded with the same or different data, but the same function is performed by all the processors at the same time; thus in Flynn's terminology [Flynn 72], CLIP4 is a single instruction stream, multiple data stream (SIMD) computer.

A simplified diagram of the CLIP4 processor is given in Figure 2.2, more detail about the processor can be found in [Duff 78]. Briefly, the processor consists of three registers A,B and C, two boolean processors P1 and P2, and 32 bits of local memory D. In the processor P1, the result for that cell is calculated and stored in D_i; processor P2 calculates the propagation and sends the result to the neighbouring cells. The propagation signals from the neighbouring cells are accepted via the neighbourhood gating. The only access of data external from the D stores is via the A register.

There are two important aspects of CLIP4 architecture which will be briefly mentioned here since they play an important part in the method by which data is represented and propagated in the array. An image in CLIP4 is represented by an array of numbers of variable precision (but usually 6 bits or 8 bits long) forming a **bit-stack** consisting of **bit-planes** (often called **d-levels** [Duff 78]) (Figure 2.3). A bit-plane D_i can be represented using a set theoretic notation

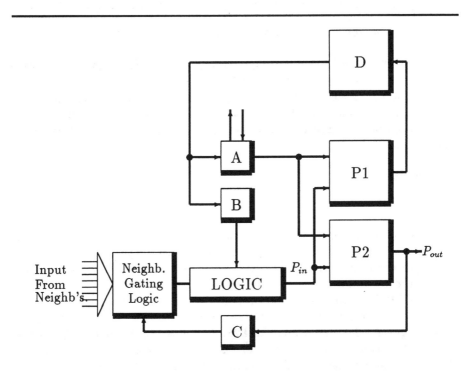

Figure 2.2: A simplified diagram of the CLIP4 processor.

as $D_i : N \times N \rightarrow \{0,1\}$; the plane has size $N \times N$. In this image representation form, each processing element (PE) holds a complete number (each bit in a binary plane). An image can then be of variable length (within the limits of storage), and not represented by an 8-bit or a 16-bit word which would be wasteful in memory allocation in CLIP4 with only 32 bits of memory per PE, since all processing is carried out in a bit-serial manner.

The local neighbourhood connectivity between the processors is very important in the way data is passed around in the array. The connectivity in CLIP4 is software controlled and there are two modes:

(1.) Square 8-connected, of which any sub-set can be chosen (Figure 2.4), and

(2.) Hexagonal 6-connected, of which again, any sub-set can be chosen.

This allows data to propagate in the array in specified directions. There are many examples of this in the following chapters.

The notation used to indicate the directions enabled for data propagation will be dirc[j] when a single direction path, j, has been enabled or dirstr("...") when more than one direction path has been enabled. Examples of these are given below:

(1.) dirc[8] ,

(2.) dirstr("026").

In (1) the data is propagated to each processor from the left (Figure 2.4) and in (2) the data from top and bottom processors are propagated; this occurs at all processors simultaneously. In all cases, a square tessellation is to be assumed, unless otherwise stated.

The CLIP4 controller (a 16-bit processor with 16 kbytes of memory) has its own assembly level language, CAP4 [Wood 79, Wood 83]. A high-level procedural language called IPC [Reynolds 82a, Reynolds 82b, Reynolds 83] has also been implemented. This language is an extension of the C language [Kernighan 78], which allows image and file handling and program development on a PDP 11/34 computer running under the UNIX operating system. A large subroutine library [Otto 82a] now exists which further facilitates ease of programming the CLIP4 computer. In 1987, the CLIP4 was upgraded to use the CLIP4A processors running at 2.5 MHz. Also, it no longer has a controller but is more directly controlled by a host, a Sun-3 workstation [Duff 89], and the array memory is mapped to the host memory giving virtual memory capabilities.

A program to be run on CLIP4 consists of (1) array instructions (these are carried out by all processors simultaneously) and (2) serial instructions (these are executed only by the controller). There are three classes of array operations [Duff 73] :

1. Pointwise – in these operations no data is transferred between the processors.

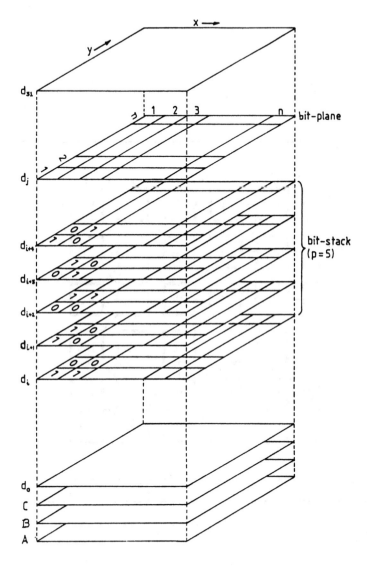

Figure 2.3: Bit-plane and bit-stack representation of an image in CLIP4. The 32 addressable d-levels are shown.

The bit-planes A,B,C are special registers used by the CLIP4 PE in processing data. D_i is the least-significant-bit and D_{i+4} is the most-significant-bit plane of the image depicted.

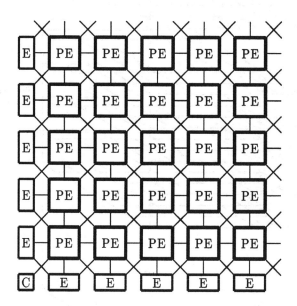

Figure 2.4: The square connectivity of the CLIP4 processors in the array.

2. Local neighbourhood – the result at each processor is dependent on the data of its 3×3 (square connectivity) neighbourhood.

3. Global – in this class of operation, the data sent to a neighbouring cell is a function of the data received from the neighbouring cells (and the data from the point itself), i.e., information can be broadcast over arbitrarily large distances in the array in one instruction. Therefore, the result (potentially) is a function of the data of the entire array.

Although CLIP4 is very fast, many applications [Ip 83a, Ip 83b, Clarke 84a, Potter 84, Potter 85] have emphasised the requirement for more local memory to prevent disc swapping (and many operations such as convolution and high-order polynomial line fitting require more memory than CLIP4 has) and for a greater resolution.

CLIP4S Hardware

CLIP4S is an SIMD computer which is software compatible with its predecessor CLIP4, but processes a 512×512 pixel image and has 64 bits of local storage per pixel. This gives a total of 2 Mbytes of data storage for the complete array. Since at present it is not economic to build a full 512×512 processor array (the biggest full coverage array currently available is the 128×128 MPP [Batcher 80]), a 512×4 processor array (256 CLIP4B chips) has been built [Fountain 83b, Fountain 87b] which scans in one dimension with edge stores over a 512×512 data array.

There are three main classes of operations which can be performed on a SIMD array such as CLIP4 and CLIP4S [Duff 73]; a brief description of how these are carried out on a one-dimensional scanning array, namely CLIP4S, is given below (for a detailed description see [Fountain 83b, Fountain 87a]).

1. Pointwise operation — at each scan position, the CLIP4S array fetches the operands from the local store, computes the result and returns the result to the local store (Figure 2.5(a)).

2. Local operation — for these operations, data is required from three sources, the local stores and the upper and lower edge stores. At each scan position, the data from the three sources needs to be accumulated, and the results computed and stored in the local stores. The array then moves on to the next scan position (Figure 2.5(b)).

3. Global operation — because the scan can occur in only one direction at a time, after 128 scan positions (from the first scan position) the direction of the scan is reversed and this process is continued until the data array stabilises or the global operation is timed out. Thus, at the first scan position, the specified data is loaded into the upper edge store and data is fetched from local stores and two results are computed; one (the current propagation path) is stored in the local array and the other (propagation signal to be passed to the next scan position) is stored in

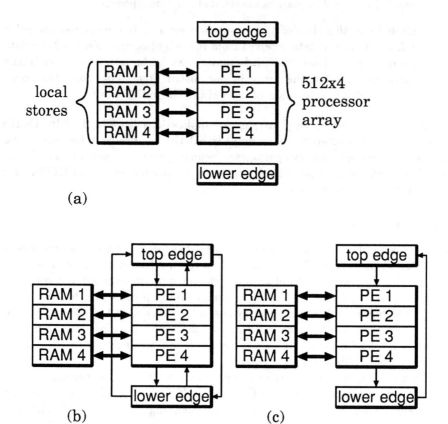

Figure 2.5: The sequencing operations on CLIP4S.

the lower edge store and is subsequently transferred to the upper edge store (Figure 2.5(c)) for propagation to begin in the next scan position.

To implement these three classes of operation, the CLIP4S controller has six different hardwired routines. The controller also has a two-level instruction pipeline. Using the three methods described above, a full coverage array of 512×512 processors is simulated in CLIP4S.

A comparison of some basic operations in CLIP4 and CLIP4S is given in Table 3.2. The CLIP4 programs were written in CAP4 [Wood 83] and the images have 6-bit precision, while for CLIP4S, the programs were written in IPC [Reynolds 83] and have 8-bit precision.

Table 2.3: Timing for some commonly used library routines.
The CLIP4 times are for 6-bit images, and the routines written in CAP [Otto 84]. The CLIP4S times are for 8-bit images, the times quoted also have significant IPC overheads. The times given are in ms.

FUNCTION	ROUTINE	CLIP4	CLIP4S
Add/Subtract two images	ADD	0.38	270
Multiply two images	MULT	4.25	3340
Divide two images	DIV	6.78	8400
Find maximum of two images	MAX	0.82	570
Shift an images one pixel	SHIFTS	0.26	215
Mask part of an image	ANDS	0.27	250
Generate X-coordinate	**RAMP_X**	4.7	650
Generate Y-coordinate	**RAMP_Y**	4.7	250
Sum grey value of an image	VOLUME		1070
Median Filter (3×3)	MEDIAN	17.5	9.5s
Local maxima (3×3)	MAXN	11.7	70.7s
Skeletonise binary image	THINR8	8.6	2.40s

Massively Parallel Processor

Massively parallel processor (MPP) is a 128×128 array of bit-serial processors, each running at 10 MHz [Batcher 80]. Each PE has six 1-bit registers A,B,C,G,P, and S. Registers ABC are used for bit-serial arithmetic; register P is used for moving data to and from the neighbouring PEs. There is also a shift-register which allows the operands to recirculate through the adder; this significantly speeds up multiplication, division and floating-point operations. While the MPP array is only four-connected, its edge connectivity is much richer than that for CLIP4 and DAP. The top and bottom and left and right edge processors can be connected to form a torus. Also the left and right edges can be connected in a *spiral* form. The I/O for the data array is carried out through the S register. The data from the S register can be transferred to RAM in a single cycle. Thus the data throughput through the array is 160 Mbytes per second. The processor array is four-connected (to its north, east, south and west neighbours). MPP exhibits local autonomy by controlling the activity of individual PEs by the state of the G register. As in CLIP4 and the Connection Machine, the state of the array can be determined by the inclusive-or operation of the all the enabled PEs. This could be used to determine whether the array is empty or finding the maximum value.

The VLSI CMOS MPP chips each contain eight PEs. An upgrade to MPP called BLITZEN is being developed by MasPar [Rosenberg 89]. The BLITZEN VLSI chip will each have 128 PEs. As with CLIP7[Fountain 83a, Fountain 88], BLITZEN will allow the local computation of addresses (as well as receiving

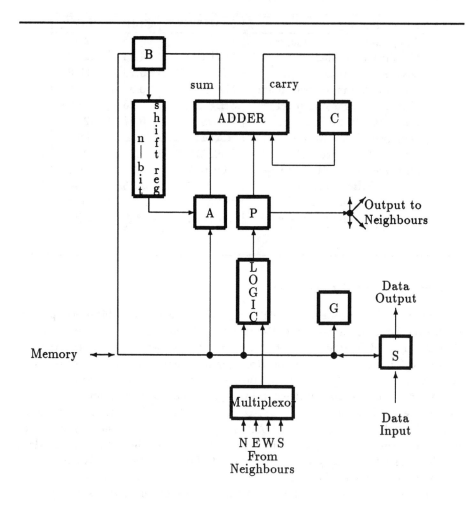

Figure 2.6: A schematic architecture of the Massively Parallel Processor (MPP)

CM chip has 16 PEs; each PE consists of:

- one proprietary arithmetic logic unit and associated latches which act in a bit-serial manner,

- four k bits of static RAMS,

- eight flag registers each of which is one bit long; one of which is a condition flag, the state of which determines whether a result is stored or not,

- one router interface which allows it to connect to the other processors, and

- a two-dimensional grid interface for local processor communications.

On CM-1, a point-wise operation takes $0.75 \mu s$, plus over-head for decoding instructions. A 32-bit add takes 24 ms; with 64k PEs working in parallel, this gives a total capacity of 2G 32-bit adds per second. As in CLIP4, the CM also has a GlobalOR instruction. This allows it to determine rapidly whether the data array is empty or acts as a check for a termination condition.

The CM has a much richer communication mechanism than the other massively parallel SIMD machines described in this book. Like most processor arrays, it supports:

(a) *Broadcast communication:* the control processor broadcasts constants to the data processors.

(b) *GlobalOR:* this function does a boolean OR of the outputs of all the data processors.

(c) *Hypercube communication:* large distance communication between data processors is carried out using the boolean 12-cube network. At the vertex of the 12-cube is one of the 4096 processor chips.

(d) *Router:* this provides data routing between data processors by message switching. It can buffer and combine data for messages being sent to a common address.

(e) *NEWS grid:* nearest-neighbour communication is carried out by a four-connected (North, East, West, South) cartesian grid system with a PE at each of the nodes. NEWS communication is about six times faster than using the hypercube because no message collision can occur.

An upgraded CM system (CM-2) was released in 1987. The CM-2 is software-compatible with the CM-1. Its enhanced features are: (a) it has larger memory, 64 kbits per PE; (b) it has optional floating-point accelerator, a floating-point processor is provided for every 32 data processors; and (c) the two-dimensional NEWS grid is replaced by an n-dimensional grid.

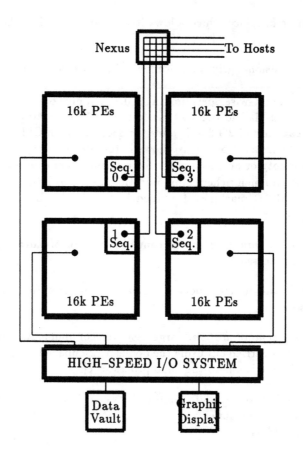

Figure 2.7: The Connection Machine.

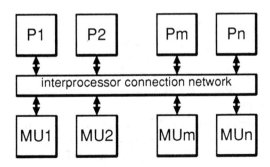

Figure 2.8: The general architecture of a MIMD machine.

2.4   MIMD Architecture

MIMD architectures have multiple processors with one or more memory units. The processors act autonomously, i.e. they execute independent instruction streams on local data streams. Synchronisation between the processors is possible by passing messages between themselves using an interconnection network (i.e. Occam [May 83] and Transputer [Homewood 87]) or by accessing data in a shared memory unit (Figure 2.8. MIMD processing units are considerably more complex than typical SIMD PEs; they are often von Neumann engines with additional support for interprocessor communication and I/O.

With an SIMD array, an important consideration in its design is the processor autonomy. In a MIMD system, an important consideration is how the processors are interconnected and its memory resources. In a system which has a global pool of memory so that it is shared between the processors, it is possible that two or more processors may require the same memory address simultaneously leading to memory contention. Therefore the memory access needs to be arbitrated through either software or hardware. Also as execution of a program consists largely of memory references, the time spent for memory access must be kept to a minimum. These requirements place demands on the processor interconnection and global memory. The cost of fast memory and the speed of the bus limits the number of processors in the system. Thus shared memory bus-systems tend to exhibit coarse-grained parallelism. The other option is to have a distributed memory system with each processor having its own local memory and the processors communicating with each other by message passing.

Most MIMD machines developed have been used in a SCMD (single code, multiple data) mode. In this paradigm, each node has a copy of the program and executes the code on its local data.

Howard Siegel [Siegel 79, Siegel 80] has studied a number of interconnection networks for a collection of processors. An interconnection network can be defined by a set of interconnection functions which maps an output of a processor to the input of another.

Multiprocessor computers have been built with primarily three types of network interconnections: *point-to-point topology* such as a hypercube (Figure 2.9), *multistage networks* (Figure 2.10) and *bus-based architectures*. The multistage networks and bus-based architectures can be linked to form a pseudo-complete graph for task-level parallelism and these are suitable for forming small to medium-sized systems.

However, the connectionist model of computation [Feldman 85] and data-parallel algorithms [Hillis 87b] requires large fine-grain parallelism. The mesh-connected processors such as CLIP4, DAP, and MPP, while very efficient for local neighbourhood operations, do not readily support global communication between processors, instead communication must occur through intermediate processors. Hypercube connectivity allows both efficient local and global communication between processors leading to its use in the Caltech Cosmic Cube [Seitz 85], CM-2 [Hillis 87a], NCUBE [Hayes 86], Intel iPSC and FPS T-series amongst others.

While for its strong connectivity, regularity and symmetry, hypercubes have been the favoured connectivity system; however, such a system is not so readily scaleable in hardware. For a hypercube consisting of N processors, each node requires $\log N$ communication channels. Therefore, different sized hypercubes require different numbers of communication channels. It is not modular; different-sized networks require a different number of communication channels, this prevents them from being fabricated onto the chips. For example, a Transputer [Homewood 87] which has four built-in channels can only support $log N > 4$ channels through extra hardware and arbitration. To overcome this problem, a hierarchical network system called *hypernet* has been proposed by Hwang and Ghost [Hwang 87]. This is achieved by providing a constant node degree, having a dense local connectivity and as the network size increases, the internode links are made more sparse to maintain the constant node degree.

A computational task may be either compute bound or I/O bound. In a computation, if the number of operations is greater than the sum of the input and output data items then it is said to be compute bound, otherwise it is I/O bound. For example, the computation of the Mandelbrot set [Mandelbrot 82]

$$\mathbf{Z}_{i+1} \leftarrow \mathbf{Z}_i^2 + \mu$$

may need the function to be iterated a thousand times to determine whether the complex value μ is within the set. The input and output are a pair of complex numbers, thus this operation is indeed compute bound, as also is matrix multiplication. However, a matrix addition is I/O bound because there are only one third as many add operations as there are elements in the input and output matrices.

Ideas for a hypercube computer are not new; they were first proposed by Squire and Palais of Michigan University [Squire 62, Squire 63]. The first hypercube computer built was the 64-node Caltech Cosmic Cube [Seitz 85]. Each

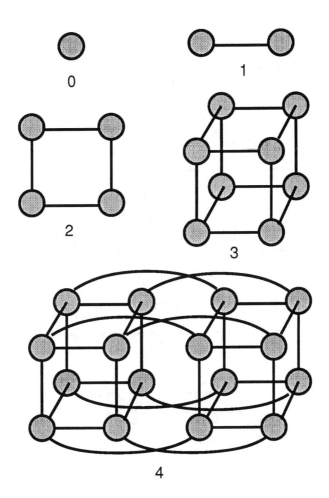

Figure 2.9: The diagram shows various degrees of hypercubes. A zero-degree hypercube is a SISD machine. A three-degree hypercube has a processor at each of the node of a cube. A higher degree hypercube consists of lower-degree hypercubes.

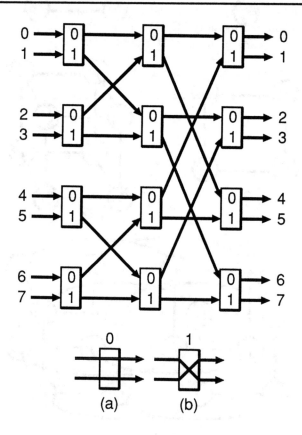

Figure 2.10: The figure shows a multistage network to connect eight processors. Each stage consists of a set of switching boxes which can be in one of four states: straight (state 0), exchange (state 1), upper broadcast, and lower broadcast.

node of the six-dimensional hypercube consists of the Intel 8086/87 processors running at 8 MHz and 128 Kbytes of RAM. Two years later in 1985, Intel launched its 128-node iPSC, Ametek a 256-node machine and NCUBE a ten-dimensional hypercube. Both the Intel and Ametek's machine had their nodes built using the Intel 80286/87 chip sets, while NCUBE'S nodes were custom designed 32-bit processors.

A tightly coupled, shared-memory system is easier to program but creates memory latency. An additional advantage of a hypercube is that it is homogeneous (i.e. all nodes can be identical) therefore the I/O problem associated with SIMD and shared-memory MIMD systems can be alleviated by attaching an I/O system at each node or at a collection of nodes [Hayes 86, Cannon 89, Reddy 90]. Furthermore, an n-dimensional hypercube is composed of hypercubes of $(n-1)$ and lower dimension. Hence, a hypercube can be subdivided into cubes of lower dimension allowing its efficient use in a multi-processing and task-level division of algorithms.

Each node of NCUBE has 128 Kbytes RAM. Each 16×22 inch board contains a 6-dimensional hypercube with a total of 8 Mbytes of memory. Thus the complete 1024-node machine is constructed on 16 boards. There are further eight boards dedicated to I/O; each has 16 NCUBE processors, each servicing eight hypercube nodes.

ZMOB [Kushner 82] consists of 256 8-bit Z80A processors running at 2.5 MHz connected to a high-speed bus. Each processor has 64 Kbytes of memory and a multiplier. The maximum rating for ZMOB is 100 MOPS. Images are partitioned and parts are sent to individual processors; a 512×512 image is partitioned into 256 32×32 sub-images.

A Transputer [Homewood 87] is not a MIMD-architecture machine, but it is a computer on a chip with a CPU, memory and four serial links which allows it to be connected to other transputers to form a MIMD machine. A T800 Transputer is a 32-bit RISC microprocessor with a 64-bit floating-point unit on the chip which is capable of 1.5 MFLOPS sustained performance. It has 4 kbytes RAM and four 20 Mbits/s serial links for communications. The serial links are capable of DMA block transfer between memory. It also has internal timers for high-priority and low-priority processes.

Transputers are typically networked to exhibit one of three types of parallelism:

1. **Pipeline:** this is temporal parallelism. The difficulty of this approach using Transputers involves extracting sufficient parallelism. In this method there is no load balancing and the throughput rate is determined by the slowest process in the chain. If a process is particularly data-dependent then it could lead to a drop in throughput for the entire network.

2. **Geometric:** this is spatial parallelism, which could lead to massive data parallelism with all units working synchronously as in a SIMD array. The problem of this approach using Transputers is the limitation of the communication bandwidth since fine grain parallelism is very expensive

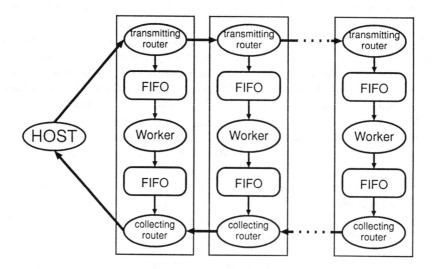

Figure 2.11: A schematic diagram of a processor farm.

because of module cost and the high cost of communication setup. Typically, a small array of Transputers is used and the images are divided between the processors in suitable chunks, balancing computation and communication.

3. **Processor Farm:** this is spatial parallelism. This involves data parallelism as with geometric parallelism but the processes are set up on the Transputers to act asynchronously. Some form of heuristics is typically used to maximise the utilisation of the network. A master processor will farm out work to worker processors when idle worker processors become available (Figure 2.11). This type of processing is used for data-dependent problems. A big assumption is that the task scheduling does not constitute a significant overhead to the problem solving and therefore can be carried out on the master processor.

Using Transputer hardware, some machines have been specifically designed and built for computer vision and image processing. They include MARVIN [Brown 89]. MARVIN consists of 25 T800 Transputers forming a 8 × 3 fully-connected mesh. The important aspect to this architecture is that the first row has busses to take I/O with industry standard MAXbus equipments. This machine was specifically designed to process stereo images.

2.5 Pipe-line Architectures

From consideration of the previous two sections on SIMD and MIMD architecture we have observed that many low-level and intermediate-level image
processing tasks fit well onto a massively parallel SIMD array. Symbolic processing tasks map well onto powerful serial computers or on a small MIMD
system [Fountain 86]. However, there are I/O bandwidth limits on moving
data out of a SIMD array. Also, the cost of SIMD arrays are generally high
because there is no volume production of these architectures. The design cost
is often much greater than component cost. Therefore, there is a need to reduce
the design cost. One way of achieving this is to make the system modular, simple and reusable. Also, there is a requirement to balance the computation rate
with the data I/O bandwidth with the host. This is the basis of a computation
model called a *systolic architecture* [Kung 82].

Pipelined processors are often specialised to carry out a task efficiently. One
of the most commonly required classes of operation required in image processing
is the *neighbourhood operation*, such as convolution, correlation or neighbourhood boolean operation. Many of these computers act with neighbourhood
parallelism in a raster operation (RasterOp) fashion. One of the earliest machines built was GLOPR [Preston 73]. It only operated upon binary images
using a hexagonal neighbourhood referred to as the Golay surround [Golay 69].
In such a neighbourhood only 14 unique (rotationally invariant) patterns are
possible. The output can be specified to be one or zero on the basis of the neighbourhood pattern. The commercial version of GLOPR is Diff3 [Danielsson 81].
It has 8 Golay processors which operate in parallel.

Another computer of a similar type is Picap I [Kruse 73]. It could process
a 64 × 64 4-bit image. However, as we observe from Table 2.4, the parallelism
exhibited by a single neighbourhood operator such as GLOPR, Picap I and even
Diff3 is rather limited. Without incurring the cost of building massively parallel
computers, a reasonable amount of parallelism and I/O balance with the host
can be achieved with pipelined processors. A very large degree of pipelined
parallelism is exhibited in Cytocomputer [Lougheed 80]. It has 115 stages in
the pipeline of two types; one had 80 stages, referred to as the two-dimensional
stage and the other is referred to as the three-dimensional stage. The 2D-
stage computes local neighbourhood boolean operations using table lookup, and
the 3D-stage computes mathematical morphology operations such as erosion,
dilation and their grey-level generalisation again using a table lookup.

In PIPE (Pipelined Image-Processing Engine) [Kent 85], the processors are
designed for local neighbourhood operations. It consists of a sequence of identical processors; on either side are specialised processors for I/O. Although all
PIPE processors are identical, they perform different operations depending on
the algorithm. Each processor can receive three input images, from the preceding processor, the processor immediately ahead and the result of the processor
itself, and output three images which need not be identical. The two main
types of operation which can be carried out are point-wise and neighbourhood
operations. By feeding results back, recursive operations such as relaxation

Table 2.4: Degree of parallelism for exhibited by various image processing architectures.

Machine	k_o	k_i	k_n	k_p	K
Picap I	1	1	9	4	36
Picap II	3	4	1	8	96
Diff3	2	8	7	1	112
Flip	16	2	17	8	4352
Cytocomputer	88	1	9	8	6336
Clip IV	1	9216	9	1	82944

algorithms and simulations of larger neighbourhood operations can be carried out.

A computer similar to PIPE is FLIP (FLexible Image Processing system) [Luetjen 80]. But instead of the pipeline carrying out RasterOp as in PIPE, FLIP has 16 general-purpose 8-bit processors running at 2 MHz which work in an image parallel mode. They work in a semisynchronous mode, passing data from one processor to another. This requires a number of very high-speed buses. Each processor has its own output bus; the combined bus is 128 bits wide.

A systolic architecture consists of either a pipeline or an array of cells, each capable of carrying out some simple computation. In such a system, the I/O only occurs at the "boundary cells". In this system the data flows from one cell to another in a pipelined fashion and operations are carried out in every cell. Compute bound problems can be implemented efficiently and cheaply on a systolic architecture.

The systolic approach has been used in the design or implementation of VLSI chips for convolution and correlation [Kung 80, Kung 81a, Kung 81b, Blackmer 81], discrete Fourier Transform [Kung 80] and median filtering [Fisher 81] types of operations which are commonly used for image processing.

However, systolic arrays are control-driven and therefore require correct timing between processors. This timing is more complex than a global clock on an SIMD machine because different cells may compute different operations. If processing times are not equal for each cell, then in a synchronous system, the clock speed will be that required by the slowest processing element. This problem of synchronisation may be overcome by using a dataflow computing principle in a systolic architecture; the resulting architecture is referred to as a *wavefront array processor* [Kung 87]. In a dataflow model, correct *sequencing* of functions is required. There have been a number of wavefront array processors constructed such as the STC-RSRE [Davie 86] and MWAP (Memory-linked Wavefront Array Processor) [Dolecek 84].

An example of a linear systolic array is WARP [Kung 84, Annaratone 87].

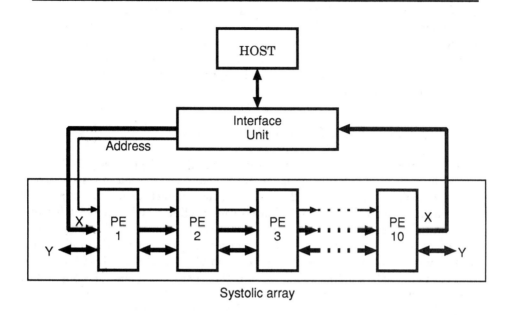

Figure 2.12: A ten cell WARP systolic array.

It consists of a ten-cell array; each cell is capable of performing at 10 mflops. Therefore, the machine has peak performance of 100 mflops. This is achieved by using two floating-point units (an adder and a multiplier) at each cell. Two prototype WARP machines were built and delivered in 1986 by General Electric and Honeywell respectively. These machines have been extensively used for image and signal processing and other scientific computation. The WARP machine consists of the ten-cell array, a host computer and an interface unit between the host and the array. Each of the cells is driven at 20 MHz by the interface unit which also handles the I/O between the array and the host. The data flows through the systolic array in two channels (X and Y), (Figure 2.12). Each of the WARP cells is a microprocessor with its own sequencer and program memory. There are two levels of parallelism in WARP; there is the processor array, but each cell also has a parallel architecture. The important aspects of the machine are that each of the processors is a powerful computing engine. There is a large amount of data memory for each cell (32 Kwords) and a high communication bandwidth (80 Mbytes/s). Because of this large communication bandwidth, WARP can operate in both a fine- and a large-grain parallelism mode.

Because of the linear array construction and high communication bandwidth between cells, the WARP machine can operate efficiently as a pipeline. The algorithm is systolic if it is partitionable across the array and each cell computes one stage of the processing as the data is pumped through the cells. Another method is to partition the data amongst the processors (each processor then has one or more lines of the image). This method of processing is typically used for neighbourhood operations. The other mode of parallelism used is where each cell has the complete (or a large part of the) image, but produces only a part of the final result; partial results from each of the cells are combined in the host. The types of operations which may use this type of parallelism are for example histogramming and Hough transformations.

2.6   Special VLSI Architectures

The VLSI technology is extensively used in digital signal processor (DSP) architectures for many real-time applications. The fast execution of instructions is accomplished by the use both of a Harvard architecture and of extensive pipelining. A typical DSP would have an instruction which would fetch two operands from memory, multiply them, add them to an accumulator, and write the result to memory. The DSPs are optimised to carry out the multiplication and accumulation (MAC) operations. For example, the AT&T DSP16A carries out integer MAC operation in 33 ns, and the Texas Instrument TMS320C30 can carry out a floating-point MAC operation in 60 ns.

Although DSPs make extensive use of internal pipelining, little support is usually provided for using them in parallel multi-processor systems. Very few DSPs have methods for synchronising on data or instructions. Exceptions are the Motorola DSP96002 and the TMS320C30 which provide instructions for "hardware interlocking". More parallel DSPs based on the dataflow paradigm have recently been made available including from NEC, the μPD7281 processor; this processor is particularly suitable for image processing.

Another problem for which VLSI solutions have been proposed is the detection of smooth curves using Montanari's figure of merit (FOM) function [Montanari 71, Clarke 84b, Guerra 86, Cheng 90]. Solutions are based upon using systolic arrays, VLSI arrays and massively parallel mesh-connected arrays. In Montanari's method, the curve cannot change direction by more than $45°$. Thus if a curve is defined by a sequence of points $P_1, P_2, \ldots, P_n$ such that P_{k-1} and P_k are neighbours, then $d(P_k, P_{k-1})$ is the orientation of the curve between these two points. This is essentially the chaincode direction for $\overline{P_{k-1}P_k}$. Then to satisfy Montanari's condition

$$\Theta(P_{i-1}, P_i, P_{i+1}) = ([d(P_i, P_{i+1}) - d(P_{i-1}, P_i)] \mod 8) \leq 1.$$

The FOM for a curve of length is

$$g_k(P_1, P_2, \ldots, P_k) = \sum_{i=1}^{k} a(P_i) - \sum_{j=2}^{2} \Theta(P_{j-1}, P_j, P_{j+1})$$

where $a(P_i)$ is the grey-level of pixel P_i. The task then is to determine the optimal FOM (G_n) by dynamic programming.

$$G_n = \max \left[g_n \text{ of all curves satisfying Montanari's condition} \right].$$

At each iteration we need to compute,

$$g_k(x,y) = \max_{-1 \leq i,j \leq 1} [g_{k-1}(x+i, y+j)] + a(x,y) - \Theta(g_{k-2}, g_{k-1}, g_k)$$

Because of the local nature of the operations, this algorithm can be efficiently implemented on an SIMD array computer.

2.7 Hybrid Architectures

From the discussions in the previous sections of this chapter, we have observed the success of SIMD processors for iconic processing, especially when the algorithms are complex. However, if only a few special functions are required, such as filtering and edge detection, then it is more economical to use specialised pipelined architectures. While iconic processing maps well on to massively parallel 2D SIMD arrays, symbolic processing maps on to a small but powerful MIMD computer. However, in many computer vision problems, both types of processing are required. Having both powerful SIMD and MIMD machines is not always useful because of the communications bottleneck between the two structures, while mapping both the iconic and symbolic processing onto one structure can also be very unsatisfactory. To address these difficulties, a number of hybrid architectures have been proposed [Fountain 86]. We will review some of these architectures in this section.

2.7.1 Trees and Pyramids

As with SIMD computers, the pyramid computers have been more influenced by the structure of the mammalian visual pathway. The pyramid cone represents a notion of the visual system hierarchy found by Hubel and Wiesel [Hubel 62], the base layer of the pyramid equating to the retinal iconic process, and each successive layer carrying out a higher and higher level of symbolic processes.

SIMD array computers such as CLIP4 [Duff 78], DAP [Reddaway 79] and MPP [Batcher 80] have shown substantial computational speed can be gained by using a large number of simple interconnected processors which carry out cellular logic operations [Preston 79]. The pyramid topology gives greater global communication and access to data than local-neighbour connected PEs in an SIMD array [Tanimoto 80, Uhr 72, Uhr 79].

From the study of commonly used image-processing operations [Cantoni 85a] because of the low cost ratio of *computation* to *communication*, the most suitable structure found for their implementation is an SIMD array. However, for many other image-processing operations, fast communication between two or more widely separated processors is also required. Pyramids provide this fast communication in $O(\log n)$, where n is the edge size of the pyramid base. Pyramids also map both the data-driven (bottom-up) and the model-driven (top-down) paradigm for image analysis and allow their execution in parallel.

For a pyramid computer: (i) The system could either be driven by a single controller so that the pyramid acts as a single SIMD machine, each layer

executing the same instruction in a lock-step fashion. (ii) Each layer could
be driven by a separate controller. This would allow the pyramid to act as
a multiple SIMD computer. This increases the efficiency of utilisation of the
computer. The problem with having a single SIMD control is that a pipeline
of operations cannot be set up in the pyramid cone. For example, the base
layer cannot compute pixel operations whilst a layer above carries out I/O operations from a previous operation. This can lead to poor utilisation of the
hardware.

A cell at layer k in a pyramid computer has typically 13 neighbours, a
parent cell in level $(k-1)$, eight neighbours in the plane k and four offspring
cells in level $(k+1)$.

While we have observed that counting is inefficient on SIMD array computers, an $(n \times n)$ mesh-connected array takes $\Omega(n)$ operations for counting, the
tree connectivity of a pyramid reduces the counting complexity to $O(\log n)$.
This affects such operations as histogramming, feature extraction [Reeves 80],
median filtering [Tanimoto 83a], object selection [Stout 83] and some forms of
object labelling [Tanimoto 82].

Although pyramid computers have been discussed as being suitable general-
purpose structures for high-speed computation of low-level image processing
and computer vision algorithms [Dyer 77] [Dyer 82, Reeves 80, Tanimoto 82,
Uhr 72] it is only recently that hardware has been developed [Tanimoto 83b,
Cantoni 87, Merigot 86]. Below we describe some hybrid architectures.

PAPIA

The designing of PAPIA (Pyramid Architecture for Parallel Image Analysis)
started in 1984. The chip consists of five PEs, a parent and its four offspring,
each with 256 bits of memory [Cantoni 85b]. Thus each chip can be regarded
as a pyramid node. The ALU is bit-serial; there are also two variable size shift
registers which allows local operations such as convolution without writing
intermediate results to the memory. There is also a boolean comparator.

The interconnection between the processors is fixed. Each processor within
a plane is connected to its four square neighbours. It is also connected to
its parents and it has connectivity to its four offspring. As in most mesh-
connected SIMD arrays (MCA), there is a neighbourhood gating which allows
simultaneous access for some boolean functions to a sub-set of the neighbours;
in this instance, the neighbours are not only the NEWS processors but also the
parent and the offspring. Also as with MCAs which can mask individual PEs,
PAPIA also provide facilities for masking pyramid layers. Using these it hopes
to provide recursive operations such as the global operation in CLIP4.

PAPIA, an eight-layer pyramid, the base layer being 128×128 square, can
be configured to act as three independent SIMD machines: the first four layers
(up to the 8×8 PEs plane) form the top section, the middle section consists
of two layers of 16×16 and 32×32 PEs respectively, and the bottom section
consists of a 64×64 plane and the base plane. Each of these sections have
their own controller and I/O busses so that images can be loaded/unloaded in

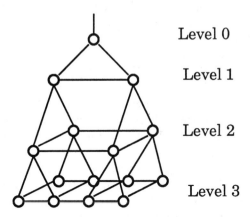

Level 0

Level 1

Level 2

Level 3

Figure 2.13: The SPHINX architecture based on binary pyramid.

a column mode as in CLIP4.

While PAPIA forms a pyramid using a quadtree structure, SPHINX's (Système Pyramidal Hiérarchise pour le traitément d'Images Numeriques) pyramid interconnection consists of a binary tree, although in each layer, the PEs are mesh-connected to their four nearest neighbours (Figure 2.13). Therefore each PE is connected to one parent and two offspring PEs. Depending on the parity of the layer, the offspring alternatively have either the same X- or Y-coordinates. Each SPHINX PE consists of a bit-serial ALU and 128 bits of RAM. The PE also has an activity register and neighbourhood gating. The hardware also has a global-OR to test result convergence, such as whether the array is empty after some iterative computation.

Other pyramid machines are being developed: one by Dave Schaefer at George Mason University and the Image Understanding Array (IUA) architecture at University of Massachusetts [Weems 89]. IUA is designed to be a real-time image processor, a 64th of which is being constructed. The pyramid is in three logical layers: the base layer consists of a 512×512 array of bit-serial processors forming the content-addressed array parallel processor (CAAPR); the intermediate layer is called the intermediate communications and associative processors (ICAP) which consists of an MIMD 64×64 array of 16-bit processors. This layer converts data from an iconic to a symbolic description. At the apex is the symbolic processing array (SPA) which is also a MIMD architecture consisting of 64 32-bit processors. Data is transferred between the layers by means of shared memory: there is 1 GByte shared memory between ICAP and CAAPR and 0.5 GByte shared between SPA and ICAP. That is, a single ICAP processor shares 262 kByte of RAM with 64 CAAPR processors.

The CAAPR layer is essentially a custom-designed bit-serial SIMD array with many of the properties of SIMD array (which we have discussed earlier in this chapter) such as local connectivity through the NEWS mesh, GlobalOR of the values of the array, and global broadcast of values to the processors. However, we know that NEWS connectivity is inefficient for passing information over a long distance. Two of the layers, CAAPR and ICAP, are controlled by a dedicated Array Control Unit (ACU), which in turn is controlled by the SPA. The switches which enable NEWS communication can be selectively disabled to form clusters of SIMD arrays. Under MIMD mode, the ICAP can run different programs on different clusters, these two layers acting as MSIMD machines. In the limit, there can be 4096 SIMD machines each of 64 PEs, each executing different code simultaneously.

Another similar pyramid architecture currently being constructed is the Warwick Pyramid Machine [Nudd 88]. Like IUA, the WPM consists of three layers, the base layer is a 256 × 256 SIMD array constructed using DAP processors [Hunt 81]. Above these are 256 processors which act as controller for a 16 × 16 array of DAP processors; thus the machine can act as a 256 MSIMD machine. The base-layer acts on iconic data and the layer above converts these to a symbolic representation. A transputer array above this carries out the high-level symbolic processing. This pyramid machine is hosted by a Sun computer on which programs are developed.

NON-VON

NON-VON is a massively parallel, fine-grained, binary tree-structured SIMD machine [Ibrahim 87]. A prototype of this, NON-VON 1, has been operational at Columbia University since 1985. The general NON-VON architecture can be classed in Flynn's terminology as a multiple SIMD (MSIMD) machine. The architecture of the NON-VON machine is depicted in Figure 2.14. It consists of three types of processors: (1) a very large number of *small processing elements* (SPEs); (2) a small number of *large processing units* (LPEs); and (3) a few *secondary processing subsystems* (SPPs) – these are intelligent devices.

Each SPE consists of an 8-bit ALU and 256 × 9-bit local RAM. At the leaf level the SPEs are interconnected to form a 2D mesh structure; it can also be configured as a linear array. Different resolution images are stored in different layers of the machine. This structure allows efficient implementation of multi-resolution algorithms [Ibrahim 84].

There are no facilities for loading images at the leaf level. This can be a bottle neck if images always have to be loaded from the root; however, this problem is alleviated somewhat by the facility to load images in parallel at some intermediate level of the tree.

The tree-structure also allows the operations to be pipelined. Thus for some functions, such as histogramming which can be highly pipelined, it performs considerably better than the SIMD array reviewed in § 2.3; its performance for point-wise operations is comparable to that for MPP, but it is considerably slower for large window convolution because data has to be moved up and

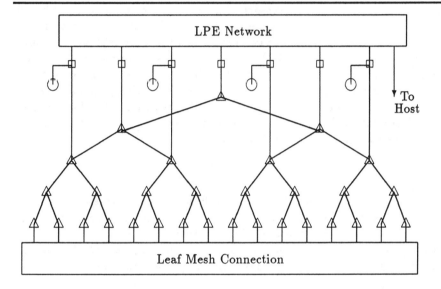

△ Small Processing Elements

□ Large Processing Elements

○ Disc

Figure 2.14: Architecture of NON-VON.

down the tree-structure. In the following chapters, we will see some typical image-processing algorithms implemented on this machine.

PASM

The PASM (PArtitionable SIMD/MIMD) [Siegel 81] system can operate as one or more SIMD and/or MIMD machines. The prototype machine, built using Motorola MC68010 processors, consists of 16 PEs and memory and 4 microcontrollers (MCs). This forms an MC-group. In the final version, there will be 1024 PEs and 32 MCs. This PASM concept is feasible (or usable) only if there are a large number of processors. The processors communicate with each other using an interconnection network [Pease 77]. This allows the system to be dynamically changed between the two modes and there are separate memories for these two modes. In the MIMD mode it acts as a normal von Neumann machine and accesses data from other PEs by network asynchronous DMA. PASM can emulate a variety of architectures such as mesh, pyramids, tree and ring. However, each processor has only one out going link. Therefore, it must communicate with other PEs one at a time; therefore, the efficiency of the emulation varies according to the architecture.

CLIP7

The array size limitation of CLIP4 gleaned from several successful image processing applications [Ip 83a, Clarke 84a, Potter 84, Hussain 88a] led to the development of CLIP4S [Fountain 87b]. But there still remained some severe limitations with memory and processing of grey-scale images. CLIP4s successor CLIP7A, (Figure 2.15), which has only a single processor per chip, has the following properties:

1. Increased amount of memory per processor. For each processor there are 64 kbytes of local RAM.

2. Improved image I/O rate. In CLIP4, the image had to be loaded/unloaded through the A register (Figure 2.2). This meant that the I/O could not be overlapped with array operations. In CLIP7, separate I/O registers are provided. There are also 8 registers for collecting data from the neighbours and a result output register.

3. Increased local control. There is a 16-bit condition register. This affects the operation of the ALU and the connectivity to the neighbourhood for data processing. There is further local autonomy provided with the CLIP7A processing element; by using two CLIP7 processors in tandem, it has the capability for the local generation of the data memory address. The main processor is concerned with result computation while the co-processor generates addresses. Compare this with CLIP4 processor (§ 2.3.1) which has two boolean processors; one computes the local result and the other the propagation result.

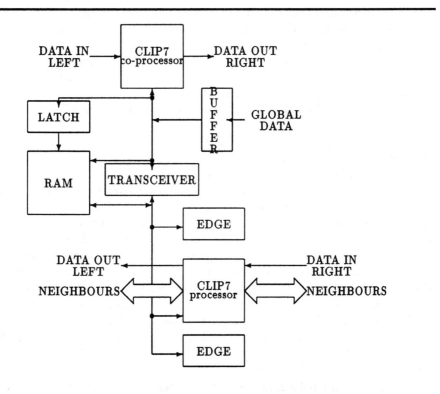

Figure 2.15: Architecture of CLIP7.

Using the CLIP7 chips, a linear array machine (CLIP7A) consisting of 256 processing elements has been constructed [Fountain 88]. There are a number of reasons for building a linear array: (1) it can emulate a 2D array for iconic processing; (2) it is a 256-element vector processor for symbolic processing; and (3) its simple and regular structure can be used to simulate more complex structures.

2.8   Concluding Remarks

This chapter has discussed the various taxonomies used for computer architectures. Despite the limitation of Flynn's taxonomy, resulting from the advent of various types of connection network and autonomy exhibited by current hardware, it still remains a popular means of classifying architectures because of its compactness and the simplicity of its description of the concurrency of processes. Three classes of computers, SIMD, MIMD, and MISD (pipeline), have been considered in this chapter and various examples in each of these classes has been presented.

We have alluded to the fact that most low-level vision algorithms maps very efficiently onto the SIMD array processors because of the iconic nature of the data and the neighbourhood operations applied to the data. However, high-level vision requires more symbolic task-parallel processing as opposed to the data-parallel processing. These problems often maps better onto the small to medium-sized networks of MIMD processors. Computer vision tasks consists of vast amounts of low- and medium-level image processing (filtering, segmentation, description and even some recognition) and some scene interpretation tasks (which involve further recognition and indeed identification of objects). These later tasks map better onto MIMD processors. It would appear, to have efficient implementation of vision algorithms, that one needs data-parallel (SIMD) processing for low-level vision and task-parallel (MIMD) processing for high-level vision. The need to move large amounts of data from the SIMD processors to the MIMD processors, for the iconic-to-symbolic conversion to occur, leads to an I/O bottleneck.

One method for balancing the I/O has been to use a pipelined architecture with specialised units running at frame rate rather than a general-purpose SIMD array processor. This approach is also more cost-effective when a production algorithm has been developed. Some have attempted to retain the general-purpose processing power of a SIMD array processor and augment it with MIMD processors through the use of a pyramid architecture.

As we have also seen, both SIMD and MIMD pyramid and tree architectures have been proposed and built because they *allow* better data I/O between the processor arrays and the host computer, or because it is thought that more efficient iconic symbolic transformation can be carried out. It has certainly been shown that operations which involve counting, histogramming and sorting are more efficiently mapped to a pyramid or tree architecture rather than to a SIMD array. However, it appears uncertain how much more efficient pyramids are at iconic-symbolic transformation. The problem with a pyramid is that, at any given level, information about an object may only be partially available at a processor because objects cross over "processor cluster boundaries."

Chapter 3

Parallel Algorithms and Languages

3.1 Introduction

Factors which affect the performance of an algorithm on a particular architecture are dependent on the degree of parallelism and the over-head incurred in scheduling and synchronising the tasks. For task-level parallelism, the finer the grain of parallelism, the greater the cost in scheduling and synchronising. This means that a linear increase in speed by increasing the number of processors is not usually possible. Many believe that algorithms can be expressed as expression evaluation in a dataflow system which then have little scheduling-and-synchronising overhead [Dennis 80, Arvind 83, Gurd 85]. Others believe that it is important to have structured data (object-orientated computation) hence a more grainy dataflow model [Gottlieb 83, Gajski 84].

The choice of an algorithm to solve a particular problem is strongly influenced by the hardware architecture and the software support tools available. Given a particular environment, a programmer will prove to be sufficiently resourceful in developing an algorithm which is "most efficient" for that architecture.

Generally, four theses have been proposed for an order of magnitude improvement in performance over conventional computers. They are:

- Develop faster circuits and augment them with methods for synchronizing parallel processes. This allows us to use the current computer architectures and hence retain the large number of algorithms and programs that have been developed.

- Develop better optimising and vectorising compilers which would allow parallel processors to be better exploited.

- Develop new algorithms better supported by new languages.

- Develop new models of computation which allow massive parallelism; these can be exploited by building large multiprocessor architectures.

However, recent developments in VLSI technology have meant none of these theses can be regarded in isolation, but they interacts with each other very much.

For image processing on serial computers, chain-coding [Freeman 74] for lines and quadtrees [Samet 80, Samet 84] for regions have been used for both data reduction and computational efficiency. On MIMD architecture, these models of computation can be efficiently mapped because of interprocessor communication and data manipulation. However, on massively parallel SIMD array processors, these processing models do not fit well and could lead to storage and execution inefficiencies.

In this chapter, we first look at various parallel paradigms such as data parallelism, task parallelism, and special (pipeline) parallelism. Data parallel processing involves distributing the data amongst the processors and running the same code on all the processors. Task parallel computing involves distributing the program amongst the processors.

We show that a new way of thinking is required if we are going to make efficient use of parallel processors for image processing and computer vision tasks. It is not sufficient just to implement the best currently available algorithms. Often, the problem has to be considered from first principles. We illustrate this with some simple examples. The need for new thinking is further emphasised when we consider parallel implementation of some commonly used techniques such as matrix multiplication, sorting and searching.

With the advent of serial computers, certain "fast" algorithms based on *divide-and-conquer* methods have been developed. These include the FFT [Brigham 74] and various methods for matrix multiplication [Strassen 69, Kronsjo 79]. The Fourier transform is an $O(n^2)$ algorithm, the FFT improves the complexity to $O(n \log n)$. However, on an array processor, both these methods have $O(n)$ complexity [Ip 80]. For matrix multiplication, the algorithm is $O(n^3)$ for the slow method. A fast method due to Strassen is $O(n^{2.81})$. We show in this chapter that a $n \times n$ matrix multiplication on an array processor can have complexity of $O(n^2)$ or better. Similar improvements in the order of the complexity for other commonly used functions such as sorting and searching are also possible.

Next we look at some image processing techniques such as relaxation as a paradigm for solving many sets of linear equations. We follow this with a discussion of the three types of operations involved in low-level computer vision. We present these techniques with particular reference to array processors and the CLIP SIMD array processor in particular. Many of the basic techniques presented will be used extensively in the following two chapters.

Finally we look at the basic principles used in the design and use of various types of parallel languages. We do not have space to consider any particular languages but we look at the general paradigm involved.

3.2   Parallel Paradigm

There are three types of parallelism: (i) data parallelism (SIMD), (ii) task parallelism (MIMD), and (iii) special parallelism (systolic). Special parallelism refers to the systolic or pipelined models. The pipeline is set up such that the algorithm is partitioned temporally and each stage computes a partial result and passes it on to the next stage. The other two types of parallelism we will describe later.

Just as there is difficulty and confusion over architecture taxonomy, so similar difficulties exist in classifying algorithms. Often, algorithms have been developed without reference to an architecture and use more than one model of parallelism and therefore are difficult to implement in practice. However, in a similar manner to hardware architecture, we can *plot* an algorithm in a space with three orthogonal axes [Kung 80]: *concurrency control, module granularity,* and *communication geometry.* In the system, there may be a single or a distributed control which may be synchronous or asynchronous. The granularity of a task is determined by the amount of work a module can carry out before it must communicate with another module. This granularity is subdivided into three levels: fine, medium and large.

Using this *space* as a basis, Kung [Kung 80] has proposed a taxonomy for parallel algorithms. Kung processes the product of

$$\{\text{concurrency control}\} \times \{\text{module granularity}\} \rightarrow \{\text{Systolic, SIMD, MIMD}\}$$

subspaces. This restricts the classification to a manageable size. Then the parallel algorithms can be classified into subspaces produced by the cross-product of {Systolic, SIMD, MIMD} × {communication geometry}.

However, it is unlikely that any one type of parallelism will solve a given problem efficiently.

The development of parallel algorithms has a long history dating back to the early 1960s [Miranker 71], although at that time no parallel computers had actually been constructed.

For fine-grain parallelism, we have seen (Chapter 2), shared memory systems suffer from memory contention leading to performance degradation. Also with large MIMD arrays which cooperate by message passing, there are communication costs and need synchronisation.

A possible efficient method for expressing fine-grain parallelism may be through dataflow machines [Watson 79, Dennis 80] and languages [Conway 63, Arvind 78, Keller 81]. The development of the dataflow computing architecture has come from representation of the program by the dataflow program graphs [Karp 66, Adams 68, Rodriguez 69]. In a dataflow paradigm, there is no concept of control flow (hence no program counters), instead an instruction is executed when all its operands are available. Consequently, many instructions may receive their operands simultaneously and hence are executed concurrently; there is no specific order for their execution.

However, there is very large investment involving current procedural languages such as FORTRAN, C and PASCAL. To see rapid acceptance of dataflow languages, the programming task has to be made easier and ready use of libraries which have been developed (possibly through language translation [Allan 80]). Further advances are being made to allow programming via the dataflow program graphs [Keller 81, Davis 81]. The obvious problem with these are the requirements for high-resolution graphics terminals. It is expected that graphical resolution better than $4K \times 4K$ may be required.

Concurrency can be further exploited by using multiprocessors as in an MIMD system with the dataflow program graph being subdivided and distributed across the processes. However the system then suffers from the communication problem in a similar fashion to other MIMD systems (similar connection network – Chapter 2). On a message-passing system, with fine-grain data parallel or systolic algorithms, communication overheads may be minimised by clustering data (and processes) together and allocating a cluster to a processor node.

Another method for expressing parallelism is to use a programming methodology such as CSP (or Occam) and develop hardware (e.g. Transputer) to act as a hardware interpretor.

3.2.1   Data-Level Parallelism

Given the number of parallel computers which have been developed both commercially and in universities, there is still a lack of software for them. Most machines are often delivered only with a compiler for a conventional language such as FORTRAN or C which have been enhanced to support the parallel model of computation such as message passing and synchronisation between processors.

The data-parallel paradigm of computation is explicitly synchronized; it maps to the SIMD [Flynn 72] model of programming where each of the processors executes the same code on its local data simultaneously (see Chapter 2 for further details). The advantages of data-parallel programming is that there is a very simple control flow, the data array changes from one state to another. Furthermore, the results are deterministic and independent of the number of physical processors in the system. This means that debugging a data-parallel program is much simpler, and it is also easier to port programs, not only between SIMD machines but also to MIMD ones. However, as we will discover later, while data-parallel algorithms map readily for low-level image processing, it is not obvious how one maps high-level reasoning tasks to this model of programming.

If there are fewer processors than the number of pixels in the image, the picture will need to be partitioned amongst the available processors. While in MPP, a 512×512 image may be "folded" so that there are 16 128×128 square arrays; in processor farms or linear arrays, it is more likely that the image will be partitioned vertically or horizontally into α-slices for the α available processors. This partitioning scheme is much simpler for the host computer in both transmitting and collecting back the images and data. For neighbourhood operations involving $k \times k$ kernels, each image slice will have to have k rows or columns added at the "top-and-bottom".

Many of the SIMD algorithms can also be implemented on systolic arrays. However, because of the pipelines, the processors execute the same operation but at different times; this is referred to as the skewed SIMD model [Annaratone 87].

3.2.2 Task-Level Parallelism

Programming a multiprocessor computer is considerably different from programming a uniprocessor computer. A program can be represented by a control graph. It consists of nodes which represent operations on data to transform it to a different state or movement of data, and arcs which represent the order in which the nodes are executed. In a serial program, one node is executed after another; there is a predefined order. A multicomputer has different programming requirements from a uniprocessor computer. A multiprocessor computer requires:

- the algorithm needs to be partitioned into sub-tasks;

- the sub-tasks and data need to be distributed amongst the processors; and

- the system must be set up to allow interprocessor communication and synchronisation.

These three steps must be carried out in the specified order because the requirement for communication and synchronisation cannot be determined in advance of the distribution of the algorithm amongst the processors. For new algorithms and architectures, the above processes have to be assessed anew.

With the MIMD message-passing hypercube type of computers, there may be excessive cost with communication between processors. Tasks may need to be partitioned (and grouped) so that communication (particularly *global* communication) is minimised. There may be a large overhead involved in initiating communication between processors. In such cases, it may be required to combine several messages into a single packet to improve efficiency. Therefore, it may be necessary to group communications and send them in larger packets, rather than in many smaller ones.

3.2.3 An Example

Consider an industrial inspection task of object recognition on a moving conveyor belt. The procedure we might typically apply on a serial computer would be:

1. Grab and digitise a picture using a video camera.

2. Threshold the image, classifying object pixels by 1s and the background pixels by 0s.

3. Raster scan until we find a unmarked non-zero point. This point represents the first point on the object.

4. The boundary is traced and stored as a chain-code; also the boundary points are marked.

5. If the object (chain-code length) is small, it is probably noise. Goto 3.

6. If the object is touching the edge of the picture, then the object is only partially visible. Goto 3.

7. If object found, then analyse the chain-code to identify it.

8. When the raster scan reaches the last point, goto 1.

There are two levels of serial computation involved here. There are the raster operations (RasterOps) to threshold the image and generate the chain-code of an object; and there is the serial computation involved for each object to be identified.

If we implement this on an MIMD computer with coarse-grain parallelism, then we will still be left with the need for the RasterOps. Our parallelism could come with processing all the objects in parallel, using the master and worker processor paradigm. It is most likely that we will use the chain-code methods for recognising the objects. Essentially, we are running serial programs on each of the processors but the tasks for recognising individual objects are now distributed amongst the available processors. This method for processing images is illustrated in Algorithm 3.1.

With SIMD array processing, this affair is reversed. We now have massive parallelism at the low level and therefore do not need the RasterOps. However, at some later stage of the processing we will have to work through each of the individual objects in the scene, attempting to recognise them. However, because the RasterOps are not used and there is no generating of Freeman chain code [Freeman 61] to represent the objects, the recognition method will be very different as a consequence.

3.3   General-Purpose Algorithms

In this section we present some general algorithms which are useful in many image-processing routines but are not exclusive to them. We explore methods of carrying out matrix multiplication using array processors. We then present some methods for sorting and searching on SIMD array computers. Finally, we consider the relaxation algorithm. It is a way of solving Laplacian equations and many of the calculus of variation methods used in *shape from X* and optical flow (see Chapter 6).

Algorithm 3.1: A processor-farming method for object recognition.

MASTER PROCESS:
begin
  for each image **do**
    PAR
      SEQ
        grab and digitise image;
        threshold image;
        delete small noise objects;
        SEQ for each object
          find non-edge connected objects;
          send object to available processor;
        collect results from workers;
    od
  end

WORKER PROCESSOR:
begin
  SEQ
    receive object data;
    trace around object boundary;
    analyse chain-code to recognise object;
    send result to master processor;
  end

3.3.1   Matrix Multiplication

An algorithm is presented for multiplying two matrices using the CLIP4 computer. The method proposed uses global propagation, it is further assumed (although not currently supported) that the global propagations are appropriately timed out. This method is significantly different from serial methods using L-U decomposition [Stoer 80], parallel versions of such algorithms have been proposed for SIMD computers and are reviewed in [Hwang 85].

Given two matrices $\mathbf{A}$ and $\mathbf{B}$ (both $n \times n$)

$$
\begin{pmatrix} a_{11} & a_{12} & \cdots & a_{1n} \\ a_{21} & a_{22} & \cdots & a_{2n} \\ \vdots & \vdots & \ddots & \vdots \\ a_{n1} & a_{n2} & \cdots & a_{nn} \end{pmatrix} \times \begin{pmatrix} b_{11} & b_{12} & \cdots & b_{1n} \\ b_{21} & b_{22} & \cdots & b_{2n} \\ \vdots & \vdots & \ddots & \vdots \\ b_{n1} & b_{n2} & \cdots & b_{nn} \end{pmatrix}
$$

$$
c_{ij} = \sum_{k=1}^{n} a_{ik} \times b_{kj} \qquad 1 \leq i,j \leq n
$$

The method consists of extracting the first column of $\mathbf{A}$, a_{i1}, and the first row of $\mathbf{B}$, b_{j1}. The values a_{i1} are spread across the array and b_{1j} down the array, the two resulting matrices are pointwise multiplied and the result stored in an accumulator. Next the a_{i2} and b_{2j} column and row are extracted, spread across and up and down the array as before, pointwise multiplied and added to the accumulator. This process is repeated for the remaining column and rows (Algorithm 3.2). This method produces an $O(n^2)$ algorithm (up to the size of the processor array) as opposed to an $O(n^3)$ algorithm on a serial computer. The parallel algorithm for an SIMD array computer has complexity $O(n^2)$ because we need to duplicate the row and column values of the matrices to the appropriate processors; this duplication requires n operations. But this is much better on CLIP4 which can use global propagation for the spread function. It would be possible to have an $O(n)$ algorithm if we were to choose a systolic approach (see Algorithm 3.3). But we will need to feed in a complete row (and column) of values in parallel. At the end of the n iterations, another n iterations are required to propagate the last inserted rows and columns through the rest of the array. At the same time, zeros will have to be inserted at the first row and column.

3.3.2   Sorting and Searching

There are well-known serial sorting algorithms [Knuth 73, Horowitz 78]. Parallel sorting algorithms are rare and have been developed only recently. Some parallel sorting algorithms are reviewed in [Lakshmivarahan 84]. Amongst the type of sorting algorithms favoured for SIMD arrays are merge sorts. One of the simplest sorting algorithm for a linear array is a type of bubble sort call odd–even transposition sort. This algorithm has $O(n)$ complexity for sorting n keys on n processors. This algorithm can be implemented using a network of *comparator* modules; various types of networks can be used [Batcher 68].

Algorithm 3.2: Parallel Matrix multiplication using an array processor.

begin
  $C(,) = 0$;
  for $i = 1$ **to** N **do**
    $D(,) = \text{SPREAD}(A(i,), \leftrightarrow)$;
    $E(,) = \text{SPREAD}(B(,i), \updownarrow)$;
    $C(,) = C(,) + D(,) * E(,)$;
  od
end

Algorithm 3.3: Parallel Matrix multiplication using a systolic approach.

begin
  $C(,) = 0$;
  for $i = 1$ **to** $2N$ **do**
    $\text{INSERT}(A(i,))$;
    $\text{INSERT}(B(,i))$;
    $C(,) = C(,) + A(,) * B(,)$;
    $\text{SHIFT}(A, \longrightarrow)$;
    $\text{SHIFT}(B, \downarrow)$;
  od
end

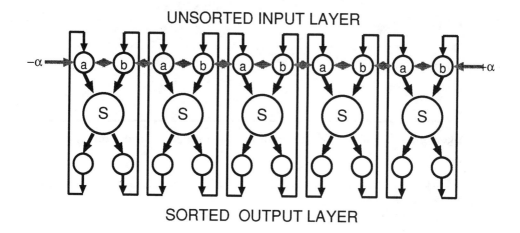

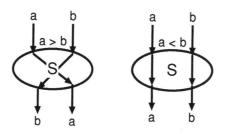

Figure 3.1: A network for sorting numbers.

The 1D odd–even transposition sort algorithm can also be implemented on a
SIMD machine (Figure 3.1). A 2D extension to this method is presented in
[Thompson 77]. This algorithm is a combination of Batcher's odd-even merge
and the odd-even transposition sort and consists of four steps for sorting an
array of 4 rows by 2 column. The steps involved are:

1. Move data at odd-processors to the left neighbouring processor.

2. Use odd–even transposition to sort each column.

3. Interchange the values at even rows.

4. Each odd-numbered processor (snake-like row major order) compares its
   value with the next processor and swaps if its value is larger.

Many 2×4 arrays can be sorted in this way in parallel. Then each of the
neighbouring 2×4 arrays can be merge-sorted to obtain larger sorted arrays (see

[Lakshmivarahan 84]). No proof is being presented here, but the complexity of parallel sorting is $O(n)$ as opposed to $O(n \log n)$ for the best serial algorithms.

Many associative search operations are particularly well suited for parallel processing. Such operations have been grouped as follows:

- Extreme searches – search for maxima, minima and median.

- Equivalence searches – i.e equal-to, not-equal-to, similar-to and approximate-to a given key.

- Threshold searches

- Adjacency searches – search for nearest record which is smaller than the key.

- Between-limits searches

- Ordered retrieval – requires ascending or descending sort.

Many of the above methods are obvious or standard and no further discussion will be given. For searches involving the smallest or largest record in objects to be determined in parallel, the reader is referred to [Otto 84].

3.3.3 Relaxation Algorithm

Begin with the consideration for solution to Poisson's equation

$$\nabla^2 u = f$$

which is Laplace's equation if f is identically zero. Assume that the domain of the problem is a square unit $0 \le x, y \ge 1$ and the domain is split into square grid mesh of size h. Then the Poisson equation can be solved by the centered difference approximation at any grid point $(\hat{x}, \hat{y})$.

$$\frac{\partial^2 u}{\partial x^2} \approx \frac{1}{h^2} \left[u(\hat{x} + h, \hat{y}) - 2u(\hat{x}, \hat{y}) + u(\hat{x} - h, \hat{y}) \right]$$

$$\frac{\partial^2 u}{\partial y^2} \approx \frac{1}{h^2} \left[u(\hat{x}, \hat{y} + h) - 2u(\hat{x}, \hat{y}) + u(\hat{x}, \hat{y} - h) \right]$$

Then the Poisson equation is approximated by

$$u(\hat{x} + h, \hat{y}) + u(\hat{x} - h, \hat{y}) + u(\hat{x}, \hat{y} + h) + u(\hat{x}, \hat{y} - h) - 4u(\hat{x}, \hat{y}) \approx h^2 f(\hat{x}, \hat{y})$$

Any interior grid point is given by

$$(x_i, y_i) = (ih, jh) \qquad i, j = 1, \ldots, N$$

and $(N + 1)h = 1$. By defining an approximation u_{ij} to an exact solution $u(x_i, y_j)$ at the N^2 interior points and requiring them to satisfy the Poisson equation exactly gives

$$u_{i+1,j} + u_{i-1,j} + u_{i,j+1} + u_{i,j-1} - 4u_{i,j} = h^2 f_{i,j} \qquad i, j = 1, \ldots, N$$

This is a linear equation in $(N+2)^2$ variables u_{ij} and can be expressed [Ortega 81] as

$$\mathbf{Ax} = \mathbf{b}$$

There are two general methods for solving this linear equation, Gaussian elimination or an iterative method. For a non-linear equation, an iterative method is required.

Consider the linear equation $\mathbf{Ax} = \mathbf{b}$, with the condition that the diagonal elements of $\mathbf{A}$ are non-zero; the simplest iterative procedure is the Jacobi method. Given an initial approximation $\mathbf{x}^0$, the next iteration is given by

$$x_i^1 = \frac{1}{a_{ij}} \left[b_i - \sum_{i \neq j} a_{ij} x_j^0 \right] \qquad i = 1, \ldots, n \tag{3.1}$$

If $\mathbf{A} = \mathbf{D} - \mathbf{B}$ and $\mathbf{D}$ is the diagonal matrix (see [Ortega 81]), the equation 3.1 may be written as

$$\mathbf{x}^{k+1} = \mathbf{D}^{-1} \left[\mathbf{b} + \mathbf{Bx}^k \right] \qquad k = 1, \ldots, n$$

However, from equation (3.1), it is seen that x_1^1 is available for use in computation of x_2^1. if the updated values are used then equation 3.1 becomes the Gauss–Seidel iterative method

$$x_i^1 = \frac{1}{a_{ij}} \left[b_i - \sum_{i \leq j} a_{ij} x_j^1 - \sum_{i \geq j} a_{ij} x_j^1 \right] \qquad i = 1, \ldots, n$$

In the matrix form, this can be expressed as

$$\mathbf{x}^{k+1} = \left[\mathbf{D} - \mathbf{L}^{-1} \right] \left[\mathbf{b} + \mathbf{Ux}^k \right] \qquad k = 1, \ldots, n$$

where $\mathbf{L}$ and $\mathbf{U}$ are the lower- and upper-triangular matrices.

The simplest form of the Poisson equation is Laplace's equation. This can be solved using the Jacobi iteration method.

$$u_{ij}^{k+1} = \frac{1}{4} \left[u_{i+1,j}^k + u_{i-1,j}^k + u_{i,j+1}^k + u_{i,j-1}^k \right]$$

The solution at each iteration is the average of the four near-neighbour values from the previous iteration. For this reason, Jacobi's method is often referred to as the method of *simultaneous displacement*. However, the order in which the computation is carried out for the Gauss–Seidel iteration method is important.

$$u_{ij}^{k+1} = \frac{1}{4} \left[u_{i+1,j}^{k+1} + u_{i-1,j}^k + u_{i,j+1}^k + u_{i,j-1}^{k+1} \right]$$

These two methods are shown schematically in Figure 3.2. The Gauss Seidel method converges considerably faster than the Jacobi method [Ortega 81]; but while the Jacobi method can be computed in parallel on a SIMD array computer, The Gauss-Seidel method can only be computed serially, in a pipeline

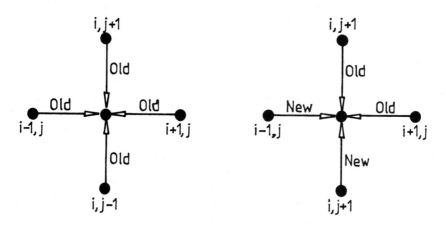

Figure 3.2: Jacobi and Gauss–Seidel iterative methods respectively.

or a MIMD computer [Barlow 82]. Many [Horn 90] choose to use the Gauss-Seidel method because commonly algorithms are implemented on von Neumann architecture and the method can be further speeded up by using successive over-relaxation. These two methods for solving linear equations are classical methods and date back to the last century; detailed discussion of the methods can be found in [Forsythe 60, Varga 62, Young 71]. The classical reference for the discretisation of the elliptic equation by a finite difference method is [Forsythe 60]. For an introduction to finite element methods, the reader is referred to [Strang 73, Mitchell 77].

3.4   Operations for Image Processing

All image-processing operations can be achieved by applying an appropriate sequence of neighbourhood operations. These neighbourhood operations can be classified into three classes of operations to be discussed later in this section.

If G is a digitised image with the pixel value at a point (i,j) of g_{ij}, then a parallel neighbourhood operation, i.e. an operation computed simultaneously at all pixels, can be expressed as

$$g_{ij}^* \; = \; \mathcal{L}(\mathcal{N}(g_{ij}))$$

where $\mathcal{L}$ is the neighbourhood operation and $\mathcal{N}$ is a set of the pixels in the neighbourhood.

There is more than one way of classifying neighbourhood operations. For example, one may use a functional classification and discuss a useful set of operations by the results they produce or classify the operations by their method of construction, i.e. the window size and computation. Using the latter method, we will divide the array instructions into three classes:

Pointwise – the result at a pixel is a function of the initial data at that pixel alone and no data is transferred between the neighbouring pixels. This is the lower limiting case of the neighbourhood operation.

Local – the result at each pixel is a function of the pixels in a (3×3) neighbourhood of that pixel. This is a special but important case of the neighbourhood operations because of the local connectivity of the processors in a SIMD computer.

Global – the data sent to the neighbouring processors is a function of the data received from the neighbouring processors (and the data at the point itself). As a consequence, information can be broadcast over arbitrarily large distances in the array in a *single instruction*. Therefore, the result (potentially) is a function of the data of the entire array. The global operation is an upper limiting case of the neighbouring operation with the special case of the neighbourhood window equal to the size of the image.

As a passing note, it should be emphasised that while all SIMD arrays (indeed any computer) can compute any of the above operations, only the CLIP4 processors support global propagation at the hardware level; in all other computers, this process has to be simulated in software.

Below, some examples of the above three classes of operations are given with especial emphasis on their implementation on the CLIP4 computer [Duff 78, Otto 84].

Pointwise Operations

Taking b_{ij} as a binary image plane, an obvious pointwise operation is image inversion,

$$b_{ij}^* = \neg b_{ij}$$

for a grey image of p-bit precision **do**
    for $n := 1$ **to** p **do**
      $b_{ij,n}^* \longleftarrow \neg b_{ij,n}$;

Amongst boolean point-wise operations involving two image planes **a** and **b** are

$$c_{ij} = a_{ij} \bigwedge b_{ij}$$
$$c_{ij} = a_{ij} \bigvee b_{ij}$$
$$c_{ij} = a_{ij} @ b_{ij}$$

where @ is the EXOR operator. Using such a set of boolean operations, integer arithmetic operations involving two grey-level images **X** and **Y** of p-bit precision can be constructed as follows [Duff 78, Otto 84]:

addition :

$$C_{-1} = 0$$

$$\left.\begin{array}{rcl} S_i & = & X_i \text{ @ } Y_i \text{ @ } C_{i-1} \\ C_i & = & (X_i \cap Y_i) \cup (X_i \cup Y_i) \cap C_{i-1} \end{array}\right] \quad 0 \le i \le p-1$$

$$S_p = C_{p-1}$$

where C is the carry and S is the sum. If the image are signed then the last step is replaced by

$$S_p = X_{p-1} \text{ @ } Y_{p-1} \text{ @ } C_{p-1}$$

subtraction :

$$C_{-1} = 0$$

$$\left.\begin{array}{rcl} D_i & = & X_i \text{ @ } \neg Y_i \text{ @ } C_{i-1} \\ C_i & = & (X_i \cap \neg Y_i) \cup (X_i \cup \neg Y_i) \cap C_{i-1} \end{array}\right] \quad 0 \le i \le p$$

where D is the difference and C is the carry.

Local Neighbourhood Operations

Before proceeding to give some examples of local neighbourhood operations, a neighbourhood gating logic function operator needs to be defined. This operator $\cup_{dir}$ ORs the data from the neighbouring pixels from the enabled directions *dir* which is a subset of the eight possible directions. As an example, $\cup_{2468}$ represents ORing of the data from the NEWS neighbouring pixels. In the CLIP4 processor, a programmable neighbourhood gating logic exists. In such a machine, local neighbourhood operations can be performed in a single instruction. However, in DAP and MPP, such neighbourhood gating logic does not exist and data has to be shifted in singularly from the required directions.

(i) shift data $\rightarrow$

$$b_{ij} \Leftarrow \bigcup_8 b_{ij}$$

(ii) delete isolated ones

$$b_{ij} \Leftarrow b_{ij} \wedge \bigcup_{1-8} b_{ij}$$

(iii) extract local 8-connected edges

$$b_{ij} \Leftarrow b_{ij} \wedge \bigcup_{2468} \neg b_{ij}$$

(iv) expand (dilation) operation

$$b_{ij} \Leftarrow b_{ij} \vee \bigcup_{dir} b_{ij}$$

where *dir* represents the enabled directions.

(v) shrink (erosion) operation

$$b_{ij} \;\Leftarrow\; b_{ij} \wedge \neg \left(\bigcup_{dir} \neg b_{ij} \right)$$

Global Propagation Operation

Global propagation operations [Duff 80, Jelinek 78, Blake 82, Otto 84] have already been defined at the beginning of §3.4. Before giving some examples of operations involving global propagation, an explanation of the notation used need to be given. Referring to the CLIP4 processor (Figure 2.2), it is observed that there are two outputs from each processor. Note also that each processor consists of two simple boolean processors, one calculates the propagation path (P2) and the other the result at that processor (P1). In a similar fashion, the operation has been divided into two; the first line calculates the propagation path and the second the result for that processor.

A more verbose explanation of the first example to delete all edge connected objects is given. The subscript to the *propagation_in$_{dir}$* indicates the direction of the propagation. If the processor is a one and receives a propagation signal (starting from the edge of the array) then that processor propagates a one else a zero (line 1). The result is that only the objects connected to the edge receive a propagation signal; no propagation occurs through the background. The edge-connected objects are deleted by noting and removing only those pixels which received a propagation signal (line 2). For the first three examples, the propagation is initiated from the edge of the array.

(i) Delete edge connected objects

$$b_{ij} \wedge propagation\_in_{1-8} \;\Rightarrow\; propagation\_out$$
$$b_{ij} \wedge \neg propagation\_in_{1-8} \;\Rightarrow\; b_{ij}$$

(ii) Fill holes in 8-connected objects

$$\neg b_{ij} \wedge propagation\_in_{2468} \;\Rightarrow\; propagation\_out$$
$$b_{ij} \vee \neg propagation\_in_{2468} \;\Rightarrow\; b_{ij}$$

(iii) Detect holes in 8-connected objects

$$\neg b_{ij} \wedge propagation\_in_{2468} \;\Rightarrow\; propagation\_out$$
$$\neg b_{ij} \wedge \neg propagation\_in_{2468} \;\Rightarrow\; b_{ij}$$

(iv) Detect the first point in the image, this requires two global propagation operations

$$b_{ij} \vee propagation\_in_{128} \;\Rightarrow\; propagation\_out$$
$$b_{ij} \wedge propagation\_in_{128} \;\Rightarrow\; b_{ij} \qquad (3.2)$$

$$b_{ij} \bigvee propagation\_in_{234} \Rightarrow propagation\_out$$
$$b_{ij} \bigwedge propagation\_in_{234} \Rightarrow b_{ij} \qquad (3.3)$$

By choosing different direction sets, other points such as the bottom-most left can be detected instead.

(v) Labelling objects - by this we mean, given an object in one field, we extract an object in another field topologically connected to it. If the label seed image is **a** and the image containing the object to be labelled is **b**, then the function is given by

$$b_{ij} \bigwedge \left(a_{ij} \bigvee propagation\_in_{1-8} \right) \Rightarrow propagation\_out$$
$$b_{ij} \bigwedge propagation\_in_{1-8} \Rightarrow b_{ij}$$

(vi) Finally an algorithm is given for generating bit-planes in a ramp image (note that the ramp is 16×4).

```
0 1 2 3 4 5 6 7 8 9 10 11 12 13 14 15
0 1 2 3 4 5 6 7 8 9 10 11 12 13 14 15
0 1 2 3 4 5 6 7 8 9 10 11 12 13 14 15
0 1 2 3 4 5 6 7 8 9 10 11 12 13 14 15
```

The least significant bit-plane of which is

```
0 1 0 1 0 1 0 1 0 1 0 1 0 1 0 1
0 1 0 1 0 1 0 1 0 1 0 1 0 1 0 1
0 1 0 1 0 1 0 1 0 1 0 1 0 1 0 1
0 1 0 1 0 1 0 1 0 1 0 1 0 1 0 1
```

This is produced by the following global propagation operation

$$\neg propagation\_in_8 \Rightarrow propagation\_out$$
$$propagation\_in_8 \Rightarrow least\_significant\_bit$$

The successive bit-planes (i.e. $1 \rightarrow 3$) are obtained by noting where the $1 \rightarrow 0$ transition occurs in the $\rightarrow$ direction in the previous bit-plane and then sending propagation left to right, "flipping" at the non-zero points. The $1 \rightarrow 0$ transition is determined using a near-neighbour operation

$$b_{ij} \Leftarrow \neg b_{ij} \bigwedge \bigcup_8 b_{ij}$$

The result for the bit-plane 1 is shown below.

```
0 0 1 0 1 0 1 0 1 0 1 0 1 0 1 0
0 0 1 0 1 0 1 0 1 0 1 0 1 0 1 0
0 0 1 0 1 0 1 0 1 0 1 0 1 0 1 0
0 0 1 0 1 0 1 0 1 0 1 0 1 0 1 0
```

The flipping of the non-zero points involves the following global propagation operations

$$b_{ij}@propagation\_in_8 \Rightarrow propagation\_out$$
$$b_{ij}@propagation\_in_8 \Rightarrow b_{ij}$$

The result of applying this operation to the previous result is given below.

$$0 \ 0 \ 1 \ 1 \ 0 \ 0 \ 1 \ 1 \ 0 \ 0 \ 1 \ 1 \ 0 \ 0 \ 1 \ 1$$
$$0 \ 0 \ 1 \ 1 \ 0 \ 0 \ 1 \ 1 \ 0 \ 0 \ 1 \ 1 \ 0 \ 0 \ 1 \ 1$$
$$0 \ 0 \ 1 \ 1 \ 0 \ 0 \ 1 \ 1 \ 0 \ 0 \ 1 \ 1 \ 0 \ 0 \ 1 \ 1$$
$$0 \ 0 \ 1 \ 1 \ 0 \ 0 \ 1 \ 1 \ 0 \ 0 \ 1 \ 1 \ 0 \ 0 \ 1 \ 1$$

3.5   Parallel Languages

Designing an image-processing language is no different from designing any other language. One must consider: (1) data structures, types of data and especially the handling of vectors and matrices; their description and representation. (2) Data operation, the various manipulations of the data, especially of matrices. (3) Program structure, control-flow and syntax.

A computer language would take one of the four forms: (1) It could be a distributed language with parts of the program running on separate autonomous processors and communicating with other processes by message passing. (2) Similar to (1), but processes *communicate* with each other by sharing resources such as memory. (3) A language may be functional in that its structure is declarative rather than imperative. (4) In logic programming, the structure of the language is represented by relationship between *statements* and not by functions. For a language to be *parallel*, it must support some concept of parallelism and communication. Parallelism could be at the level of processes (Ada), objects (Emerald), statements (Occam), expressions (functional languages), or clauses (logic language – PARLOG, Concurrent PROLOG). Either the communication is *point-to-point*, such as in Occam, or it may be broadcast.

Of the vast number of programs in existence very few are written in a parallel language. Even on supercomputers people are running non-parallel code and using these machines as very fast serial computers. There continues to be an increase in the processing power of conventional serial processors and improvements in compiler technology which act as a disincentive to port these codes to "novel" parallel architectures. It would appear that, the only method of having these programs running on parallel hardware would be to provide automatic translation of codes from conventional languages to parallel languages. Efforts are continuing at translating serial codes into parallel ones by *recognising* the granularity of the algorithm [McCreary 89].

At its simplest, the assignment operations which are independent of each other can be evaluated in parallel. It is also possible that many conditional operations can also be executed in parallel. However, there are problems with history sensitivity (see [Backus 78]). It is possible also to parallelise loop operations. Loop operations can be history-sensitive, where the data produced in one execution of the loop is used in a later execution of the same loop; this type of loop operation is said to be *cyclic*. If the loop operation is not history-sensitive (such as assigning an array to a constant value), then the loop operation is said to be *acyclic*. Both of these types of loop operations can be (to a large extent) regarded as being vector operations and can be executed in parallel; either as point-wise or as neighbourhood operations. Indeed, much of this book discusses these types of operations and their implementation on various types of array processors. However, certain loop operations cannot be executed in an SIMD mode (such as the Gaussian elimination method; see earlier in this chapter).

Translation of algorithms to data-parallel processing on array processors is often difficult because they are not suited for random access types of operations. Solutions often need to be reformulated to make use of synchronous operations of SIMD processing. In a MIMD multi-computer, each processor can execute very different tasks. However, they are very often used for running the same code on each of the processors but with different data (SCMD – single code, multiple data mode). Although there is no explicit synchronisation, as in SIMD mode, synchronisation occurs through communication and the need to share data by message passing.

Computational models have been characterised into three groups by Backus [Backus 78] based on the following criteria: program foundation, history sensitivity, semantics and program structure.

Simple operational model: They have a simple and concise mathematical foundation (such as for the Turing machine). Since the model only allows simple state transitions, the program structure is poorly defined. Also, programs are history-sensitive because of the method used for information storage.

Applicative model: They have a concise foundation and use reduction semantics for program structure. They have no concept of stage, hence are not history-sensitive. An example of this model is Lisp.

von Neumann model: These are in most common usage and form the basis of all Algol-based languages such as FORTRAN, Pascal, Modula-2 and C. They have program foundation and are history-sensitive. The program structure is variable and the state transitions are complex.

A program can be represented by a control graph consisting of *nodes* and *arcs*. The nodes represent operations and data and the arcs represent the data flow (or control flow). For a multiprocessor environment, Gajski and Peir define two models of computation based on execution sequencing [Gajski 85]. In a sequential computational model, only one node is executed at a time and the current node execution starts only after its predecessor has finished. A parallel model of computation is better suited on a multiprocessor architecture. However, parallel computation must accommodate the program being partitioned into tasks (each represented by a node) which may be distributed on different processors; this distribution of tasks will require the task to be scheduled as processors become available and also needs synchronisation between tasks so that correct data is operated upon.

Scheduling is commonly used by operating systems involving multi-tasking or multi-user systems. Similarly, in a parallel machine, where it is not economical for there to be one processor per task and each task is of different size, scheduling is necessary to assign a task to one or more processors. The scheduling of tasks can either be static or dynamic. For the static case, the algorithm designer allocates the required processors for the task. The communication needs to be set up to synchronise the execution of tasks and passing of data between them. In the CSP model [Hoare 85], the synchronisation is achieved through message passing. However, in general, it is not possible to know the "run-time" of a task; this is often highly dependent on the data. In dynamic scheduling, the run-time utilisation of processors leads to task scheduling. This involves a master processor and a number of worker processors. The master processor determines which worker processors are idle and then partitions the task amongst these processors. However, this involves some further scheduling"overheads". For example, in a computer vision problem, each processor may have the complete program but is supplied with only a portion of the data. Initially, each processor may be given an equal portion of the image slices. By monitoring the time taken by each processor to complete its task it is possible to determine which were the maximally and minimally utilised. Assuming that the image content does not change significantly in successive frames, the processor which was most heavily-utilised may be given a smaller amount of image, and those under-utilised given more of the image. In this way better over-all utilisation of the processors is possible.

What language are we to use? This would depend on what language is suitable for the type of knowledge or data representation we require and on the language being reasonably efficient efficient on the hardware that we plan to use.

The control structure of an SIMD array processor is *serial*. However, we do need a special control structure, the *parallel-if* operator. This would allow different operations to be carried out in different parts of the array. With this provision, the overall control is as for a serial computer. While the use of the parallel-if operator is semantically clear, it is not usually necessary because this construct can be carried out by use of masking operations as shown in the example below:

<u>where</u> (image$_1$ < image$_2$) <u>do</u>
      image$_3$ = image$_1$ + *thr*;
      <u>elsewhere</u>
            image$_3$ = image$_2$; <u>od</u>

This operation can be implemented using the following masking operations:

mask = *compare*(image$_2$, image$_1$, >);

image$_3$ = (**image$_1$** + *thr*)&**mask**;
image$_3$ = **image$_3$** + (**image$_2$**&¬**mask**);

The major difference is that an SIMD processor can act upon multiple objects in a uniform fashion in parallel. The consequence of this is that we no longer have to use RasterOps; this fact leads to the development and use of algorithms for image processing diffferent to that one would normally use in an SISD computer (see §3.2.3).

However, we need many more new data types on an SIMD architecture as well as the various scalar and array types supported in a conventional serial language (such as C, Pascal, and FORTRAN). There are particular needs for matrices (2D array) and vectors (1D array) of various data types which can be treated as a single object. A language to be extensively used for image processing must provide means of handling the *object IMAGE* and also data types for convolution masks, binary masks and direction lists. An example of this is CLIP4, the basic memory consists of the a D-level. A D-level can be used to represent a binary image or ninety-six 96-bit element array, and a stack of D-levels can be used to represent a grey-image.

The language chosen to program an SIMD array processor must not hinder the speed of processing. This usually means that the program must be compiled and not interpreted. However, interaction is necessary in debugging programs. The language must come with a debugging tool on the host. Further tools must then be provided for examining the array memory. However, with SIMD programming we will not suffer from communication or memory deadlocks. The language must make efficient use of memory. SIMD processing is at an iconic level and not symbolic; therefore an AI language such as LISP is not really suitable. With all this in mind the IPC (Image Processing C) [Reynolds 82b, Reynolds 82a] was developed for high-level programming of CLIP based on the C programming language [Kernighan 78]. Some other languages for programming SIMD arrays are mentioned below.

C* [Rose 87] used for programing the Connection Machine is also an extension to C. In C*, there are primarily two data types: scalar and vector, referred to as mono and poly respectively. C* also support aspects of object-orientated programming such as abstract data types, and a parallel class type called a domain. A domain is mapped to individual processing elements and can be acted upon in parallel.

NON-VON [Ibrahim 87] is programmed in a high-level Pascal-based language called N-PASCAL [Bacon 82]. This language has a new data-type *vector* which allows parallelism to be expressed at the data-level, with a data element on an individual SPE. The data elements in the SPEs are operated upon an SIMD mode. Parallel conditional statements are executed through the use of the **where** statement.

where <conditional expression> **do** <statement>
    [**elsewhere** <statement>]

Parallel Pascal [Reeves 84] was designed for the Massively Parallel Processor (MPP) [Batcher 80]. It is an extension to Pascal [Jensen 74], providing such keywords as **parallel** to specify the array memory. As in Pascal, the arrays must also be conformable in Parallel Pascal. Also, Parallel Pascal also allows subrange indexing. An example of this is:

$$a[5,] := b[3,];$$

This assigns the value in the 5th row of b to 3rd row of a. In IPC, this statement has to be fully written out as follows: **mask** is a vertical *line* marking the 3rd row.

tmp$ = ands(b, mask);
a$shiftr$(tmp, $dirc$[4], 2);
a$shiftr$(mask, $dirc$[4], 2);
a$ = ad$d$tmp, ands(a, $nots$(mask)));

The danger of using statements such as $a[5,] := b[3,]$ in a high-level language is that the underlying parallelism gets deeply hidden from the programmer. It could encourage the use of random-access of memory in programs (such as the use of chaincodes) which is grossly inefficient of SIMD array processors. It is important that the programmer is not completely sanitized from the parallel architecture by the language; rather the language must reflect the underlying architecture and encourage the programmer to use parallel operations. For example, in CLIP4 individual pixels cannot be accessed without special efforts which cannot be said of Parallel Pascal.

In the past decade, a very large number of languages have been developed for distributed programming [Bal 89]. For multicomputers, we could have sequential processes running on each of the processors and the processes connected by message passing; this is the basis for CSP [Hoare 78, Hoare 85] and Occam [May 83, Inmos 84]. Communicating Sequential Processes (CSP) was designed by Hoare [Hoare 78] to be a simple language allowing implementation on a variety of architectures. In the CSP model, a fixed number of sequential processes are created, these can communicate with each other through synchronous message passing.

However, it is believed by many that imperative (algorithmic) languages do not express parallelism more explicitly at a finer level and are inherently sequential. Hence, they suffer from the von Neumann bottleneck [Backus 78]. Functional languages are not history-sensitive and therefore can be evaluated in any order, hence more in parallel. A logic program can be both declarative and procedural. Examples of parallel logic languages are Concurrent PROLOG [Shapiro 87] and PARLOG [Clark 86]. The parallelism in logic programming arises from the clauses and sub-clauses being evaluated in parallel. This is called *Or/AND*-parallelism. An example of this is given below:

grandfather(X,Y) :- father(X,A) , father(A,Y)
grandfather(X,Y) :- father(X,A) , mother(A,Y)

Linda

Linda is a programming paradigm based on shared memory called a *tuple space* [Ahuja 86]. For its efficient execution, Linda requires distributed data structures. Linda has been implemented on a variety of computers including Ethernet networked Micro-Vaxes and AT&T S/Net multi-computers.

Linda provides parallelism through shared memory using a *tuple space* (TS). Conventionally, memory is accessed using addresses; however in Linda, they are accessed by *logical names*. In Linda, memory can be accessed in three ways **read**, **add**, and **remove**. For memory content to be changed, it must first be removed from the TS, updated and then reinstated in to the TS. Using this model, it is possible for many processes to access Linda's memory concurrently. There is no point-to-point message passing between processes as in CSP. All interaction and communication between processes occurs through the TS; one process writes to it and another reads it. There are four functions associated with the TS: **out()**, **in()**, **read()**, and **eval**. The function **out(a)** adds a tuple *a* to the TS and the process continues executing. The function **in(b)** withdraws a tuple and executes it. If no matching tuple *b* is found, then the process suspends until a match is found. If more than one match exists, then one is chosen arbitrarily. The function **read(c)** is similar to **in(b)** except only a copy of the matching tuple is taken. The function **eval(d)** is similar to **out(a)** except that it adds an unevaluated tuple to the TS.

While in conventional multiprocessing the program is partitioned into parallel tasks which are executed on *n* different processors, in the Linda paradigm, *n* copies of the program are executed on the *n* worker processors. There need not be any partitioning of the program. Each of the worker processors seeks work to be done from the TS. This paradigm has a number of useful properties: (1) A program can be developed on a single processor and then when debugged executed on *n* processors. In developing the parallel program we need only be aware of the size of the task which determines the granularity of the parallelism. (2) The load balancing is automatic. Each worker process seeks work, executes it, and searches for another task.

3.6   Concluding Remarks

To make effective use of multicomputers, algorithms have to be partitioned into subtasks and mapped onto the available processors and set up the network for the communication between the processors for a given network topology. However, very few tools exist which can help the programmer to carry out these tasks automatically. The efficiency of the programs created is largely the result of skill and the craft of the programmer and the task is very difficult and the methods used involve considerable experimentation. An algorithm written for a given machine in a given language may need to be reimplemented if the processor network topology is changed such as adding more processors.

It should be noted that many parallel algorithms actually run faster than conventional programs even on serial computers. Many believe that one should start with the best available algorithms (whether parallel or serial) and then think of ways of making them run more efficiently on the available architecture. We believe that one needs to go one step further back from this. One needs to start with the problem and analyse it to discover what is required for a solution. Only then should we think of algorithms to solve the problem in hand.

Writing data-parallel programs on an SIMD array processor is generally simple (although devising an algorithm often requires lateral thinking) because of the sequential control involved. It is generally more difficult to write a program for an MIMD processor because one should be aware of the need to partition both code and data across the processors and to set up communication channels between the processors. This leads to a complex control structure. Debugging of programs is also difficult because one has to keep a tag on communication.

Chapter 4

Low-Level Image Processing

4.1 Introduction

Computer vision is a process of extracting, characterising, and interpreting information from real world images. This process may be subdivided into: (a) sensing, (b) preprocessing, (c) segmentation, (d) description, (e) recognition, and (f) interpretation. Sensing is the process of acquiring images. Preprocessing includes the topics of image enhancement and noise suppression. The segmentation process partitions the image into a useful set of objects and the background. These techniques usually work by extracting certain features from an image by emphasising the desired properties and suppressing the unfavourable ones and by balancing one desired property against another until favourable results are obtained. Description labels the objects with information such as their texture, size or shape. Recognition then identifies the labelled objects. Finally, interpretation puts the recognised objects into real-world context.

This is a general introduction to the following three chapters.

Computer vision is usually subdivided into three levels: low-level, medium-level, and high-level. This classification has little to do with the way the human vision system subdivides its vision tasks; but it is convenient when describing various algorithms and the types of processing and representation involved. Low-level vision involves sensing and pre-processing of image data; medium-level involves segmentation and description of the image data. This level of processing may involve some iconic–symbolic transformations. High-level vision consists of recognition of objects and interpretation of the scene. While object recognition could involve both iconic and symbolic processing, scene interpretation is very much the domain of symbolic processing. This chapter primarily deals with preprocessing of digital images. Sensing of images is beyond the scope of this book; an introduction can be found in [Fu 87]. The next two chapters deal primarily with image segmentation and description.

This chapter looks primarily at image quality, image enhancement, filtering and edge detection. We discuss image enhancement through histogram modification and transformation. Image filtering to remove noise or improve image fidelity can involve the use of both linear and non-linear filters. Linear filtering involves convolving the image with a template such as local averaging or convolving it with a Gaussian mask. Linear filters, while they remove noise, they also blur the edge. for this reason, non-linear filters such as morphological or rank filters are more commonly used. In §4.4, we give an introduction to mathematical morphology. This is extended to a description of grey-scale mathematical morphology functions and rank filters. This section concludes with their use in spatial low-, high-, and bandpass filters. In the section following, we describe the parallel implementation of the Fourier transform which is used both for filtering and feature detection. In the final section of this chapter we describe various edge detectors and their parallel implementation.

4.2   Image Quality

The measures such as point-wise standard deviation (SD) and signal-to-noise ratio (SNR) give a quantitative result of noise in the imaging system. A measurement of the width of edges in an image gives a quantitative value of the separation distance of two identical objects before they can be identified with any certainty.

There are various measures of noise, which may be classified as being global or local. Global measures involve the entire image, for instance, the root-mean-square deviation between the image and the actual object or between the image and its ensemble average. Local measures are, for example, local signal-to-noise ratio (SNR).

One can define the signal as being the image that would be obtainable in the absence of noise. Computationally, the signal may then be estimated by taking an ensemble average of the noisy image. The noise can then be defined as being the root-mean-square deviation of the noisy image from the mean or signal image [Pratt 78].

The signal-to-noise ratio of a series of images using the RCA 1" vidicon camera and CLIP4S was measured using the profile method (described later in this chapter). The result is shown in Figure 4.1. At low grey-values (region 1), the noise tends to dominate the signal. At high grey-values (region 3), there is signal saturation. The region found for the signal to be free from significant noise and signal saturation is approximately the grey-value range of 60–230 for an 8-bit input image.

4.2.1   Point-wise mean and standard deviation

A quantitative measure of the temporal noise (and the temporal noise is not independent of small area spatial noise, so the measure being discussed is a combination of the two) can be made by measuring the point-wise standard deviation (SD) of a test image. The test image being used is shown in Figure 4.2.

With the camera automatic gain control (agc) on, the measure (SD/mean) is about 4%. The SD is largest, as would be expected, at the edges and the brighter part of the image. The reasons for the larger SD in the brighter

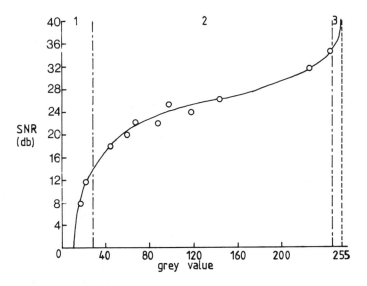

Figure 4.1: The signal-to-noise ratio plotted against grey-value for images input using a Vidicon camera.

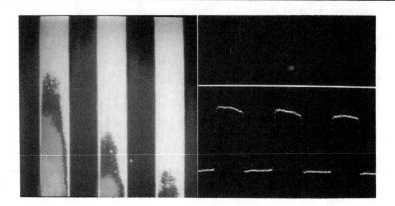

Figure 4.2: The Ranche grating pattern and the edge profile.

regions of an image are twofold. Firstly, the action of the agc is to lower the
camera sensitivity to bright light. Secondly, even with the agc on, the camera
tube suffers from severe burns in the bright regions, and in these regions the
sensitivity of the camera falls and measures a lower grey-value than previously.
This second problem has been found to be more severe than the first. This
problem can be graphically illustrated, if instead of inputting all the images
before conducting any calculations, some intermediate calculations are carried
out such as adding successive images into an accumulator and squaring the
images and adding into another accumulator, then in this second method, the
camera tube is exposed to light for about 3-4 times longer than in the first
method, and the (SD/mean) measure is about 17%.

A conclusion which may be drawn is, the image quality can be improved
by a factor of $\sqrt{n}$ if n images of the same field are averaged. But n needs to
be small, $n = 8$ being most suitable, to prevent tube burn by long exposure to
bright light.

4.3   Enhancement and Filtering

Image filtering techniques can be subdivided into two main categories: ones
which act on the whole or large-section of the image and the others involving
small neighbourhood windows.

Global techniques include Wiener or least-squares filtering [Andrewds 77,
Rosenfeld 82] and Kalman filters. Wiener filter techniques require a statistical
model of the signal (i.e. uncorrupted image) and the noise. However, statistical
models for most images are unknown or too complex to be described by simple random processes. The filtered images produced by these techniques have
blurred edges and loose details. Therefore, the results are often unsatisfactory
and computationally very expensive.

Local methods are generally computationally more efficient. However, their greater advantage comes from their ability to process several windows in parallel. In this section, we will explore the local filter techniques.

4.3.1 Histogram Transforms

A histogram of an image is used to display the distribution of grey-values in the image. An 8-bit image has 256 grey-levels ranging from absolute black to absolute white. Another important method of measuring image quality is to generate the profile of the image. If an image is regarded as a terrain in Cartesian 3-space, then the z-axis represents its grey-level at a given (x,y) position in the array. The profile of the image is a cross-section through the image, plotted for grey-level (z) against x-coordinate for a given y-value (or z versus y for given x). The profile of the image can be used to determine the width of the edges or the signal to noise ratio.

There are a very large number of applications for histograms, such as in segmentation [Weszka 78], enhancement [Gonzalez 84], filtering [Ballard 82], and image encoding [Pavlidis 82] amongst many others.

Enhancement is a method which attempts to improve the appearance of an image, and can provide additional visual information [Andrews 74]. This enhancement can be achieved by various means; amongst the most common are (1) intensity mapping, (2) edge modelling, (3) edge sharpening and (4) the use of pseudocolor.

The intensity mapping is a non-linear operation, it involves pointwise mapping of intensity value to another globally.

$$\hat{g}(x,y) = f(g(x,y))$$

such a mapping function $f(\cdot)$ can correct for film and display nonlinearities.

Edge modelling involves simulating the mechanism of the psychophysics of the human perception system [Stockman 72]. It attempts to precompensate for the visual system for Mach banding, intensity response, spectral (chromatic) response and temporal response. For example, the eye differentiates at lower spectral frequencies and emphasises the higher spectral frequencies [Cornsweet 70]. Edge sharpening involves subtracting low-frequency signals, this gives an enhanced image. This property is useful because humans have a desire to see sharp edges.

Often the foreground and background intensities are quite close and it is difficult to differentiate them, hence there is a need to separate them spectrally. At other times, details are lost because a large percentage of pixels have a similar value. By rescaling these values, the dynamic range can be increased. The method for histogram rescaling with different functions is discussed in [Billingley 70, Prewitt 70, Hall 74, Troxel 78]. Some of the common functions used are given in Figure 4.3

Algorithm 4.1 give the basic structure for histogram generation. The method we employ for (1) distributing the task, (2) comparing pixel value, and (3) counting them will be determined by the architecture of the computer. In

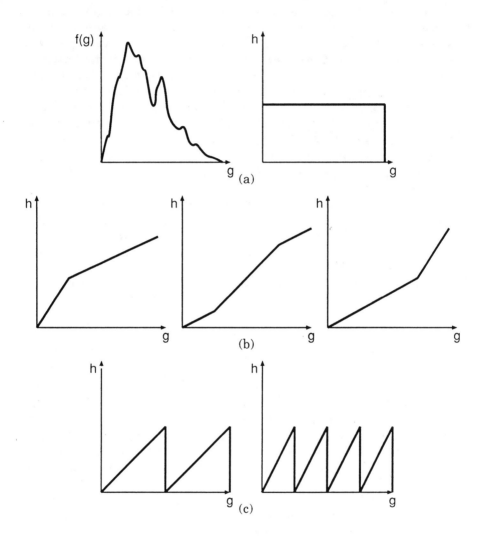

Figure 4.3: Common functions used in histogram transforms: (a) Histogram equalisation, (b) grey-level stretching, and (c) bit-level truncation.

Algorithm 4.1: Histogram generation.

<u>for</u> $i = 0$ <u>to</u> $2^p - 1$ <u>do</u>   // for each possible grey-level
    find all pixels in the input image equal to i
    count these pixels
    put result in i^{th} bin of histogram <u>od</u>

---

many SIMD array computers such as CLIP4, DAP and MPP, there is special hardware for counting. Also the processing elements are bit-serial. In CLIP4, the COUNT instruction shifts data to the left edge of the array where the counting occurs [Otto 81]. Also in CLIP4, the most efficient method for searching all the grey-levels present in an image is to use a tree search and "divide and conquer" strategy [Otto 84]. The time taken to histogram a 6-bit image on the prototype CLIP4A was 11 ms [Otto 84]; on the new CLIP4, this timing have improved by a factor 2.5.

On NON-VON [Ibrahim 87], which is a binary tree structure SIMD computer, the histogramming is carried out in a slightly different manner. For every grey-level in the image, it needs to execute the following steps: (1) Mark all pixels with a label 1 if it is in the bin-range, otherwise mark it with a 0. This is computed in parallel in the leaf of all the PEs. (2) Count the number of 1s using the tree connectivity of the architecture. This needs $O(\log n)$ operations, where n is the number of PEs. The searching and counting tasks can be pipelined giving greater efficiency. The complete task takes $O(b + \log n)$ operations. For an 128×128 image and 256 bins, histogram computation on NON-VON takes 2.16 ms.

Histogram equalisation involves mapping the original grey-levels to new values in a "monotonic increasing" fashion (to maintain light–dark relationship) such that in the new histogram, each bin contain approximately the same number of pixels, (see Figure 4.4).

To enhance detailed information in a narrow range, the use of histogram techniques has been suggested [Hummel 75, Hummel 77, Frei 77]. The techniques expand the dynamic range of the narrow peak areas. To equalise the histogram, the observed cumulative distribution function needs to be a linear one. The mapping function for this is [Pratt 78]

$$\hat{g}(x,y) = [g_{max} - g_{min}] P(g) + g_{min}$$

the human visual system also carries out non-linear transformations [Frei 77, Stockman 72], the mapping function is histogram hyperbolization [Frei 77, Pratt 78, Wang 83]. A parallel algorithm for histogram equalization (adapted from [Otto 84]) is given in Algorithm 4.2.

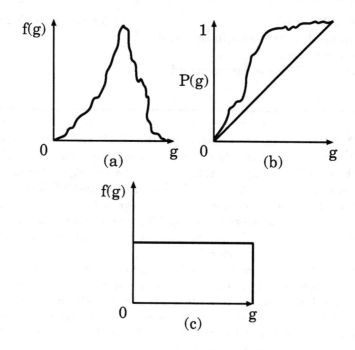

Figure 4.4: Histogram equalisation.

Algorithm 4.2: Histogram equalisation.

begin
   result $\leftarrow$ 0;
  for each grey-level **do**
      Find points at level i in the input image;
      Apply remapping function to find new value.
      Set the points to the new value in the **result** image **od**
  where

$$\text{new value} = \frac{\#(points<i)+\frac{1}{2}\#(points=i)}{\text{size of image}} \cdot 2^p$$

end

6-bit / 8-bit image

The precision of an image is important in determining the speed of an operation if carried out on a bit-serial computer. If an operation has complexity $O(p^2)$, where p is the bit precision, such that an computation on an 8-bit image takes 4 times longer than on a 6-bit image, should one still use an 8-bit image?

The answer is very much dependent on the application. For example, it may be reasonable to truncate an 8-bit input image to a 6-bit image for subsequent processing if the segmentation process involves extracting very strong edges which lie between the object and the background. But truncating to 6-bit is not suitable in segmenting a ridge-like feature. Any truncation has the effect of flattening out the ridge and induces problems in determining local maxima.

The time taken by an 8-bit operation is often prohibitive, and it is important to try to work with either a 6-bit or a 7-bit image. However, it is often not possible to truncate an 8-bit image to either a 6-bit or a 7-bit image whilst maintaining sufficient information for subsequent segmentation of fine features. But more often than not, one is interested in segmenting a single class of features in an image, and the grey-range of these objects is likely to be only a small part of the 256 grey-levels available. The objects are likely to be surrounded by a pedestal of background and other features could be brighter. The grey-range of the object can be determined by observing the profile across the object. The background level (z_1) can then be set to zero, and the maximum grey-value of the object (z_2) is set to either 63 or 127 as required. either a 6-bit or a 7-bit image); and the range between z_2 and z_1 is linearly remapped onto either 64 or 128 values (Algorithm 4.3).

The grey-scale transformation that is commonly used is

$$
g' = \begin{cases} \frac{(g-z_1)\cdot(2^p-1)}{z_2-z_1} & z_1 < g < z_2 \\ 2^p - 1 & g \geq z_2 \\ 0 & g \leq z_1 \end{cases} \tag{4.1}
$$

where g is the original grey-value, g' is the transformed grey-value and the precision of the image is p bits.

The routine GAIN_CONTROL requires four parameters: an 8-bit input image to be transformed, **im**, the direction of the profile, sdir, the profile sampling, stype, and the precision of the output image, p. Three types of profile sampling are catered for: (i) choosing a region in the image and generating five lines at an interval of 10 pixels, (ii) generating 16 lines at an interval of 32 pixels to cover the whole array and (iii) generating 16 lines at a random interval.

The LINE_CURSOR routine generates a vertical or a horizontal cursor line which can be moved about the array and once in position (determined by the operator), returns the position of that vertical or horizontal strip in the array.

Depending on whether $((z_2 - z_1) < (2^p - 1))$ is true, two options are available; if true, the images **res** and **mask** are added together and their histogram equalised, otherwise a grey-level transformation based on equation (4.1) is computed (step (12)). The routine SCALE(**im**,p,q), scales the image by multiplying it by p and dividing by q, both p and q are integers.

Algorithm 4.3: An algorithm for grey-scale transformation.

GAIN_CONTROL(im, *sdir*, *stype*, *p*)
begin
  length ← $2^p - 1$;
  switch (*sdir*)
       then
             case 1 : **then** *dirn0* ← 8; *dirn1* ← *H*; **esac**
             case 2 : **then** *dirn0* ← 6; *dirn1* ← *V*; **esac** **break**
  switch (*stype*)
       then
             case 0 : **then**
                  stval ← *LINE_CURSOR*(*dirn0*);
                  stline ← 0;
                  rxy ← *RAMP*(*dirc*[*dirn0*], 9);
                  for *i* := −20 **to** 20 **step** 10 **do**
                      tmp ← *EQUAL*(rxy, *stval* + *i*);
                      stline ← *OR*(stline, tmp);
                  od esac
             case 1 : **then**
                  rxy ← *RAMP*(*dirc*[*dirn0*], 5);
                  stline ← *EQUAL*(rxy, 31); **esac**
             case 2 : **then**
                  stline ← 0;
                  rxy ← *RAMP*(*dirc*[*dirn0*], 9);
                  for *i* := 1 **to** 16 **do**
                      stval ← rand() **mod** 512;
                      tmp ← *EQUAL*(rxy, *stval* + *i*);
                      stline ← *OR*(stline, tmp);
                  od esac
  break
  prof ← *PROFILE*(im, stline, *dirn1*);
  small ← *LINE_CURSOR*(*dirn0*);
  big ← *LINE_CURSOR*(*dirn0*);

Algorithm 4.4: continued...

mask ← *plane* ← *THR*(**im**, *big*);
plane ← *OR*(**plane**, *NOT*(*THR*(**im**, *small*)));
res ← *AND*(*SUB*(**im**, *small*), *NOT*(**plane**));
mask ← *MULT*(**mask**, *length*);
<u>if</u> ((*big* − *small*) < *length*) <u>then</u>
                        res ← *ADD*(**res**, **mask**);
                        res ← *HISTEQ*(**res**, *p*); <u>fi</u>
<u>else</u>
     res ← *SCALE*(**res**, *length*, *big* − *small*);
     res ← *ADD*(**res**, **mask**); <u>elsf</u>

return(*res*);
<u>where</u>
        <u>proc</u> *PROFILE*(**im**, **line**, *dir1*) ≡
            <u>if</u> (*dir1* = 8) <u>then</u> *dir2* ← *EW*;
            <u>elsf</u> (*dir1* = 6) <u>then</u> *dir2* ← *NS*; <u>fi</u>
            r ← *RAMP*(*dirc*[*dir1*], *log*(*Arraysize*));
            tmp ← *SPREADS*(*ANDS*(**im**, **line**), *dir2*);
            prof ← *EQUALtmp*, **r**);
            return(**prof**); .
<u>end</u>

4.3.2   Linear Filters

Image enhancement involves noise removal to deblur the edges of objects and to highlight specified features. Many of the image-enhancement techniques currently being employed are heuristic and problem orientated. Consequently, the suitability of a particular technique applied to a particular image can only be subjectively judged [Andrewds 77].

If the image is corrupted by random impulse noise, then averaging over n image frames will reduce the noise by a factor of $\sqrt{n}$. However, this may not be possible or suitable. In such circumstances, linear or non-linear local window operations may be employed. The simplest of the linear operations is equal-weighted averaging [Prewitt 70, Rosenfeld 82]. While this method is efficient in removing noise it will blur edges and blurring is more severe for larger windows. The blurring effect can be reduced slightly by using a weighted averaging technique.

$$g'(x,y) = \sum_{i=-m}^{m} \sum_{j=-n}^{n} w(i,j)g(x-i,y-j)$$

where $w(i,j)$ are normalised weights and are often binomial coefficients, g is the noisy image and g' is the filtered image. Other weighting factors have been explored and these are surveyed in [Wang 83].

A local smoothing operator based on the σ probability of the Gaussian distribution has been proposed by Lee [Lee 83]. Noise in images is generally assumed as spatially uncorrelated, as additive and as having a Gaussian distribution. The principle is to calculate the average of the pixels within the window which are within a fixed sigma range of the centre pixel.

$$\delta_{k,l} = \begin{cases} 1 & if\ (g_{i,j} - t\sigma) \le g_{k,l} \le (g_{i,j} + t\sigma) \\ 0 & otherwise \end{cases} \tag{4.2}$$

Then

$$\hat{g}_{i,j} = \sum_{k=i-n}^{n+1} \sum_{l=j-m}^{m+j} \delta_{k,l} g_{k,l} \Big/ \sum \sum \delta_{k,l}$$

Lee sets $t = 2$. However, this has the problem that it allows through small noise areas. To prevent this problem, it is specified that the number of pixels averaged should be greater than 4 for a 7×7 and greater than 3 for a 5×5 window; otherwise, the average of the neighbourhood is taken. A simpler technique has been used [Graham 66, Prewitt 70], the centre pixel is replaced by the average of the neighbourhood if the absolute difference between the average and the centre pixel is smaller than a threshold.

There are a number of other local filters which are based on similar statistical properties. For example, Nagao suggests subdividing a window into sub-windows. For each sub-window, the variance is computed. The value of the centre pixel is replaced by the average of the pixels with the lowest variance [Nagao 79].

4.4 Mathematical Morphology

Mathematical morphology is an approach to image processing based on shape and geometry. Many 2D SIMD cellular array computers such as CLIP4, DAP and MPP, with their neighbourhood connectivity, are ideally suited to such operations; and others such as Golay [Golay 69], Diff3 [Graham 80], and PICAP [Kruse 77] have special hardware for these purposes.

Mathematical morphology is based upon set theory. The set consists of the grey-values in the image; thus in a binary image, the set of all white pixels gives a complete description of the image. This set in a Euclidian 3-space could represent either a solid binary object, a spatio-temporal binary object, or a grey image (the grey image consisting of stacked binary planes).

Basic mathematical morphology operators are the *dilation* and *erosion* proposed by Minkowski [Minkowski 03] based upon set theoretic operations. If we have two sets $\mathcal{A}$ and $\mathcal{B}$ in N-space with elements **a** and **b** respectively, $\mathbf{a} = (a_1, ..., a_n)$ and $\mathbf{b} = (b_1, ..., b_n)$ being N-tuple of element coordinates, then the *dilation* function combines the sets by vector addition of the set elements; and *erosion* combines the two sets $\mathcal{A}$ and $\mathcal{B}$ by vector subtraction of the set elements. Thus image $\mathcal{A}$ is transformed by the structuring elements $\mathcal{B}$ into a new image; $\mathcal{B}$ is analogous to a convolution kernel. We observe the duality of the dilation and erosion operation. $\mathcal{A}^c$ is the complement of $\mathcal{A}$.

$$\mathcal{A} \oplus \mathcal{B} = \{c \in E^N \mid c = a + b : a \in \mathcal{A}, b \in \mathcal{B}\} = \bigcup_{b \in \mathcal{B}} \mathcal{A}_b$$

$$\mathcal{A} \ominus \mathcal{B} = \{c \in E^N \mid c + b \in \mathcal{A} : \text{for every } b \in \mathcal{B}\} = (\mathcal{A}^c \oplus \mathcal{B})^c = \bigcap_{b \in \mathcal{B}} \mathcal{A}_b$$

We will note here some of the properties of mathematical morphology; details and proofs may be found in [Matheron 75, Serra 82b, Haralick 87]

1. $\mathcal{A} \oplus \mathcal{B} = \mathcal{B} \oplus \mathcal{A}$

2. $\mathcal{A} \oplus (\mathcal{B} \oplus \mathcal{C}) = (\mathcal{A} \oplus \mathcal{B}) \oplus \mathcal{C}$

3. $\mathcal{A} \cup (\mathcal{B} \oplus \mathcal{C}) = (\mathcal{A} \oplus \mathcal{B}) \cup (\mathcal{B} \oplus \mathcal{C})$

Properties 2 and 3 are important because it allows large structuring elements to be decomposed into a number of smaller structuring elements. This would make their implementation on neighbourhood-connected processor arrays such as CLIP4, DAP and MPP very efficient.

The *dilation* operation is often used as a smoothing operator [Kirsh 57, Unger 58, Preston 61, Golay 69]. Two special form of the erosion/dilation duality is the *expand/shrink* duality [Golay 69, Duff 73]. We express these operations as

expand: (at all pixels, simultaneously do) if the centre pixel is 1, then set the nearest neighbouring pixels also to 1.

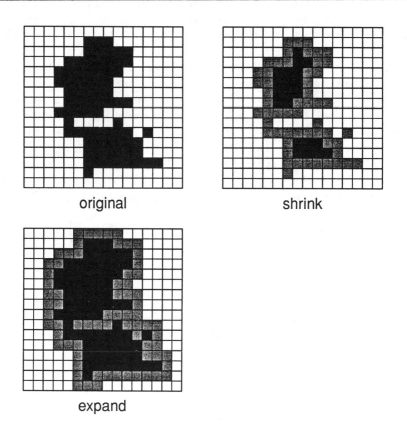

original   shrink

expand

Figure 4.5: Expand and shrink operation. For the shrink operation, the grey-pixels have been removed, and for expand operation, the grey-pixels have been added.

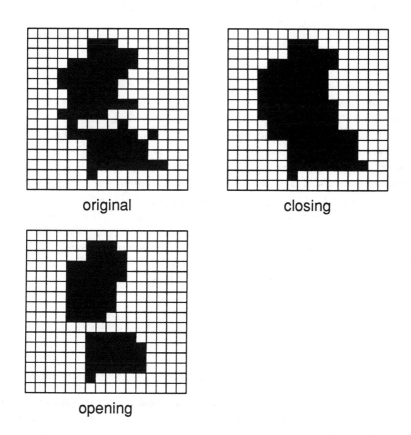

Figure 4.6: Opening and closing operation.

shrink: (at all pixels, simultaneously do) if the centre pixel is 1 but any of its
nearest neighbouring pixels are 0, then replace the centre pixel by a 0.

However, because of the built-in neighbourhood gating in array processors, it
is simpler to compute for an expand operation: *if any neighbourhood pixel is a
1, then become a 1.* this is accomplished by taking the OR of the neighbouring
pixels. For the shrink operation, the AND of the neighbours is taken. If the
neighbours are all 1s and the centre pixel is a 1, then it remains a 1; otherwise
it transforms to a 0. The dilation and erosion operations are usually operated
in pairs. The function $\mathcal{I} \circ \mathcal{A}$ and $\mathcal{I} \bullet \mathcal{A}$ are referred to as the *opening* and *closing*
of I by the structuring element A.

$$\mathcal{I} \circ \mathcal{A} = (\mathcal{I} \ominus \mathcal{A}) \oplus \mathcal{A}$$

$$\mathcal{I} \bullet \mathcal{A} = (\mathcal{I} \oplus \mathcal{A}) \ominus \mathcal{A}$$

These properties of mathematical morphology were first studied by Matheron
[Matheron 65, Matheron 75]. Opening an image $\mathcal{I}$ by the structuring elements
$\mathcal{A}$ has the effect on $\mathcal{I}$ of smoothing the contours of the objects, eliminating
islands, sharp peaks and capes smaller than the structuring elements $\mathcal{A}$ and
breaking narrow isthmuses. Closing an image $\mathcal{I}$ by $\mathcal{A}$ again smooths contours,
joins narrow breaks, fills in coves and eliminates small holes.

A significant property of the iterative application of the dilation and erosion
functions is the idempotent transform; further application of these operators
has no change on the transformed image. These form a complete closed set for
the objects in the image, allowing description of shapes through the opening
or closing operations. Like dilation and erosion operations, the opening and
closing functions also form a dual transform pair.

$$(\mathcal{A} \bullet B)^c = \mathcal{A}^c \circ \breve{B}$$

$$(\mathcal{A}^c \bullet B)^c = \mathcal{A} \circ \breve{B}$$

4.4.1   Grey-Scale Morphology

Concepts of the dilation, erosion, opening and closing operations are extended
for use on grey-scale images through the use of neighbourhood maximum
($\max_n$) and minimum ($\min_n$) functions. The use of ($\max_n$) and ($\min_n$) operations was first suggested by Nakagawa and Rosenfeld [Nakagawa 78]. These
operators are based on fuzzy logic and fuzzy set theory introduced by Zadeh
in 1965 [Zadeh 73, Zadeh 83].

The fuzzy logic equivalence of the $\cup$ and $\cap$ functions are the $\max_p$ and $\min_p$
operations. In image operations, these would be the pointwise functions. Given
two grey images $\mathcal{A}$ and $\mathcal{B}$,

$$\mathcal{A} \cap \mathcal{B} \iff \min_p(\mathcal{A}, \mathcal{B})$$
$$\mathcal{A} \cup \mathcal{B} \iff \max_p(\mathcal{A}, \mathcal{B})$$
$$\neg \mathcal{A} \iff 1 - \mathcal{A}$$

The fuzzy logic equivalence of the erosion and dilation dual is the $\min_n$ and $\max_n$ dual. These two functions compare the values within the given filter window with the value of the central pixel; $\min_n$ replaces the centre pixel by the minimum value in the neighbourhood, and conversely for $\max_n$.

$$\min_n(\mathcal{A}) = 1 - \max_n(1 - \mathcal{A})$$

$$\max_n(\mathcal{A}) = 1 - \min_n(1 - \mathcal{A})$$

The grey-level equivalent of the *opening* and *closing* operations are $\mathcal{O}(\mathcal{A})$ and $\mathcal{C}(\mathcal{A})$ respectively.

$$\mathcal{O}(\mathcal{A}) = \max_n(\min_n(\mathcal{A}))$$

$$\mathcal{C}(\mathcal{A}) = \min_n(\max_n(\mathcal{A}))$$

Some of the properties of $\min_n$ and $\max_n$ filters can be summarised as follows:

$$\min_n(\max_n(\min_n(\mathcal{G}))) = \min_n(\mathcal{G})$$

$$\max_n(\min_n(\max_n(\mathcal{G}))) = \max_n(\mathcal{G})$$

$$\min_n(\max_n(\mathcal{G})) \geq \mathcal{G} \geq \max_n(\min_n(\mathcal{G}))$$

The $\max_n$ and $\min_n$ operations have been used in designing high-, low-, and band-pass filters [Goetcherian 80, Hodgson 85]. Peleg *et al.* have used grey-scale mathematical morphology for medial-axis transform [Peleg 81] and texture analysis [Peleg 84, Werman 84]. Other usage of mathematical morphology we will explore throughout this book.

4.4.2 Rank Filters

As we have seen, for a local linear filter, the output is a linear combination (i.e. weighted average) of the neighbouring pixels. Although, from discussion in the previous section, we have observed that edges are blurred by their application. To improve upon them, trimmed (i.e. values far from the average or centre pixel are rejected from computation) or non-linear combinations of linear filters [Nagao 79, Lee 83] have been suggested.

Salt-and-pepper noise in images which are created by bit-error can be removed by use of median filters [Justusson 81, Ataman 80, Nodes 82, Rosenfeld 82]. In a small window, the pixels are nearly homogeneous; only a small portion of these pixels are noise pixels. These noise pixels tend to have values in the extreme rank positions; hence they are not selected by the median filter. The edge restoration or sharpening can be achieved by iterative application of extremum filters [Kramer 75]. Below, a brief review of these rank filters is given before presenting results of applying rank filters (median and extremum filters) to enhance and restore sharp edges. A detailed review of the properties of rank filters is given in [Hodgson 85]. Both Ip [Ip 84] and Levy-Mandel [Levy-Mandel 86] have used rank filters for noise suppression and to enhance their cephalometric images.

The rank filters are a class of non-linear filters. With a rank filter, for a window about a pixel ij, the N pixels within the neighbourhood are ranked according to value:

$$g_{ij}^1 \leq g_{ij}^2 \leq \cdots \leq g_{ij}^N$$

Then the output value is chosen.

$$f_{ij} \;=\; R_k(g_{ij}) \tag{4.3}$$

where g is the input image, f is the output image and the kth rank value within the window is chosen.

Three special rank filters are the $\min_n$, $\max_n$ and median filters.

$$\min_n(g) \;=\; R_1(g)$$

$$\max_n(g) \;=\; R_N(g)$$

$$median(g) \;=\; R_{\lceil \frac{N}{2} \rceil}(g)$$

The effect of rank filters is to reduce the variance in the image indicated by narrowing of the histogram (Figure 4.7). Note, other than for the median filter, equation (4.3) is only true when $\mathcal{G}$ is a constant.

Repeated application of max or min filter will continue to change the image, until eventually the histogram is a delta function (i.e. the image is single-valued). The speed of convergence to a delta function is increased by increasing the neighbourhood size. Thus edges are propagated by max and min filters. Repeated application of a median filter to an image also leads to a root image, which consists of regions of locally monotonic slopes, edges and certain saddle points. The root properties of the median filter are discussed in [Nodes 82]. The net effect of all three of these rank filters is to reduce the variance in the image (see Figure 4.7).

Well-known properties of local median filters are (1) they preserve sharp edges and (2) they eliminate spike (salt-and-pepper) noise [Scoller 84]. More detailed descriptions of the properties of the median filter can be found in [Huang 82]. Salt-and-pepper noise can also be removed by the application of combination of the $\max_n$ and $\min_n$ filters.

$$\mathcal{G} = \min_n(\max_n(\max_n(\min_n(\mathcal{F}))))$$

These form a family of spatial filters. We can express iterative application of $\max_n$ filter as

$$\max_n^i(\mathcal{F}) = \max_n(\max_n(...\max_n(\mathcal{F})...))$$

and similarly for $\min_n^i(\mathcal{F})$. The application of $\max_n^i(\mathcal{F})$ and $\min_n^i(\mathcal{F})$ also filters spike noise.

$$\mathcal{G} \;=\; \min_n^k(\max_n^k(\mathcal{F})) \tag{4.4}$$

$$\mathcal{G} \;=\; \max_n^k(\min_n^k(\mathcal{F})) \tag{4.5}$$

$$\mathcal{G} \;=\; \max_n^k(\min_n^{2k}(\max_n^k(\mathcal{F}))) \tag{4.6}$$

$$\mathcal{G} \;=\; \min_n^k(\max_n^{2k}(\min_n^k(\mathcal{F}))) \tag{4.7}$$

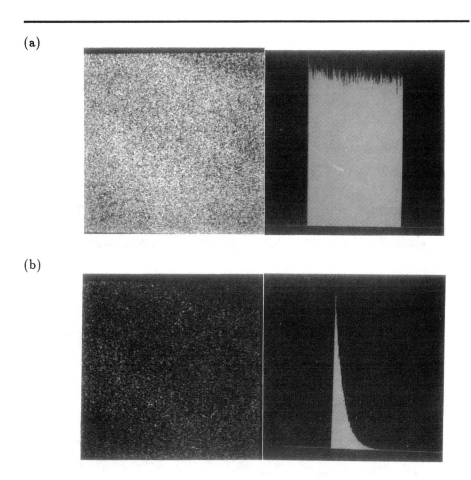

Figure 4.7: Properties of rank filters in reducing the variance in a noise image. (a) The original noise image and its histogram, (b) the image after a local-neighbourhood minimum operation and its histogram.

(c)

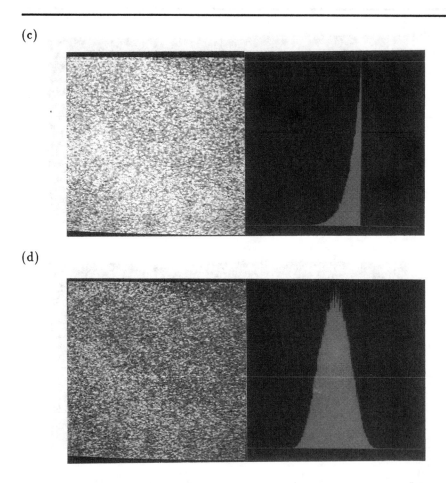

(d)

Figure 4.7: (c) The result of applying a local neighbourhood maximum operation and (d) the result of a 3×3 median filter and its histogram.

(e)

(f)

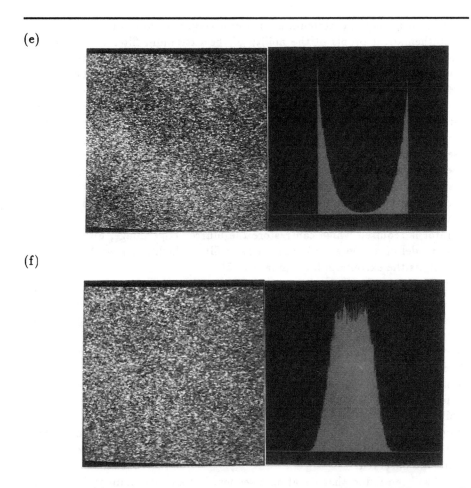

Figure 4.7: (e) The result of applying a 3×3 extremum filter and (f) the result of applying a median and an extremum filter. The narrowing of the histogram indicates the reduction of variance.

A rank filter which sharpens the edges in an image, especially after several iterations, is the extremum filter [Kramer 75]. The filter works by replacing the central pixel value within the filter window by the extreme of the grey-values within the window which is nearest in value to the current pixel value.

$$f_{ij} = Ext(g_{ij}) = \begin{cases} max_n(g_{ij}) & if \quad max_n(g_{ij}) - g_{ij} \leq g_{ij} - min_n(g_{ij}) \\ min_n(g_{ij}) & otherwise \end{cases}$$

where $Ext(g_{ij})$ is the extremum filter over a neighbourhood window.

Mathematical proofs of the ability of the extremum filter to reconstruct edges, i.e. sharpen blurred edges to step edges, are given in [Kramer 75]. In this thesis, the arguments of the properties of the extremum filter will be totally qualitative and some results are presented.

Consider the example in Figure 4.8. The original image consisting of black and white stripes (and its profile) is given in Figure 4.8(a). Consider the case when the edges have been blurred by a low-pass filter, Figure 4.8(b). In this case, a spline was fitted to the noisy image. This image could also have been generated by fitting a spline to an image of black and white stripes which is noise free and has ideal step edges. However, it would have been very difficult to generate the different shades within the stripes which are required if the convergence property of the filter is to be considered for real images. An operator is required which can restore the blurred edge image, Figure 4.8(b), to its model of a step edge image. A rank filter which is known to have this property is the extremum filter [Kramer 75].

Applying a 5×5 neighbourhood extremum filter to the image in Figure 4.8(b), the results are observed in Figure 4.8(c) after 11 iterations. It can be observed that the blurred edges have been sharpened to step edges. However, noise points are also enhanced by the filter and dendritic formations are observed. The consequence of this is that iteratively applying the extremum filter does not produce a simple convergence of the image to its root image consisting of step edge stripes. The convergence appears as a lightly damped oscillation, i.e. the difference between successive iterated images decreases to a minimum and then "explodes" and then converges again. Kramer [Kramer 75] reported that 20–50 iterations were required to observe complete convergence for 8-bit 27×33 pixel images. However, after only a few iterations reasonable results are observed.

It has been found that iterating a sequence of an extremum and a median filter, rather than an extremum filter alone, increases the convergence rate. A root image is almost observable after 11 iterations when a sequence of a 5×5 window extremum and median filter is applied to the 8-bit 512×512 image in Figure 4.8(b). The result is shown in Figure 4.8(d). The restoration of the blurred image to the original, Figure 4.8(a), is better than that obtained using iterated extremum filter alone, Figure 4.8(c).

The conventional method for applying the median filter is to use the histogram modification scheme [Garibotto 79, Huang 79]. In this method, the histogram is generated for the first window position and the median value determined. For subsequent window positions, a number of pixels will move in

(a)

(b)

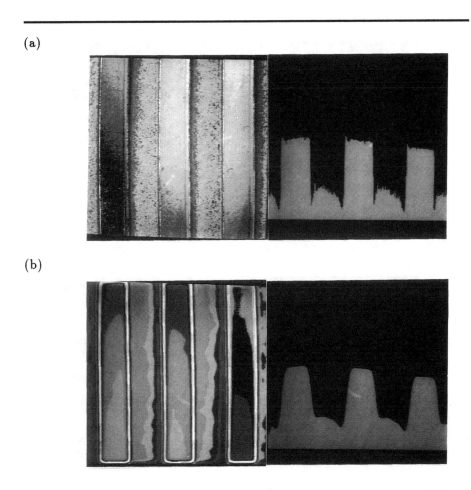

Figure 4.8: The result of sharpening blurred edges using rank filters. (a) The original noisy stripe image and its profile. (b) The result of blurring the step edges using a spline function.

(c)

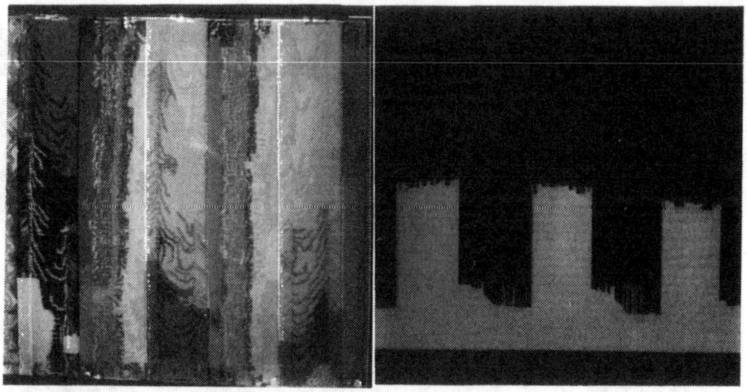

(d)

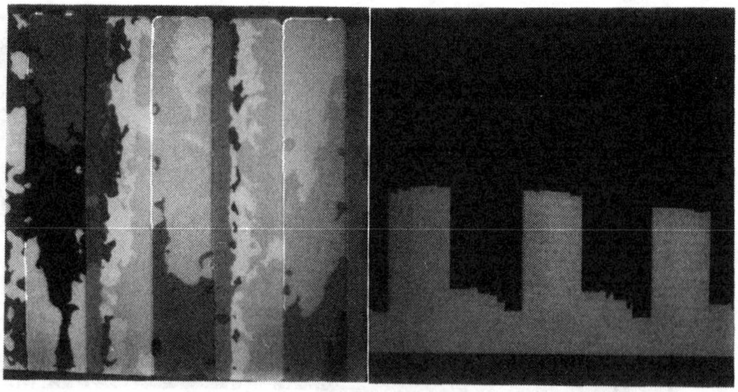

Figure 4.8: (c) The result of applying 5×5 extremum filter, iterated 11 times. (d) The result of iterating 11 times a sequence of 5×5 extremum and median filters. In both cases the step edges are restored.

and an equal number move out; for those moving out of the window, their values are removed from the histogram and new values are inserted for those moving in. Again the median value is determined. If a piano scan is adopted then the histogram need only be initialised once. This method for computing the median within a window is considerably faster than sorting for all but the smallest of windows. For a $M \times M$ window, the operation has complexity $O(M)$ while the best sorting algorithm is $O(M \log M)$. The memory requirement for this method is reasonable for images up to 8-bits; the histogram need only have 256 bins. For images with a greater precision, it is more memory efficient to use a sorted list [Bedner 84].

The above methods for computing the median are *serial*. However, median filtering is essentially SIMD parallel; the value at each pixel is to be replaced with the median of its neighbouring pixels. The value computed at one window position is not dependent on the value computed at another window position. Parallel algorithms for computing any rank value using a "bit" sorting method has been proposed (along with hardware implementation) by both Ataman and Danielsson [Ataman 80, Danielsson 81]. The method starts by looking at the *most significant bit* (msb) of all the pixels in its neighbourhood. It counts the number of zeros, if this number, n_0, is greater than the rank value R, then the msb of the rank value must be a 0; otherwise a 1. If msb is 0, the other numbers with msb 1 are discarded. We examine the *next significant bit* for number of 1s, n_{01} in the remaining pixels. If $(n_0 - n_{01} > R)$, then the next significant bit is also a 0. Other numbers with 1s as their significant bit are discarded (see Figure 4.9). The discarded pixels are noted in a register, **S**. Remaining pixels are examined in this way using a binary search to find the median (or any other rank value). However, note only one rank value can be determined at a time. This algorithm is also particularly suitable for bit-serial processors because of the way the median value is built up one bit at a time (Algorithm 4.5).

4.4.3 Non-linear Spatial Filters

Equations (4.4)–(4.7) represents types of spatial low-pass filter. The "cut-off" frequency is determined by the window shape and size; the window size is $(2k+1) \times (2k+1)$, the larger the k, the lower is the "cut-off" frequency. However, these are not true low-pass filters because they preserve edge information. But for most image processing purposes, these form a good approximation to low-pass filters.

The spatial band-pass and high-pass filters form the non-linear spatial low-pass filters [Goetcherian 80, Preston 83] in the same way as for linear band-pass and high-pass filters. A band-pass filter is the difference between two low-pass filters with different cut-off frequencies determined by r and s.

$$\text{BAND PASS} = \max_n^r(\min_n^{2r}(\max_n^r(\mathcal{F}))) - \max_n^s(\min_n^{2s}(\max_n^s(\mathcal{F}))), \quad r > s.$$

A high-pass filtered image is obtained by subtracting a low-pass filtered image from the original.

$$\text{HIGH PASS} = \mathcal{F} - \max_n^r(\min_n^{2r}(\max_n^r(\mathcal{F}))).$$

Algorithm 4.5: Median filter using a bit searching method.

begin
  RANK → R;
  for $i := 0$ **to** N **do**
    $s[i] \leftarrow 1$; **od**
  for each bit-plane j **do**
    $k \leftarrow \sim_i \left(s[i] \cdot B_i^j \right)$;
    if $(k > R)$ **then**
                  $R \leftarrow R - k$;
                  $A^j \leftarrow 0$;
              else
                  $R \leftarrow R + k$;
                  $A^j \leftarrow 1$; **elsf**
      for $i := 0$ **to** N **do**
        $s[i] \leftarrow (s[i] \cdot A^j \equiv B^j$; **od od**
  end

---

numbers	msb		$\longrightarrow$			lsb
23	0	1	0	1	1	1
57	1 0	1 0	1 0	0 0	0 0	1 0
18	0	1	0	0 0	1 0	0 0
43	1 0	0 0	1 0	0 0	1 0	1 0
26	0	1	1	0 0	1 0	0 0
3	0	0 0	0 0	0 0	1 0	1 0
9	0	0 0	1 0	0 0	0 0	1 0
25	0	1	1	0 0	0 0	1 0
35	1 0	0 0	0 0	0 0	1 0	1 0
median	0	1	0	1	1	1
N	6	3	5	4	4	4

Figure 4.9: Median filtering by binary search for 6-bit images.

Algorithm 4.6: Choosing a High-pass filter cut-off frequency.

<u>**begin**</u>
    $I^* \leftarrow$ *Original_Image*;
    brightest_object $\leftarrow MAX\_VALUE(I^*)$;
    counter $\leftarrow 0$;
    <u>**while**</u> (*new_brightest_object* $\geq (p * brightest\_object)$) <u>**do**</u>
        $I^* \leftarrow S(I^*)$;
        new_brightest_object $\leftarrow MAX\_VALUE(I^*)$;
        counter $\leftarrow$ *counter* $+ 1$; <u>**od**</u>
    $I^* \leftarrow E^{counter}(I^*)$;
    $I_{hp} \leftarrow$ *Original_Image* $- I^*$;
<u>**end**</u>

Note, we could have used any of the definitions for the low-pass filters given in equations (4.4)–(4.7) as required.

4.4.4 Automatically Choosing a Cut-off Frequency for a High-Pass Filter

An automatic means to determine the cut-off frequency for a high-pass filter when processing real images is not a trivial task. Real images do not contain only a single class of objects but a great variety of them. If there was only one class of object in the image, i.e. similar shapes or sizes or equally bright objects, then an automatic means of choosing a cut-off frequency would be easy. A suitable algorithm is given below (Algorithm 4.6).

The algorithm determines the value of the brightest object in the image using the MAX_VALUE operation. This determines the maximum in an image by inspecting each of the bit-planes in the image, starting with the most-significant bit plane. This process is made clear in Algorithm 4.6.

Having found the largest value in the image (Algorithm 4.7), *brightest_object*, local neighbourhood minimum operation is iteratively applied until the value of the new brightest object in the image is a fraction p of the original; p is chosen to be typically between $1/2$ and $1/3$. The same number of iterations of the local neighbourhood maximum operation is then applied; the result is that the image I^* is essentially low-pass filtered. The high-pass filtered image is obtained by subtracting this from the original. The cut-off frequency is then a function of the width of the ridges and therefore is also a function of p.

Algorithm 4.7: Computing the maximum value in an image.

<u>**begin**</u>
    mask ← *set a bit_plane to*1;
    max_value ← 0;
    msb := *most_significant_bit_plane*;
    lsb := *least_significant_bit_plane*;
    <u>**for**</u> *i* := *msb* <u>**down**</u> <u>to</u> *lsb* <u>**do**</u>
        tmask ← *bit_plane*(*i*)*AND mask*;
        <u>**if**</u> *tmask* = ¬*empty* <u>**then**</u>
                    max_value ← 2^i + *max_value*;
                    mask ← *tmask*; <u>**fi od**</u>
<u>**end**</u>

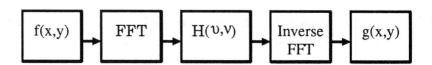

Figure 4.10: A scheme for image filtering in the Fourier domain.

4.5 Fast Fourier Transform

We have regarded an image as a matrix with each pixel represented by an intensity value. But an image can also be regarded as being combinations of two-dimensional Fourier series of different frequencies. Duda and Hart [Duda 73] have noted that homogeneous regions in an image correspond to low-frequency Fourier components, while edges and noise are represented by the higher frequencies. Thus, image filtering can be carried out in the Fourier domain. A scheme for this is given in Figure 4.10.

If $f(x)$ is a continuous real function, then the *Fourier transform* of $f(x)$ is $\mathcal{F}\{f(x)\}$

$$\mathcal{F}\{f(x)\} = F(u) = \int_{-\infty}^{\infty} f(x) \exp\left[-\text{j}2\pi ux\right] \, \mathrm{d}x$$

where j = $\sqrt{-1}$. Given $F(u)$, $f(x)$ can be obtained by *inverse Fourier transform*

$$\mathcal{F}\{F(u)\} = f(x) = \int_{-\infty}^{\infty} F(u) \exp\left[\text{j}2\pi ux\right] \, \mathrm{d}u$$

Note that $F(u)$ is a complex value.

$$F(u) = \Re(u) + \Im(u) = |\,F(u)\,|\,e^{j\phi(u)}$$

where

$$|\,F(u)\,| = \sqrt{\Re^2(u) + \Im^2(u)}$$

is the *Fourier spectrum* of $f(x)$, and

$$\phi(u) = \tan^{-1}\left[\frac{\Im(u)}{\Re(u)}\right]$$

is the phase angle. The square of the Fourier spectrum is called the *power spectrum* of $f(x)$.

However, for many problems such as image processing, the function $f(x)$ is not continuous, but is quantised spatially and have discrete value. A discrete *Fourier transform pair* [Blackman 58, Cooley 67b, Cooley 67a, Brigham 74] is

$$F(u) = \frac{1}{N}\sum_{x=0}^{N-1} f(x)\exp\left[-j2\pi ux/N\right] \tag{4.8}$$

and

$$f(x) = \sum_{u=0}^{N-1} F(u)\exp\left[j2\pi ux/N\right]$$

An important property of DFT for a 2D function is, it is separable.

$$F(u,v) = \frac{1}{N}\sum_{x=0}^{N-1}\sum_{y=0}^{N-1} f(x,y)\exp\left[-j2\pi(ux+vy)/N\right]$$

$$F(u,v) = \frac{1}{N}\sum_{x=0}^{N-1}\exp\left[-j2\pi ux/N\right]\sum_{y=0}^{N-1} f(x,y)\exp\left[-j2\pi vy/N\right]$$

This can be written in the form

$$F(u,v) = \frac{1}{N}\sum_{x=0}^{N-1} F(x,v)\exp\left[-j2\pi ux/N\right]$$

where

$$F(u,v) = N\left[\frac{1}{N}\sum_{y=0}^{N-1} f(x,y)\exp\left[-j2\pi vy/N\right]\right]$$

The complexity of FFT is $O(N^2)$. However, it can be decomposed into an $O(N\log_2 N)$ problem a *fast Fourier transform* (FFT). Let us express

$$W_N = \exp\left[-j2\pi/N\right]$$

and assume

$$N = 2^n$$

where n is a positive integer. Then we can put $N = 2M$, where M is also a positive integer. Then equation (4.8) can be written as

$$
\begin{aligned}
F(u) &= \frac{1}{2M} \sum_{x=0}^{2M-1} f(x) W_{2M}^{ux} \\
&= \frac{1}{2} \left\{ \frac{1}{M} \sum_{x=0}^{M-1} f(2x) W_{2M}^{u(2x)} + \frac{1}{M} \sum_{x=0}^{M-1} f(2x+1) W_{2M}^{u(2x+1)} \right\} \\
&= \frac{1}{2} \left\{ \frac{1}{M} \sum_{x=0}^{M-1} f(2x) W_{M}^{ux} + \frac{1}{M} \sum_{x=0}^{M-1} f(2x+1) W_{M}^{ux} W_{2M}^{u} \right\}
\end{aligned}
$$

since $W_{2M}^{2ux} = W_{M}^{ux}$. This can be further expressed as

$$
\begin{aligned}
F(u) &= \left\{ F_{\text{even}}(u) + F_{\text{odd}}(u) W_{2M}^{u} \right\} / 2 \\
F(u+M) &= \left\{ F_{\text{even}}(u) - F_{\text{odd}}(u) W_{2M}^{u} \right\} / 2
\end{aligned}
\qquad 0 \leq u < M
$$

Special hardware has been proposed [Bergland 69, Stone 71, Pease 77]. On a 1D SIMD array with M PEs with $n-$cube connectivity, Jamieson [Jamieson 86] describes an algorithm for the computation of $N-$point FFT using the radix-2 method and higher-radix methods for fewer PEs. At each stage, M "butterfly" operations are carried out and $\log_2 N$ stages are required (Figure 4.11). This algorithm has been implemented on PASM [Bronson 90]. On CLIP4, an SIMD processor array, the complexity is $O(N)$ for both the DFT and DFFT [Ip 80]; however, it is better to compute DFFT because fewer operations per PE are required; this gives it a more accurate result.

A real-valued Hartley transform [Hartley 42] has been proposed as an alternative to FFT [Bracewell 84]. Although it is much debated whether fast Hartley transform (FHT) is faster [Zakhor 87], it does have advantages over FFT in that (i) it has the same formula for both the forward and inverse transform, and (ii) it is more memory efficient since it does not use complex numbers.

Given 2D real image data, the DHT $H(u,v)$ is defined as

$$
H(u,v) = \frac{1}{MN} \sum_{x=0}^{M-1} \sum_{y=0}^{N-1} f(x,y) \text{cas} \left[2\pi \left(ux/M + vy/N \right) \right]
$$

where $\text{cas}\,\alpha = \cos\alpha + \sin\alpha$.

4.6   Edge Detection

At its simplest, an edge is a sharp discontinuity in grey-level profile. However, the situation is complicated by the presence of noise and image resolution. An edge is specified by its magnitude and its direction. A number of linked edge points may be better approximated by a linear segment called an *edgel*. Edges and 'edgels' form an important step in image segmentation and object recognition (see Chapters 5, 6 and 7). There are many varieties of edges;

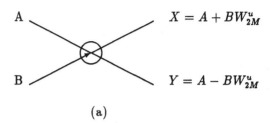

(a)

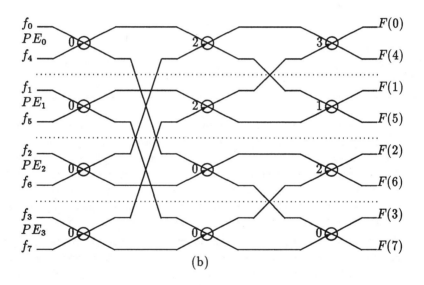

(b)

Figure 4.11: (a) The butterfly operation and (b) dataflow for 8-point FFT on 4 processors.

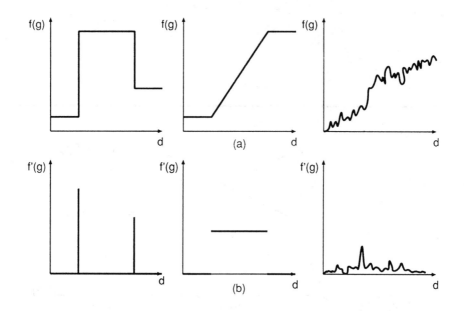

Figure 4.12: Three types of edges are shown. Along with the edge profiles, the first derivates of the edges are given.

they may be classified into three major classes: step-edge, roof-edge and line-edge (see Figure 4.12). A large edge can be regarded as being constructed of many short linear line segments called *edgels*; each edgel has a position and orientation. Edgels correspond to local discontinuities in the intensity surface. A function having a discontinuity of nth order has a delta function in its nth derivative. A linear edge has a zero order discontinuity; a step edge has a first order discontinuity; and a roof edge has a second order discontinuity.

The simplest edge detector is a difference operation [Roberts 65, Prewitt 70]. Edge detectors have been extensively reviewed in several books and surveys and will not be discussed in any detail here. For a more complete discussion, the reader is referred to the following: [Davis 76, Rosenfeld 82, Pratt 78]. Edge detection is often carried out by spatial differentiation and thresholding.

$$g_x = \frac{\partial G(x,y)}{\partial x}$$
$$g_y = \frac{\partial G(x,y)}{\partial y}$$

Prewitt directional derivative of vector direction z subtending an angle Φ [Prewitt 70]

$$\nabla\{G(x,y)\} = \frac{\partial G(x,y)}{\partial z} = g_x \cos\Phi + g_y \sin\Phi$$

and the edge magnitude is given by

$$| \nabla \{G(x,y)\} | = \sqrt{g_x^2 + g_y^2}$$

The second derivatives are defined as

$$g_{xx} = \frac{\partial^2 G(x,y)}{\partial x^2} \quad \text{etc.}$$

$$\nabla^2 \{G(x,y)\} = g_{xx} + 2g_x g_y + g_{yy}$$

and the nth derivatives given by

$$\nabla^n \{G(x,y)\} = \sum_{j=0}^{n} \frac{n!}{(n-j)!j!} \frac{\partial^j G}{\partial x^j} \frac{\partial G^{n-j}}{\partial y^{n-j}}$$

The Laplacian

$$L(G(x,y)) = g_{xx} + g_{yy} = \frac{\partial^2 G}{\partial x^2} + \frac{\partial^2 G}{\partial y^2}$$

This is a second derivative method.

A good edge operator must have the following properties; it must (1) operate locally, (2) be efficient, (3) be sensitive to the orientation and magnitude of an edge, (4) work in the presence of noise, (5) be insensitive to threshold values. A further condition imposed by Canny [Canny 86] is that the operator must not have multiple responses to a single edge.

Common edge operators used as due to Sobel [Prewitt 70], Roberts [Roberts 65], Kirsh [Kirsh 71], Marr and Hildreth [Marr 80], Haralick [Haralick 80, Haralick 84], and Nalwa and Binford [Nalwa 86].

The difference operators are local neighbourhood operations and they are efficiently implemented on parallel array computers such as CLIP4 because of local neighbourhood connectivity. Amongst the most popularly used 3×3 edge operators are the Prewitt and the Sobel [Rosenfeld 82]. Davies [Davies 84a] has shown that of the 3×3 edge detectors, the Sobel is the most accurate. The commonality between these two operators is that they compute the edge magnitude and edge direction. A local neighbourhood edge operator which is direction-invariant is the Laplacian. These three operators are illustrated in Figure 4.13. The Laplacian is less popular because it responds strongly to noise and corners [Rosenfeld 82] and because it gives a response on both sides of an edge. Because of the weighting of the central pixel, while the Laplacian will respond to edge points, it will enhance more strongly isolated points, corners, lines and line ends if these have a higher contrast than the surroundings. Furthermore, the Laplacian returns a signed image and for an 8-bit image, it may return a 10-bit result.

Using such 3×3 gradient operators, a possible edge detection scheme has been suggested in [Robinson 77]. This method is illustrated in Figure 4.14. Sobel operation gives the $x-$ and $y-$components of the edge magnitude and direction. Therefore, trigonometry would give the direction and magnitude of the edge. On an SIMD array such as CLIP4, computation of the square-root function using bit-serial arithmetic is expensive and small amount of local memory available.

-1	0	1
-1	0	1
-1	0	1

Δ_x

1	1	1
0	0	0
-1	-1	-1

Δ_y

Prewitt Operator

-1	0	1
-2	0	2
-1	0	1

Δ_x

1	2	1
0	0	0
-1	-2	-1

Δ_y

Sobel Operator

0	1	0
1	-4	1
0	1	0

Laplacian Operator

Figure 4.13: 3×3 neighbourhood gradient operators.

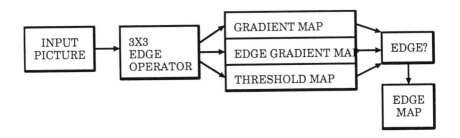

Figure 4.14: An edge detection scheme due to Robinson.

A_0	A_1	A_2
A_7	A_c	A_3
A_6	A_5	A_4

$$A_c = \max_{i=0}^{3} \mid S_i - T - i \mid$$

where $S_i = (A_{0+i} + 2A_{1+i} + A_{2+i})$ and $T_i = (A_{4+i} + 2A_{5+i} + A_{6+i})$ and the subscripts are evaluated in modulo 8.

A similar method is proposed by Kirsh [Kirsh 71]

$$A_c = \max \left(1, \max_{i=0}^{7} \mid 5S_i - 3T_i \mid \right)$$

where $S_i = A_i + A_{i+1} + A_{i+2}$ and $T_i = A_{i+3} + A_{i+4} + A_{i+5} + A_{i+6} + A_{i+7}$.

Apart from the local operations, there are regional operations to detect edges. Amongst this class are the Heuckel operators [Hueckel 71] and the Marr-Hildreth Difference of Gaussian (DOG) [Marr 80]. These edge operators involve large windows and often needs floating-point calculations to maintain accuracy.

If we are not interested in the direction of the edge but only in its presence, we should use, where possible, direction-invariant edge detectors. One such edge detector is the Laplacian. Better direction invariant edge operators, Rosenfeld suggests [Rosenfeld 82], are the maximum of the absolute second differences and the absolute difference between the mean and the median.

This brings us to consider a grey-level version of a binary edge operator. Although boolean (AND, OR) operators do not form an isomorphism with corresponding grey-level (MIN, MAX) operators, some boolean operations such as edge finding, spatial filtering and skeletonising have equivalent grey-level image operations. An edge detector of this type has been suggested [Goetcherian 80] using the neighbourhood minimum operation (see Algorithm 4.8). The reason why the operator works is because the edges between regions shift when the

---

Algorithm 4.8: Edge operator using mathematical morphology.

$$\mathbf{EDGE}^- \longleftarrow \mathcal{G} - \min_n(\mathcal{G})$$

$$\mathbf{EDGE}^+ \longleftarrow \max_n(\mathcal{G}) - \mathcal{G}$$

$$\mathbf{EDGE} \longleftarrow \min_p(\mathbf{EDGE}^+, \mathbf{EDGE}^-)$$

---

neighbourhood minimum operator is applied, giving rise to high-intensity regions in the difference image where the edges have moved. A modification to this edge detector is the "Lee edge detector" [Lee 86] (see Algorithm 4.8). To detect ramp edges, this is modified [Verbeek 88] to,

$$\mathbf{EDGE} \longleftarrow \min_p(\max_n(\mathcal{G}) - \min_n(\max_n(\mathcal{G})), \max_n(\min_n(\mathcal{G})) - \min_n(\mathcal{G}))$$

The non-ramp (or "texture edges") are detected by,

$$\mathbf{EDGE} \longleftarrow \min_p(\min_n(\max_n(\mathcal{G})) - \mathcal{G}, \mathcal{G} - \max_n(\min_n(\mathcal{G}))$$

In this algorithm $\mathcal{G}$ is the grey image and $\min_n$ is the near-neighbour minimum operator and $\max_n$ is the near neighbourhood maximum operator. The result is always positive since $\mathcal{G} \geq \min_n(\mathcal{G})$ and $\mathcal{G} \leq \max_n(\mathcal{G})$. The results obtained using this edge operators are better than the Laplacian, being less sensitive to spike noise because the rank filter reduces the variance.

The first-order differential edge operators are efficient since they operate on the difference of the neighbouring pixels; however, they have their limitation because they are sensitive to changes in illumination. It is better to use a second-order differential operator; however, they are sensitive to local maxima. They also produce faint edges in featureless regions due to the presence of noise, therefore one has to apply a threshold to remove these edges which often occur in the shadow regions. The illumination effect can of course be removed by filtering. We know that irradiance is a function of the product of reflectance and illumination. Since these two components of irradiance have different spatial frequency characteristics, they can be separated by spatial high-pass filtering. However, Horn [Horn 86a] points out that filtering yields only an approximation of the reflectance, this is referred to as image "lightness" [Land 71]. To overcome the problem of uneven illumination, Johnson [Johnson 90] has developed a contrast-based Sobel operator, using the absolute form of the Sobel [Abdou 79].

$$S_{i,j} = \mid N_{i,j} \otimes \Delta_x^3 \mid + \mid N_{i,j} \otimes \Delta_y^3 \mid$$

where the pixels in the neighbourhood of $N_{i,j}$ are convolved with 3×3 Sobel kernels Δ_x and Δ_y. Assuming that local brightness is averaged over the scale of the operator, Johnson's contrast-based edge detector is

$$S_{i,j} = \frac{|\; N_{i,j} \otimes \Delta_x^3 \;| \; + |\; N_{i,j} \otimes \Delta_y^3 \;|}{\frac{1}{9}(N_{i,j} \otimes A) + d}$$

where A is the averaging operator (a 3×3 kernel whose coefficients are all 1), and d is chosen for noise-sensitivity and dynamic range for a given sensor and scene. The method is easily extended to use larger edge detection kernels [Johnson 90].

Marr and Hildreth [Marr 80] note that in a natural scene, changes in intensity occur over different range of scales; therefore, there is no single operator which is universally applicable for edge detection. Things which give rise to intensity changes are: (1) changes in illumination, these could be due to shadows, visible light sources and illumination gradient; (2) changes in orientation and the position of the observer from the visible surface; and (3) changes in surface reflectance. All these physical properties (with the exception of diffraction pattern, but these are not common in nature) are spatially localised in their own scale; examples of them are contours, creases, scratches, marks, shadows and shadings. As a consequence of this a filter is required which is smooth and localised in the spatial domain and has a small spatial variance. There is only one function with this property, namely the Gaussian [Leipnik 60].

The Marr–Hildreth edge operator consists of (1) convolution with $\nabla^2 G$, (2) localisation of the zero-crossings, and (3) checking the alignment and orientation of local segments of zero-crossings.

$$f(x,y) = \nabla^2 \left[G(r) \otimes I(x,y) \right]$$

using the derivative rule for convolution, we have

$$f(x,y) = \nabla^2 G \otimes I(x,y)$$

This resembles the difference of two Gaussians (DOG) [Wilson 77].

The Laplacian of Gaussian, $\nabla^2 G$ is given by

$$\nabla^2 G(r) = -\frac{1}{\pi \sigma^4} \left[1 - \frac{r^2}{2\sigma^2} \right] \exp \left[-\frac{r^2}{2\sigma^2} \right]$$

Different σ gives different channels. If for a set of independent $\nabla^2 G$ channels over a contiguous range of scales, and if we have zero-crossings present which have the same position and orientation, then the presence of intensity change due to some single physical phenomena (such as reflectance, illumination, depth or surface orientation) can be inferred; otherwise, the zero-crossings could have been due to some distinct surface or physical phenomena. The minimum number of zero-crossings required are two, these must be well separated in frequency.

Wilson and Giese's DOG function in 1D is given by

$$DOG(\sigma_e, \sigma_i) = \frac{1}{\sqrt{2\pi}\sigma_e} \exp \left[-\frac{x^2}{2\sigma_e^2} \right] - \frac{1}{\sqrt{2\pi}\sigma_i} \exp \left[-\frac{x^2}{2\sigma_i^2} \right]$$

where σ_e and σ_i are the excitatory and inhibitory space constants. We can write $\sigma_e = \sigma$ and $\sigma_i = \sigma + \delta\sigma$ then

$$DOG(\sigma_e, \sigma_i) = \delta\sigma \frac{\partial}{\partial\sigma}\left[\left(\frac{1}{\sigma}\right)\left(\exp\left[-\frac{x^2}{2\sigma^2}\right]\right)\right]$$

$$= -\left(\frac{1}{\sigma^2} - \frac{x^2}{\sigma^4}\right)\exp\left[-\frac{x^2}{2\sigma^2}\right]$$

Since the requirement is to create a narrow bandpass differential operator, Marr–Hildreth shows [Marr 80] that we need to choose $\sigma_i/\sigma_e = 1.6$.

Huertas and Medioni [Huertas 86] show that the Laplacian of Gaussian can be written as a sum of two filters

$$\nabla^2 G = K\left(2 - \frac{x^2 + y^2}{\sigma^2}\right)\exp\left[-\left(\frac{x^2 + y^2}{2\sigma^2}\right)\right]$$

$$= h_1(x)h_2(y) + h_2(x)h_1(y)$$

where

$$h_1(\xi) = -\sqrt{K}\left(1 - \frac{\xi^2}{\sigma^2}\right)\exp\left[-\left(\frac{\xi^2}{2\sigma^2}\right)\right]$$

$$h_1(\xi) = \sqrt{K}\exp\left[-\left(\frac{\xi^2}{2\sigma^2}\right)\right]$$

Seit *et al.* [Seit 88] use these to detect the zero-crossings:

$$\text{direction} = \begin{cases} \text{horizontal} & \text{if } |\,x\,| > |\,2y\,| \\ \text{vertical} & \text{if } |\,y\,| > |\,2x\,| \\ \text{diagonal} & \text{otherwise} \end{cases}$$

and

$$\text{magnitude} = \max(|\,x\,|, |\,y\,|).$$

Advantages of Marr–Hildreth's method are (1) the Gaussian smoothing reduces the effect of noise, (2) the second derivative detects localised edge contours, the loci of which are the gradient maxima; also looking for zero-crossing avoids arbitrary thresholding and thinning, (3) it guarantees closed curves (this is very important for many segmentation tasks), and (4) it does not give a response in area where the intensity is changing smoothly. The disadvantage is that it is computationally expensive because very large convolution masks are required. Wiejak, Buxton and Buxton discuss method for speeding up its computation [Wiejak 85, Forshaw 88]. A method for speedup is that many 2D convolutions are separable to two 1D convolution. For example consider the Sobel kernel

$$\begin{bmatrix} 1 & 2 & 1 \\ 0 & 0 & 0 \\ -1 & -2 & -1 \end{bmatrix} \equiv \begin{pmatrix} 1 \\ 0 \\ -1 \end{pmatrix}(1\ 2\ 1)$$

In this example the gain is slight, but could be significant for larger kernels such as the Gaussian.

$$\exp\left[-\left(\frac{r^2}{2\sigma^2}\right)\right] = \exp\left[-\left(\frac{x^2}{2\sigma^2}\right)\right]\exp\left[-\left(\frac{y^2}{2\sigma^2}\right)\right]$$

Canny [Canny 86] developed an edge operator which extracts not only step edges but also ridge and roof edges. A ridge is a two closely spaced step edge. These are often too small to be dealt effectively by even the narrowest of edge operators. A roof edge is a concave junction of two planar surfaces of polyhedral objects.

Canny derives an edge operator which is approximated by the first derivative of a Gaussian $G'(x)$, where

$$G(x) = \exp\left[-\left(\frac{x^2}{2\sigma^2}\right)\right]$$

The impulse response of the first derivative filter is

$$f(x) = -\frac{x}{\chi^2}\exp\left[-\left(\frac{x^2}{2\sigma^2}\right)\right]$$

Canny notes that his edge operator is almost identical to the 1D Marr–Hildreth operator. While Marr–Hildreth needs no thresholding, an adaptive thresholding technique can be advantageous for a first derivative operator. However, first derivative operators respond to smooth shading, therefore both Canny and Nevatia–Babu [Nevatia 80] detects false edges on smoothly shaded surfaces.

The noise characteristics of an operator are determined by its size; bigger operators have the effect of larger smoothing of random noise. However, a large operator can overlap several features and hence it reduces resolution and the detectability and localisation of high-curvature edges is also reduced. Directional operators introduce averaging which is along the edge; however, isotropic operators, such as the Marr–Hildreth, smooth across the edge.

Many of these algorithms have been evaluated by Chen *et al.* for edge detection [Chen 88] on brain images obtained from various sources such as computer tomography (CT), magnetic resonance imaging (MRI), and positron emission tomography (PET). These results indicate that none of the edge detectors mentioned are universally applicable. Despite these numerous efforts, the problem of edge detection remains largely unsolved. Some work better than others for particular examples. Hence, the user must choose an operator which best suits his needs.

Chapter 5

Intermediate Level Image Processing

5.1   Introduction

In this chapter we explore various methods for image segmentation. We consider three methods in some detail, these involve (i) thresholding, (ii) region-based methods such as *split-and-merge* and (ii) use of the texture information to separate out the *regions of interest* from the background. Then we look at feature detection by use of the Hough transformation and corner detection. Once features have been detected they need to be described symbolically. Possible methods for object or shape description include "medial-axis" by using shape *skeleton* or use a number of boundary descriptors which are covered in §5.6. Some of these methods can be used in recognising partially occluded objects. We next consider object description by fitting lines and splines to a set of points. Finally we present methods for describing shapes by computing their convex hull.

5.2   Segmentation

As already stated, segmentation is a form of pixel classification, i.e. we are assigning each pixel into a class of pixels based on some property of the pixel. Segmentation algorithms can be divided into two classes: (1) based on pixel similarities, or (2) based on surface discontinuities. The latter method involves edge detection (see Section 4.6). In this section, we will consider the former method.

As a result of segmentation, the regions should be homogeneous with respect to the segmentation criterion, such as uniformity in grey-level intensity or texture; the regions should be simple, should not contain many small holes; and adjacent regions should be significantly different.

Haralick and Shapiro [Haralick 85] states: "As there is no theory of clustering, there is no theory of image segmentation." The result is that segmentation techniques have been developed on an *ad hoc* basis. A segmentation may have

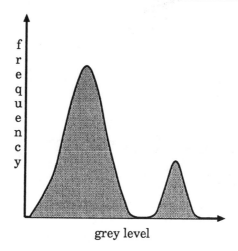

Figure 5.1: A bimodal histogram.

two types of error: (1) extra regions may be generated or (2) regions may be missing from the segmented image. These two errors may be combined such that shapes of segmented regions do not match accurately with the object boundaries.

5.2.1   Thresholding

One of the earliest method for image segmentation was based on knowledge or expectation of object occupancy [Doyle 62] called the *p_tile* method. If the object occupied $p\%$ of the image, then the image should be thresholded such that at least $p\%$ of the brightest pixels are mapped to the object. Another simple approach to segmentation is to segment the grey-level image $\mathcal{G}$ at a value T to produce a binary image $\mathcal{B}$. A threshold operation is defined as

$$\mathcal{B}_{ij} = \begin{cases} 1 & \mathcal{G}_{ij} \geq T \\ 0 & \text{otherwise} \end{cases}$$

How do we select a suitable value for T? Usually, T is obtained from the histogram of the image. If we have a simple image such that the objects have a mean intensity value μ_o with a small variance and similarly for the background μ_b. If μ_o and μ_b are sufficiently far apart such that they do not overlap, the histogram is said to be bimodal (Figure 5.1). In this case, the threshold value can be chosen to be the mid-value between these two peaks. This is often called the *mode method* [Prewitt 66].

If there are N objects with distinct grey-levels, the image may be segmented by multiple thresholds

$$
\mathcal{R}_{ij} = \begin{cases} N & \text{if } \mathcal{G}_{ij} \geq T_{N-1} \\ N-1 & \text{if } T_{N-2} \leq \mathcal{G}_{ij} \leq T_{N-1} \\ \vdots & \vdots \\ 1 & \text{if } T_0 \leq \mathcal{G}_{ij} \leq T_1 \\ 0 & \text{otherwise} \end{cases}
$$

The desired objects can then be thresholded out of image $\mathcal{R}$. However, because of reflectance, shadows and uneven illumination, a global threshold often gives unsatisfactory results.

Several methods have been developed for automatically selecting threshold values, these have been reviewed in [Weszka 78, Rosenfeld 82, Sahoo 88].

A threshold selection method based on entropy has been proposed by Pun [Pun 80, Pun 81] and extended by Kapur *et al.* [Kapur 85a]. Pun defines two *a posteriori* entropies [Abramson 63] of the background (H_b) and foreground (H_w) pixels after thresholding at a value t.

$$
H_b(t) = -\sum_{i=0}^{t} p_i \ln p_i
$$

$$
H_w(t) = -\sum_{i=t+1}^{l-1} p_i \ln p_i
$$

where t is the threshold values and p_i is the probability at intensity i ($p_i = f_i/N^2$). The threshold is defined such that $H = H_b + H_w$ is a maximum. Kapur *et al.* have extended the definition of the *a posteriori* entropy to involve two probability distributions (object distribution and background distribution)

$$
H_b(t) = -\sum_{i=0}^{t} \frac{p_i}{P_t} \ln \left(\frac{p_i}{P_t} \right)
$$

$$
H_w(t) = -\sum_{i=t+1}^{l-1} \frac{p_i}{1-P_t} \ln \left(\frac{p_i}{1-P_t} \right)
$$

where $P_t = \sum_{i=0}^{t} p_i$. A threshold T is chosen such that

$$
T = \mathbf{Arg} \max_{t \in \mathcal{G}} \{ H_b(t) + H_w(t) \}
$$

A similar method has been proposed by Johannson and Bille [Johannsen 82] which attempts to minimise the interdependence between the two *a posteriori* entropies.

$$
T = \mathbf{Arg} \min_{t \in \mathcal{G}} \{ S(t) + \overline{S}(t) \}
$$

where

$$
S(t) = \ln \left(\sum_{i=0}^{t} p_i \right) - \frac{1}{\sum_{i=0}^{t} p_i} \left[p_t \ln p_t + \left(\sum_{i=0}^{t-1} p_i \right) \ln \left(\sum_{i=0}^{t-1} p_i \right) \right]
$$

and

$$\overline{S}(t) = \ln\left(\sum_{i=t}^{l-1} p_i\right) - \frac{1}{\sum_{i=t}^{l-1} p_i}\left[p_t \ln p_t + \left(\sum_{i=t+1}^{l-1} p_i\right)\ln\left(\sum_{i=t+1}^{l-1} p_i\right)\right]$$

However, histogram modality is not often exhibited. For these, it has been suggested that the histogram prior to threshold selection may be enhanced by reducing the contribution to the histogram of the edge pixels [Weszka 74, Mason 75]. This is referred to as the *gradient-weighted histogram* method.

In many instances, global thresholding for image segmentation does not work because the histogram may be unimodal and narrow. A possible solution may be to choose a recursive thresholding scheme [Ohlander 75]. The first threshold may generate two or more regions. For each region, histograms are generated and if the histogram is not unimodal, then thresholding is carried out. This recursive process terminates for a region when its histogram is unimodal. However, recursion is very difficult to implement on an SIMD machine. It could of course be implemented on a MIMD machine but there is the overhead of data transfer between processors or memory contention in a shared memory system.

The local thresholding method proposed by Chow and Kenko [Chow 72] holds more promise. An image is divided into a number of subimages. For each subimage, a histogram is generated. If the histogram is bimodal, a thresholding level is selected for that subimage. This method has been extended [Nakagawa 78] to use trimodal histogram. For the other subimages, a threshold is defined by a local weighted averaging process. On an MIMD Transputer-based array [Homewood 87], each subimage can be placed on a single Transputer. On the processors for which no threshold value is computed, it can demand the values from its neighbours and take a weighted mean. Subimages are created with local threshold values and the complete image is smoothed. Then the threshold is carried out by comparing the local pixel value with the local defined threshold value.

$$B_{ij} = \begin{cases} 1 & \text{if } \mathcal{G}_{ij} \geq T_{ij} \\ 0 & \text{otherwise} \end{cases}$$

This last process is of course very efficient on SIMD arrays. Hardware has been proposed for local histogram generation for an SIMD array, but no so such hardware has been built. Chow and Kenko's method is more promising on augmented pyramids or MIMD machines. Also the threshold selection algorithm can be substituted by many of the other methods which have been discussed here.

If an image $\mathcal{G}$ is a function $f : \mathcal{G} \rightarrow N \times N$ of 2^p grey-levels, then generating a histogram on a serial machine is an $O(N^2)$ operation and on an SIMD array it requires 2^p threshold and counting operations. Histogramming techniques on parallel architectures have been presented in Section 4.3.1. Its efficiency varies considerably depending on the computer architecture. On a machine such as CLIP4, avoiding histogramming may be an advantage. There are a number of methods for threshold selection which do not involve histogram generation. We will explore some of these techniques below.

Edge pixels have values which lie between the foreground and background pixel values. The average of the edge pixel values could then form a suitable threshold value [Katz 65]. This is called the *high-gradient pixel average* method. Weszka developed a similar method called *high-gradient histogram* method, it chooses the threshold from the peak in the histogram [Weszka 75]. There are variation on these methods which are discussed in detail in [Kohler 81].

The Ridler and Calvard method chooses the threshold value by a process of relaxation. The algorithm first calculates the mean grey-value, t_0, and thresholds the image at this value.

$$t_0 = \sum_{x,y} f(x,y)/N$$

where f(x,y) is the grey-value at that coordinate and N is the total number of pixels in the image. For the nth iteration, the threshold value, t_n, is calculated as

$$t_n = \frac{1}{2}\left[\frac{\sum_{x,y} f(x,y) * T_{n-1}(x,y)}{M_{n-1}} + \frac{\sum_{x,y} f(x,y) * \overline{T_{n-1}(x,y)}}{N - M_{n-1}}\right]$$

where $T_{n-1}(x,y)$ is a binary image resulting from thresholding the image at t_{n-1}. The algorithm stops when the threshold value converges or after a predetermined number of iterations. If large amounts of background are present which we wish to ignore, the algorithm needs to be modified to read

$$t_n = \frac{1}{2}\left[\frac{\sum_{x,y} f(x,y) * T_{n-1}(x,y)}{M_{n-1}} + \frac{\sum_{x,y} f(x,y) * \overline{T_{n-1}(x,y)}}{N_B - M_{n-1}}\right]$$

where N_B is the number of pixels below the t_{n-1} threshold value not counting the pixels with grey-level 0. The initial condition is also modified:

$$t_0 = \frac{\sum_{x,y} f(x,y) * T_{>0}}{N_B}$$

where $T_{>0}$ are the non-zero pixels.

Another adaptive thresholding technique which does not involve histogramming is due to Reeves [Reeves 82]. This is based on the neighbourhood max-min operation,

$$MMT(i,j) \leftarrow 1 \text{ iff } 2P(i,j) > max_L(P(i,j)) + min_L(P(i,j))$$

where max_L and min_L are neighbourhood maximum and minimum operations calculated over a $L \times L$ size window. These are large window mathematical morphology operations which are the result of iterative application of the local-neighbourhood maximum and minimum operations; which are efficient on cellular logic arrays such as CLIP4, DAP and MPP. This method is particularly good at segmenting edges in the image and small features. The window size found to give good results is 17×17 pixels. Since the method is sensitive

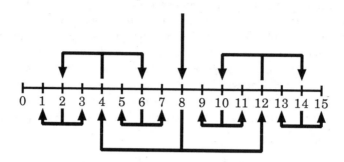

Figure 5.2: Traversing a binary tree to converge to a threshold value.

to noise, the noise in the image needs to be reduced by (1) averaging several images and (2) applying a large window median filter.

The Ridler–Calvard method fails primarily because the histogram distribution of many images is not bimodal. Kapur's method suffers from the same problem, the entropy map has no visible maximum. Furthermore, both these methods are *slow*. For the Ridler–Calvard method, if the histogram is bimodal, then convergence occurs after a single iteration. However, when the histogram is a noisy monotonic version, the relaxation is badly behaved. The slowness of the Kapur's algorithm for an SIMD array arises from the need to calculate the histogram and logarithms.

We present a fast thresholding algorithm (Algorithm 5.1) which does not require to calculate a histogram. Furthermore, the algorithm has a worst case complexity of $O(p)$, where p is the number of bit-planes in the image. Presented with an image **im** and the percentage of the edgels (edge elements), p_tile, above a given threshold, a suitable threshold value for the image may be calculated. It is assumed that the image noise has essentially been removed by some filtering process. Knowing the total number of pixels in the image (total_no_edge) and number of the strongest edge pixels that needs to be kept (p_edge_to_keep), a binary tree can be used to choose the threshold value (Figure 5.2).

We begin by initially thresholding the edge image (**edge_im**) of im at a level 2^{p-1}. The threshold value is increased if the number of edgels is found to be significantly greater than p_edge_to_keep; otherwise, if significantly fewer, we decrease the threshold level based on a binary-tree (see Figure 5.2 for an example). This process is continued until the number of edgels resulting from the threshold is equal to p_edge_to_keep (within a given error limit). The details of this algorithm are given in Algorithm 5.1.

If **edge_im** has p-bit precision, then to generate its histogram, 2^r threshold and count operations would be required, as opposed to only $(p + 2)$ threshold and count operations using the binary tree. The time taken to calculate the

Algorithm 5.1: Threshold selection based on edge strength.

THR_EDGE(**im** , p_tile);
edge_im ⟵ $EDGE$(**im**);
thr_val ⟵ 2^{p-1};
$step$ ⟵ 2^{p-2};
$total\_no\_edge$ ⟵ $\sum THR$(**edge_im**, 0);
$p\_edge\_to\_keep$ ⟵ $total\_no\_edge * p\_tile$;
new_edge_value ⟵ $\sum THR$(**edge_im**, thr_val);
$test$ ⟵ $(|new\_edge\_value - p\_edge\_to\_keep| > LIMIT) \wedge step > 1$;
<u>while</u> ($test$)<u>do</u>
        if ($new\_edge\_value > p\_edge\_to\_keep$)
           then
                thr_val ⟵ $thr\_val + step$; **fi**
      else
           thr_val ⟵ $thr\_val - step$; **elsf**
      tmp_im ⟵ THR(**edge_im**, thr_val);
      new_edge_value ⟵ $\sum$ **tmp_im**;
      $step$ ⟵ $step/2$;
      $test$ ⟵ $|new\_edge\_value - p\_edge\_to\_keep|$;
      $test$ ⟵ $(test > LIMIT) \wedge step > 1$;
<u>od</u>
$grey\_edge\_value$ ⟵ $\sum ANDS$(**im**, **tmp_im**);
thr_val ⟵ $\lceil grey\_edge\_value/new\_edge\_value \rceil$;
im_result ⟵ THR(**im**, thr_val);
$return$(**im_result**);

number of pixels in a given level is approximately equal to the time taken for a single threshold.

The LIMIT is set at 5% of total_no_edge, then the threshold value the algorithm converges to has p_tile ± 5% of the total edge pixels. Having found the p_tile ± 5% of the strongest edge pixels, these pixels are used to calculate the threshold value to apply to the original image **im**.

5.2.2 Region-Based Segmentation

Apart from multiple thresholding, there are other methods for region-based segmentation. Regions, which are contiguous, simple connected clusters of pixels, are mutually exclusive and exhaustive (i.e. a pixel can only belong to a single region and all pixels have to belong to some region). A region may support a set of predicates; however, an adjacent region cannot support the same set of predicates.

The advantage of using a region-based segmentation are: (1) there are far fewer regions than pixels in an image, thus allowing data compression; and (2) regions are connected and unique. The disadvantages of the method are: (1) assumptions are made about the uniformity of image features; (2) a region could be erroneously considered to be a single surface; and (3) surface properties or reflection could produce noise regions.

There are two principal approaches to region-based segmentation:

Region growing: Initially each pixel could be considered to be a separate region; these are referred to as the *atomic regions*. Adjacent regions are merged if they have similar properties (such as grey-level). This merging process continues until no two adjacent regions are similar. The similarity between two regions is often based upon simple statistics such as the variance measure or the range of grey-level within the regions. Region-based segmentations are described in [Brice 70, Yakimovsky 73, Feldman 74, Harlow 73]

Region splitting: Initially the image is regarded as being a single region. Each region is recursively subdivided into subregions if the region is not homogeneous. The measure for homogeneity is similar to that for region growing. Robertson *et al.* subdivided a region either horizontally or vertically if the pixel variance in the region was large [Robertson 73]. Others [Prewitt 70] have used the bimodality of histograms to split regions. This is referred to as the *mode method.* This method has been extended [Ohlander 75, Ohlander 78] to use multiple thresholds from the histogram of the region. However, as discussed above, a histogram gives global information about a region. Using this method, some pixel could have a wrong label; therefore pre- and post-filtering is required [Ohlander 75].

Growing regions is a more difficult task then region splitting. However, region splitting, using the method described in [Robertson 73, Prewitt 70], can lead to the generation of more regions than required. Some region merging at the end of the splitting phase is required. However, given some suitable growing points, regions can be grown using mathematical morphology operations very efficiently on SIMD array processors.

The region-splitting and merging methods have been combined by Horowitz and Pavlidis [Horowitz 74]. Given a predicate $\mathcal{H}$ which operates upon a region $\mathcal{R}$ (which could be a match for similarity of colour, i.e. calculates the mean and variance of the pixel intensity in the region), we could use the result of the predicate to either split or merge regions. Three tests are required:

1.

$$\mathcal{H}(\mathcal{R}) = \text{false}$$

Split the region into four sub-regions using the quadtree method (see Figure 5.3).

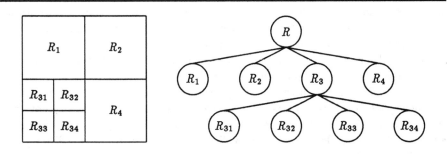

Figure 5.3: Splitting and merging of regions.

2.
$$\mathcal{H}(\mathcal{R}_{k1} \cup \mathcal{R}_{k2} \cup \mathcal{R}_{k3} \cup \mathcal{R}_{k4}) = \text{true}$$

Merge the regions of the quadtree leafs k into the region k.

3.
$$\mathcal{H}(\mathcal{R}_i \cup \mathcal{R}_j) = \text{true}$$

If two adjacent regions are similar, then merge these two regions.

5.2.3   Texture

In the previous section, we described methods for image segmentation based on the grey-level properties of objects. These methods generally work well for man-made objects which usually have a smooth grey-level surface. We observe a textured region as being *homogeneous*, although the intensity across the region may be non-uniform. This leads to the intensity-based segmentation methods to produce results which do not match with our perception of the scene.

Texture is important not only for distinguishing different objects but also because the texture gradient contains information describing the objects depth and orientation [Gibson 50]. Using texture to extract the shape of objects is discussed in Chapter 6. Texture can be described by its their statistical or structural properties [Lipkin 70]. A texture surface having no definite pattern is said to be *stochastic*, while texture with a definite array of subpatterns is said to be *deterministic*. These textured surfaces can then be described by some placement rule for the pattern primitives. In reality, deterministic texture is corrupted by noise so that it is no longer ideal; this is referred to as the *observable texture*. If the pattern making up the deterministic texture itself has subpatterns, then these are called *microtextures* and the larger patterns are called *macrotextures*.

The simplest statistical description for texture involves the use of first-order statistics of individual pixels such as their intensity level. Also for a set of pixels, the mean, variance, skewness and kurtosis (the first, second, third and fourth moments respectively) can be used as texture features [Nagao 76].

Julesz observed that the human eye is also sensitive to second-order statistics [Julesz 75]. The second-order statistic measurements involve the probability of a pair of pixels having some grey-level value. A set of such statistics are measured for pairs of pixels at different distances and rotations; these are summarised in a table called a *co-occurrence matrix*. $P(I_1, I_2, | d, \theta)$ is the probability that two pixels separated by a distance d at orientation θ will have the intensity value I_1 and I_2. Co-occurrence matrices were first used by Haralick *et al.*, computing the matrices for $d = 1$ and $\theta = \{0, 45, 90, 135\}$ degrees [Haralick 73]. Fourteen texture discrimination features were defined by Haralick *et al.* for each matrix. However, in practice, only a subset of these are used (such as energy, entropy, correlation, local homogeneity and inertia [Conners 80]). A similar method is to measure the difference rather than pixel similarity [Weszka 78].

A two-step approach to texture detection using statistical measurements is described by Laws [Laws 79, Laws 80]. The first step consists of computing microstatistic features using either 3×3 or 5×5 convolution masks. In the second step, macrostatistic features are computed over larger windows using such techniques as sum of squares or absolute value of the microfeature image.

All the methods described above involve operations over small neighbourhoods and a small number of directions. The computation is local and independent of the data; the methods are data parallel and map well to the SIMD model of processing. There are other methods for texture measurement which do not conform so well to the SIMD model. Among these are the grey-level run-length method [Galloway 75]. This method consists of computing the run length for a given intensity value at different orientation. Other methods (while being essentially SIMD) involve global or large window computation and therefore tend to be inefficient for computation on locally connected array processors. These latter methods could involve the computation of power spectrum [Ekhlund 79, Chen 82] and autocorrelation [Kaizer 55].

In the structural, as opposed to statistical, approach to texture feature extraction, first the texture primitives have to be located, then the placement rule has to be found, and finally, the primitives are characterised by their regularity. Various methods have been developed for both detecting and representing the primitives; these include the use of tree grammar [Lu 78] and histograms of the separation between primitives [Davis 79]. Leu and Wee extracted primitives by thresholding edge images using a technique similar to that described in §5.2.1 (page 118). Then the edge connected regions are removed. The shapes of the remaining regions are characterised by computing their area, perimeter length, Euler number, orientation (major axis), spreadness and elongation.

Many structured textures consist of line features, such as walls, fabrics, wire mesh and rows of windows. Line features can be very effectively extracted using the Hough transform technique. The pattern in the Hough space obtained from

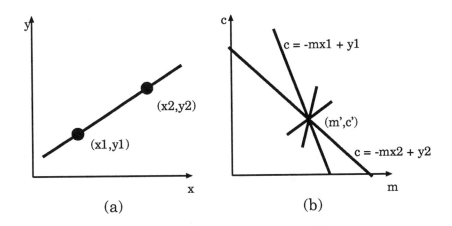

Figure 5.4: A line in the (a) image space and (b) parameter space.

the transform would be a regular set of peaks. From this, the line separation
and orientation structure information can be obtained [Eichmann 88]. This
method for texture analysis is invariant to scale, translation and orientation of
the texture pattern.

5.3   Hough Transform

The Hough transform provides a method for detecting straight lines in digital
images [Hough 62]. The method is equivalent to both template matching and
to spatial matched filtering [Sklansky 78]. The method proposed by Hough
has been extended to detect many analytically defined shapes such as circles
[Kimme 75] and ellipses [Tsuji 78].

Given a point in the Euclidian space (x_i, y_i) and an equation for a line in
the form $y_i = m \cdot x_i + c$, there are an infinite number of lines that pass through
the point (x_i, y_i). However, consider the same equation in the *parameter space*,
$c = -m \cdot x_i + y_i$, then we have a straight line for a given point (x_i, y_i). Another
point (x_j, y_j) will have a line associated with it and this line will intersect the
line associated with (x_i, y_i) at (m', c'). The method is illustrated in Figure 5.4.
In the original formulation of the Hough transform, for every point in the image
plane, the intercept value c was plotted for a range of gradient (m_{max}, m_{min})
values. However, this method suffers from singularities when the lines are
near vertical, i.e. $m \rightarrow \infty$. To overcome this problem, a parametric definition,
$x \cos \theta + y \sin \theta = \rho$, for a line is used [Duda 73]. Furthermore, using the
gradient information [Kimme 75] of the edgels in the image, the computation
can be considerably reduced as shown in the example algorithm given below.

Algorithm 5.2: Line detection using Hough transform.

1. Set accumulator A(c,m) = 0.

2. Differentiate the image using Sobel Operators

   (a) x−gradient (g_x) and y−gradient (g_y).

3. If the edge magnitude is greater then a threshold, calculate the gradient $m = (g_y/g_x)$.

4. Calculate $c = -mx_i + y_i$.

5. Increment accumulator, A(c,m) = A(c,m) + 1.

6. Repeat (3) to (5) for all edge points.

7. Peak in A(c,m) gives the line gradient and intercept.

However, this method suffers from singularities when the lines are near vertical, i.e. $m \rightarrow \infty$. To overcome this problem, a parametric definition, $x \cos \theta + y \sin \theta = \rho$, for a line is used [Duda 72]. The method has been subsequently developed to detect both analytically defined curves and lines.

The Hough transform has been proved to work in the presence of noise or even when parts of an object are obscured or not present. Consequently, it has found successful use in many fields. Among its uses have been the detection of human haemoglobin fingerprints [Ballard 75], the detection of tumours in chest radiographs [Kimme 75], the detection of chest ribs [Wechsler 77], the detection of straight edges in industrial images [Davies 86a] and food product inspection [Davies 84b]. However, whilst a Hough transform will locate an object in an image scene, the product will still need to be scrutinised. One fast method for checking food products is using a *radial histogram* [Davies 85] when for example biscuits have circular symmetry. The radial histogram is a plot of average intensity value, I, of an annular ring at a distance r from the centre located by the Hough transform. For a homogeneous object, the function $I(r)$ is a constant within the body and discontinuous at the boundary. If $I(r : r \mapsto 0, R)$ is not a constant, or a discontinuity at $I(R)$ is not observed then the object is either only partially present (i.e. parts of it are missing or partially occluded) or is non-homogeneous.

Another example of biscuit inspection is to be found in [Edmonds 88]. In this instance, rectangular biscuits are located and inspected. The system is capable of inspecting only a single object in an image scene. A Hough transform is employed to extract the boundary of the object. The boundary sides determined are matched against a model to check size disparity. Two other faults may arise with the biscuits, (i) overflow (i.e. too much chocolate) or (ii) show-through (i.e. insufficient chocolate covering. The former by computing the size of the biscuit area touching but on the outside of the Hough generated boundary; and the latter by a simple threshold.

Although the Hough transform is a fast method [Davies 86b] for detecting straight lines and circles, results have been computed, until recently, in the parameter or Hough space. However, recently, it has been shown that for a straight line, the foot-of-normal can be mapped into the image space [Davies 86a]. The corresponding situation is more obvious for the circle [Kimme 75] and the ellipse [Ballard 81]. Using these notations, parallel Hough transforms for parametric curves have been presented [Hussain 88a] using two one dimensional accumulators.

The Hough transform can also be used to recognise an arbitrary shape; this is referred to as the generalised Hough transform (GHT) [Ballard 81]. This method is essentially a form of boundary description. The method consists of choosing a point within the object, say the centroid. As the object is traced around the boundary, at the edge point, the distance to the centre point, r, is calculated along with the angle, ϕ, pointing to the centre point. The ϕ, r data set is referred to as the R-table [Ballard 81]. Figure 5.5 shows the ϕ, r geometry used for the generalised Hough transform.

The Generalised Hough Transform

Earlier in this chapter, we gave a description of the Hough transform used to locate analytically described curves such as straight lines, arcs or circles. This method works even in the presence of noise or even if only partially viewed. We have also discussed the generalised Hough transform (GHT) for describing arbitrary shape at any orientation [Ballard 81]. A derivation of the GHT is used by [Rives 85] to recognise occluded objects.

Rives *et al.* are attempting to recognise knives in an image scene by locating the handle and the tips using a GHT model. In a similar fashion to Ballard's R-table, an S-table is used to represent objects and models. The boundaries of the segmented objects are transformed into a set of piecewise linear contours. The S-table consists of:

(i) the length L_i of segment S_i,

(ii) the orientation of the segment S_i,

(iii) the orthogonal distance D_{i1} from S_i to a chosen centre C.

(iv) the distance D_{i2} between the projection of C parallel to S_i and the midpoint of S_i.

Given a model description as above, the idea is to locate in the image all the point which can possibly be the centres of the model. Details of the algorithms used for prediction-verification and the 2D Hough accumulation are described

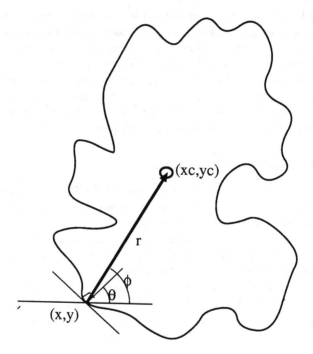

Figure 5.5: The geometry used by the generalised Hough transform.

in [Rives 86]. The system has no mirror symmetry, all objects are stored as two possible models.

The Hough transform has been proved to work in the presence of noise or even when parts of an object are obscured or not present. Consequently, it has found successful use in many fields. Among its uses has been the detection of human haemoglobin fingerprints [Ballard 75], the detection of tumours in chest radiographs [Kimme 75], the detection of chest ribs [Wechsler 77], the detection of straight edges in industrial images [Davies 86a] and food product inspection [Davies 84b].

5.3.1 Parallel Hough Transform Operations

A variety of specialised hardware and software for various architectures has been proposed for the computation of the Hough transform in an effort to speed up the computation. The proposed systems include sequential engines with pipelining [Hanahara 88], systolic arrays [Chung 85, Li 89], VLSI architectures [Rhodes 88], SIMD trees [Ibrahim 87], SIMD arrays [Silberberg 85, Cypher 87, Guerra 87, Rosenfeld 88, Hussain 88a], pipelined [Sanz 87], and scanned arrays [Fisher 87]. Most of these systems are for computing the Hough transform for straight line segments.

Using specialised sequential hardware with limited pipelining, Hanrahan has reported the time required to accumulate vote in a $512 \times 512(\rho, \theta)$ Hough space for 1024 points for a single θ value is 0.26s [Hanahara 88]. Assuming that only 3% of the image are edge points and θ is sampled at five degrees, the time taken to accumulate the vote for the image would be 12.9s. In this system, the raster addresses of the feature points (H_i, V_i) is converted to centre coordinate system, $x_i = V_i - N/2, y_i = H_i - N/2$, where $N = 256$, image space is $256 \times 256 \times 8$-bits, and the parameter space used is $512 \times 512 \times 8$-bits. The computation is in three stages. First the edge points are detected using a pipeline which involves convolution, thinning, local-maxima detection and computation of zero-crossings. Next the ρ values are computed and accumulated for a series of θ values. Finally, peak detection is carried out in the 2D histogram. The ρ computation and histogramming occur in parallel (see Figure 5.6).

Rhodes describe a wafer circuit, which is restructurable VLSI (RVLSI), for computing the Hough transform. The RVLSI consists of component logic cells which are uncommitted two-level metal interconnectivity matrix. This can be used to build customized circuits by using a laser to complete the logic circuit; the custom chip formed is then non-volatile and the process is irreversible [Raffel 85a, Raffel 85b]. The chip consists of an array of pipelined MACs (multiplier and accumulators) and parallel-serial converter (PS). The PS carries out data conversion such that the input is bit-serial and the output is bit-parallel. The wafer chip consists of 352 MAC cells in the middle and on each side there are a column of PS and SP (serial-parallel) cells. The Hough array used is 64 θ and 256 ρ values. For each edge point (x, y), it chooses 64 θ values and multiplexes them to each of the cells. The ρ computation for each θ is computed by a pair of MAC cells in parallel (see Figure 5.7). It uses two of the

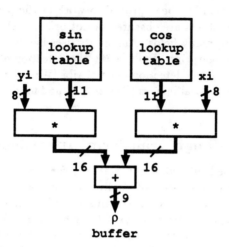

Figure 5.6: The computation of the ρ values in a pipeline.

PS cells to get the (x, y) coordinates in. The Hough transform for 64 θ values are carried out in parallel by 128 MAC cells and there are 4 SP cells to get the ρ values out. The input data is clocked in at 5 MHz. Therefore, we get a new edge point in about every 3.2 μs, and four ρ values are produced every 200 ns; a complete set of output for the 64 θ values every 3.2 μs. The Hough space is partitioned into four sections to deal with the high data output rate. The four sectors can be written to in parallel. A typical 256×256 image could have 10% edge structure. For such an image, the Hough transform can be produced in about 20 ms. This is close to frame rate.

For mesh-connected SIMD array computers such as GAPP, Silberberg has described a parallel Hough transform method which which works through each θ value at a time [Silberberg 85]. The edge points in the image are computed using the Sobel edge operator. Therefore each processor has the direction of the line. Then working through each θ value between 0 and π, the host computes the $\cos \theta$ and $\sin \theta$ values and broadcasts these to all the PEs. If at any PE, the θ value it has for the line is not in the range of the values broadcasted, it does nothing and waits for the next values to be broadcast. Otherwise, it computes the ρ value, $\rho = x \cos \theta + y \sin \theta$, where (x, y) are the coordinates of the pixels and the PEs. The transform is now carried out in two dimensions by moving the (ρ, θ) value computed to the appropriate PE with address (ρ, θ) so that it can be accumulated. This moving of data is the most time-consuming part of the algorithm since the operation essentially involves carrying out a series of 2D histogramming; a global operation. This operation is carried out by cyclically rotating the ρ values through the PEs vertically. At each stage, the ρ value is compared with the PE's row number. If the ρ value matches

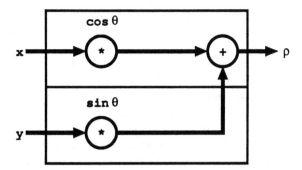

Figure 5.7: A cell in a RVLSI chip for the Hough transform.

the row number, then the counter at that PE is incremented. This achieves histogramming for a particular θ value. These values are then shifted cyclically in a horizontal fashion to the appropriate θ column and the ρ-count value is added to the counter. This accomplishes the 2D histogramming for a single θ value. We see that the order computation using this method is $O(\theta NM)$, where the Hough space is quantised to $M \times N$.

Rosenfeld *et al.* have looked at extending this method by working through different θs at each PE [Rosenfeld 88]. Because each PE has already computed a θ value, the edge gradient at that PE, it computes its own range of θ and the corresponding $\sin \theta$ and $\cos \theta$ values. This is done by the host sending an offset number to each PE and adding this to the PEs θ value. If k different offsets are used, then each PE has k (ρ, θ) pairs. Again one needs to compute the histogram for the (ρ, θ) values. This could be done by first moving the (ρ, θ) pair into the correct θ column, then into the correct ρ row. The computational cost of this method is $O(kMN)$ and k is smaller than the number of θs used. However, the method is expensive in its use of memory because for a given θ value, any PE might have to hold a complete set values equal to the number in the peak.

A systolic method for computing the Hough transform has been presented by Li *et al.* [Li 89]. The advantage of the systolic approach is that it minimises the external I/O. The system consists of a linear array of compute cells, a routing network and accumulator cells. If there are N processing cells, and ρ accumulator is quantised to M_r cells, then the routing network is a $N \times M_r$ array. The concurrency in the method arises from processing a row (or column) of an image together. All the cells compute ρ for a given θ. The first cell computes

$$\rho_{ij} = (i \cos \theta_k + j \sin \theta_k)/\Delta \rho$$

The next cell then computes

$$\rho_{i+1,j} = \left((i+1) \cos \theta_k + j \sin \theta_k \right) / \Delta \rho$$

The difference in the destination between the two adjacent cells is then

$$\rho_{i+1,j} - \rho_{ij} = \cos \theta_k / \Delta \rho$$

The destination registers are preloaded with the value $j \times \sin \theta_k / \Delta \rho$; then in every time unit, $\cos \theta_k / \Delta \rho$ is added to the destination register. As a packet moves down the routing network, its value is decremented. When it becomes zero, it is moved across to the accumulator.

To begin with we will look at the implementation of the Hough transform to determine the centre of a circle. If we have an image $\mathcal{F}$ and its pixels have a grey-level value $\mathcal{F}(x, y)$ then we extract edge gradients using Sobel operators such that $\mathcal{F}_x$ and $\mathcal{F}_y$ are the Sobel gradients at each pixel in the $x-$ and $y-$directions. If the edge gradient magnitude, $\|\mathcal{F}\|$, at a given pixel (x, y) is greater than zero, the centre coordinate of a circle belonging to that point, for a given radius of a circle, can be computed. A method for computing the Hough transform on a SIMD array processor is given in Algorithm 5.3.

The result is, we have two images, $\mathbf{x_c}$ and $\mathbf{y_c}$, containing the $x-$centre coordinates and the $y-$centre coordinates respectively. To calculate and plot the centre coordinates in the image space, we require to find some correspondence between the two images. A method for this is given in Algorithm 5.4.

Algorithm 5.4 Getting correspondence between x_c and y_c.

1. Plot histogram for image $\mathbf{x_c}$ and store result in **hist**

2. Clean histogram image **hist** *for noise by*

   (a) Removing impulse noise using the following two 3×3 templates. If the template fits, delete the central pixel. The first template deletes isolated lines and the second deletes isolated points.

0	1	0
0	1	0
0	1	0

1

0	0	0
0	1	0
0	0	0

2

   (b) Deleting histogram bins below a given threshold.

   (c) Recovering lost peaks. The combined effect of steps (a) and (b), referred to as function $\mathcal{G}$, is that some 'correct' peaks will also be removed.

$$\text{temphist} \longleftarrow \mathcal{G}(\text{hist});$$

   These lost peaks can be recovered by a labelling process. Using **temphist** *as a seed image and propagating signals vertically from the seeds, only the objects connected topologically in a vertical sense in image* **hist** *are extracted.*

Algorithm 5.3: Calculate x_c and y_c in parallel

1. Store image containing circles in $\mathcal{F}$

2. Calculate the edge gradient in $x-$ and $y-$direction using the Sobel operator and store the result in $\mathcal{F}_x$ and $\mathcal{F}_y$ respectively.

3. Calculate the edge magnitude image.

$$\|\mathcal{F}\| \longleftarrow \sqrt{\mathcal{F}_x^2 + \mathcal{F}_y^2};$$

4. Set a threshold for an edge magnitude.

$$\mathbf{mask} \longleftarrow THRESHOLD(\|\mathcal{F}\|, thr);$$

5. Modify $\|\mathcal{F}\|$ to take care of division by zero. The modification required is, where $\|\mathcal{F}\|_{x,y} = 0$, set it to some non-zero value.

6. Calculate x_c, y_c :

$$\mathbf{x_c} \longleftarrow AND((\mathbf{xramp} - \lceil R * \mathcal{F}_x / \|\mathcal{F}\| \rceil), \mathbf{mask});$$

$$\mathbf{y_c} \longleftarrow AND((\mathbf{yramp} - \lceil R * \mathcal{F}_y / \|\mathcal{F}\| \rceil), \mathbf{mask});$$

The AND function acts as a masking operation, removing the results calculated for the modified $\|\mathcal{F}\|$ pixels.

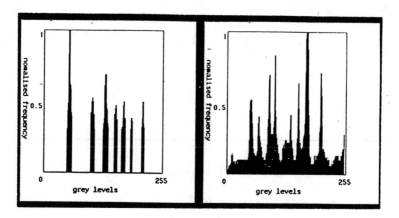

Figure 5.8: The two histograms containing the centre coordinates.

3. *The above steps may not have separated the multiply joined maxima. They can be separated using a spatial high-pass filter on the cleaned histogram image* **hist**.

$$\textbf{blob} \longleftarrow EXOR(S^n E^n(\textbf{hist}), \textbf{hist});$$

4. *Spreading* **blob** *in the vertical plane.*

5. *Removing the histogram bins which are marked by the image* **blob**.

$$\textbf{hist} \longleftarrow AND(\textbf{hist}, NOT(\textbf{blob}));$$

6. *Skeletonising the image* **hist**.

7. *Repeat to 10 for each existing bin* x *corresponding to each* x_c *(see text for detail),*

8. *Extracting the* y_c *for a given* x_c *by masking operations.*

$$\textbf{mask} \longleftarrow THRESHOLD(\textbf{x}_c, x_c);$$

$$\textbf{yim} \longleftarrow AND(\textbf{y}_c, \textbf{mask});$$

9. *Generating a clean histogram for* **yim** *using steps (2)-(6).*

10. *Getting the y-coordinate and plot* (x_c, y_c).

The Hough transform is a very robust technique as is demonstrated with some O ring objects in an image simulating the sella turcica with known radius. Figure 5.8 show the two histograms containing the centre coordinate data and the result for the given image is shown in Figure 5.9.

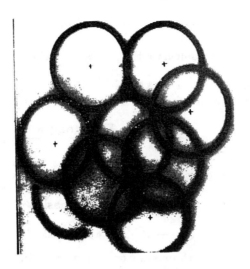

Figure 5.9: The result of the Hough transform for detecting circles.

The parallelism is obvious when computing the centre of a circle using the Hough transform because the result does not require to be plotted in the Hough space but can be plotted directly back into the real or *image* space. For line Hough transforms this has been different. However, Davies' method [Davies 86a] of computing the *foot-of-normal* of a straight line gives us an opportunity for parallel computation in image space. Furthermore, a modification to Algorithm 5.4 can be used for its computation, Algorithm 5.5.

The correspondence between x_0 and y_0 is found using Algorithm 5.4. This method has been used to segment the mandible image successfully as shown in Figure 5.10.

Merlin has described a form of generalised Hough transform (GHT) which has a parallel implementation through match filtering [Merlin 75]. The method, which uses a segmented edge image, has the following set of steps:

(a) Mark a point on a template curve; call this the reference point A.

(b) Rotate the template by 180° about the point A; call this the template E.

(c) Make a trace of the template E for each edge point in the image. If the template E consists of a set of vectors (i.e. each point in the template is described by a vector), then the trace for each point is constructed by taking a copy of the edge image and shifting by the template vector and accumulating.

(d) Search for peaks in the accumulator. A peak gives the point A.

Algorithm 5.5: Calculate the *foot-of-normal* of a straight line.

• Carry out steps (1)–(5) as in **Algorithm 7.1.1**.
(6) Calculate the common factor to be used in calculating x_0 and y_0, ν :

$$\nu \longleftarrow \frac{\mathbf{xramp} * \mathcal{F}_x + \mathbf{yramp} * \mathcal{F}_y}{\|\mathcal{F}\|}$$

(7) Calculate x_0 and y_0 and store result in **xim** and **yim** respectively.

$$\mathbf{x_0} \longleftarrow AND(\nu * \mathcal{F}_x, \mathbf{mask})$$

$$\mathbf{y_0} \longleftarrow AND(\nu * \mathcal{F}_y, \mathbf{mask})$$

The AND function acts as a masking operation, removing the results calculated
for the modified $\|\mathcal{F}\|$ pixels.

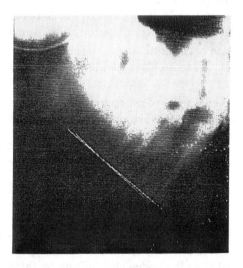

Figure 5.10: Segmenting the mandible line using Hough transform.

In summary, there are various way of implementing the Hough transform based on how the Hough space and the image are distributed amongst the processors. One option is for all the processors to have a copy or access to all feature points in the image and each processor computes the Hough transform for a range of theta values [Olson 87]. The Hough space is local to its memory. Finally, when the transformation has been computed, the peaks are detected in parallel for its θ range in each of the processors and only the peak data is finally sent to a master processor. This reduces the communication between the various processors. However, we have the initial expense of sharing or copying the image data. Another method is to distribute portions of the image to various processors but to have a global (shared) Hough space. The argument being that the communication is lower because there are fewer edge points than pixels in the image and hence the communication involved in having a common Hough space is smaller than if complete images were shared or copies sent to each of the processors. Finally, we can choose to distribute both the image and the Hough space between the processors. Partial peaks can be detected in each of the processors in parallel. Only after this, are the partially computed peaks merged. The first two methods are essentially MIMD processing and can be implemented on Transputer-based MIMD processors. The problem with these implementation is there could be memory contention in either sharing the image or the Hough space and the data transfer overhead. The third method is usually implemented on SIMD processors [Silberberg 85, Rosenfeld 88, Li 89] or can be implemented on shared-memory MIMD processors. The problem with this method is that a "multi-dimensional" histogram is generated using shift and summation operations. Such histogram generation is generally inefficient on SIMD array processors as we have discussed in this section.

5.4   Corner Detection

The usual method for "L-shaped" corner detection has been the use of 2nd derivatives to measure the rate of change of the gradient direction ("CORNERNESS") with gradient magnitude ("EDGENESS") [Beaudet 78, Dreschler 81, Kitchen 82, Zuniga 83]. A corner is declared to exist if

$$
\begin{aligned}
CORNER \quad &\leftarrow \quad (CORNERNESS > thr_c) \\
&\cap \quad (EDGENESS > thr_e)
\end{aligned}
$$

An example of such a corner detector is due to Kitchen and Rosenfeld:

$$
C = \frac{g_{xx}g_y^2 + g_{yy}g_x^2 - 2g_{xy}g_x g_y}{(g_x^2 + g_y^2)^{3/2}} \geq \text{thr}
$$

The equivalence proof for these corner detectors is provided in [Nagel 81, Shah 84]. For reasons of computational expense and also the memory usage, we wish to avoid the computation of the second derivatives such as g_{xx}, g_{yy} and g_{xy}. Fortunately, there are methods for corner detection without using the second derivatives. We will consider two such methods: Plessey corner detector and a heuristic corner detector.

One such corner detector is the Plessey algorithm [Harris 87, Noble 88]. This requires the computation of the eigenvalues λ_1 and λ_2 of the matrix $\mathbf{A}$.

$$\mathbf{A} = \left[\begin{array}{cc} < I_x^2 > & < I_x I_y > \\ < I_x I_y > & < I_y^2 > \end{array} \right]$$

The values $< I_x^2 >, < I_y^2 >$ and $< I_x I_y >$ are computed over a neighbourhood. A corner is declared if both the eigenvalues are 'large'. A major drawback to using this algorithm is the large computation involved (three neighbourhood operations using Gaussian weighting).

The above methods are data parallel in that they involve computation within small neighbourhoods. The result at any pixel is not dependant on the computation of corner at any other pixel. The next method we will discuss is more serial in its computation (but can be task parallel in that each edge segments can be distributed over different processors. But because chain coding is used, the computation time will depend on the size of the edge segments and hence not fixed.

The heuristic corner detection method is illustrated in Figure 5.11. As we raster scan the edge gradient magnitude image, we find a point $\mathbf{P0}$ which is above a threshold. We center a circular annulus operator about $\mathbf{P0}$. As we track around the annulus of radius r, we find a point at $\mathbf{P1}$ above a threshold. A reasonable assumption to make is that most edge points lie on a straight point and corners are rare; this speeds up our algorithm. Therefore, we search at position $\mathbf{P3}$ to see if a point exists; $\mathbf{P3} = \mathbf{P2} + 180°$. If such a point exists then $\mathbf{P0},\mathbf{P1}$ and $\mathbf{P3}$ are collinear and no corner exists at $\mathbf{P0}$. However, if $\mathbf{P3}$ is empty, then we search for another point on the annulus; which may be found at position $\mathbf{P2}$. A corner of angle θ then exists at $\mathbf{P0}$.

$$\theta = 180° - \alpha \parallel \mathbf{P1} - \mathbf{P2} \parallel;$$

where $\alpha = 13°$. The direction of the corner is given by the vector $\vec{n}$ and

$$r\vec{n} = (X_n - X_0)\mathbf{i} + (Y_n - Y_0)\mathbf{j}$$

where (X_0, Y_0) is the coordinates of the $\mathbf{P0}$ cell and (X_n, Y_n) is the $\lceil (\mathbf{P1} + \mathbf{P2})/2 \rceil$ cell on the annulus.

There are a few advantages to using this method: (i) it is easy to calculate the angle of the corner, its accurate to approximately $\pm 13°$; (ii) from (i) the direction of the corner can be determined; and (iii) determine whether the corner is acute or obtuse. However, the method will detect false corners if there two or more independent lines appears within the window because no line connectivity is checked.

5.5   Thinning Algorithms

Thinning is the process of successively removing layer by layer the "outer shell" of an object, leaving behind a single-pixel-wide connected frame. This

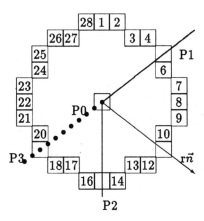

Figure 5.11: A heuristic corner detector.

"thinned" object is called the *skeleton* or the *medial axis* of the body. Sherman was the first to develop the concept of using thinned objects to recognise patterns [Sherman 59]. Skeletons have mostly found their use in data compression [Freeman 61, Deeker 72, Udupa 75, Judd 79] and shape description [Freeman 61, Hilditch 69].

If we assume that S is the set of pixels in the object, and B is the set of boundary pixels, then for each point p in S we can compute its shortest distance ρ_{min} to a boundary point. If two or more boundary points exist with the same ρ_{min}, then p is a *medial axis transformation* (MAT) point. However, determining the medial axis of a body in this fashion is computationally prohibitive. If there are n points in S and k points in B, then on a serial computer it would require $O(kn)$ operations to determine the MAT for a body. This method can be parallelised; but not SIMD or data parallel because the operations needed involve random access and depend on the number of points in S and B. We can partition S into a number of smaller sets: $S = \{S_1, S_2, \ldots, S_n\}$ and place these on separate processors with each of the processors having a copy of B. If we have N processors, then $O(nk/N)$ operations are required; the best we can achieve is when $n = N$.

To exploit a greater amount of parallelism, the same operations need to be carried out at each pixel, determining whether it is part of the skeleton; retaining these points and deleting the rest. However, in deleting any pixels we must ensure that the *connectedness* of the figure is maintained. A pixel which can be deleted such that its neighbourhood of pixels is still connected is called a *simple* pixel. An *east border* pixel of an object is one for which the pixel immediately to its right is not in the object. Other *border* pixels can be defined in a similar fashion. Rosenfeld has shown that removing a set of border pixels

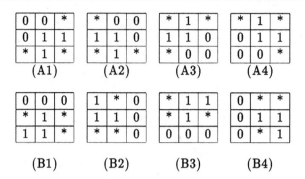

Figure 5.12: The Arcelli masks used in thinning.

in parallel preserves connectedness [Rosenfeld 75]. However, we cannot remove all border pixels in parallel. The order in which they are removed has to be chosen such that the symmetry of the shape is preserved.

An algorithm based on this principle is due to Arcelli *et al.* [Arcelli 75] and has been implemented on the CLIP4 SIMD processor array [Hilditch 83]. Arcelli's algorithm uses eight masks (see Figure 5.12). For all the pixels (in parallel), its neighbourhood is compared to mask [A1]. If there is a match then the central pixel of the neighbourhood is deleted. Because of the powerful neighbourhood gating operations in CLIP4, only two instructions are required to test whether the mask matches the data; once to match the ones and then to match the zeros. This process is then repeated for the masks [B1],[A2],[B2],[A3],[B3],[A4], and [B4] in the prescribed order. This forms a cycle of operations. This cycle has to be repeated until no pixels are removed by a cycle of operations.

Instead of computing which pixels are simple, pixels can be marked for deletion by computing their *crossing number*. The crossing number is defined by [Davies 81, Hilditch 83]

$$\mathcal{X} = \sum_{k=1}^{4} \left(n_{2k} \bigwedge \overline{n_{2(k+1)}} \right) + \left(n_{2k+1} \bigwedge \overline{n_{2k}} \bigwedge \overline{n_{2(k+1)}} \right)$$

where the neighbourhood used is defined to be

n1	n2	n3
n8	n0	n4
n7	n6	n5

Any pixel with $\mathcal{X} = 2$ may be deleted in parallel.

Another method based on crossing numbers is due to Zhang and Suen [Zhang 84]. The method consists of two subiterations. For both subiterations:

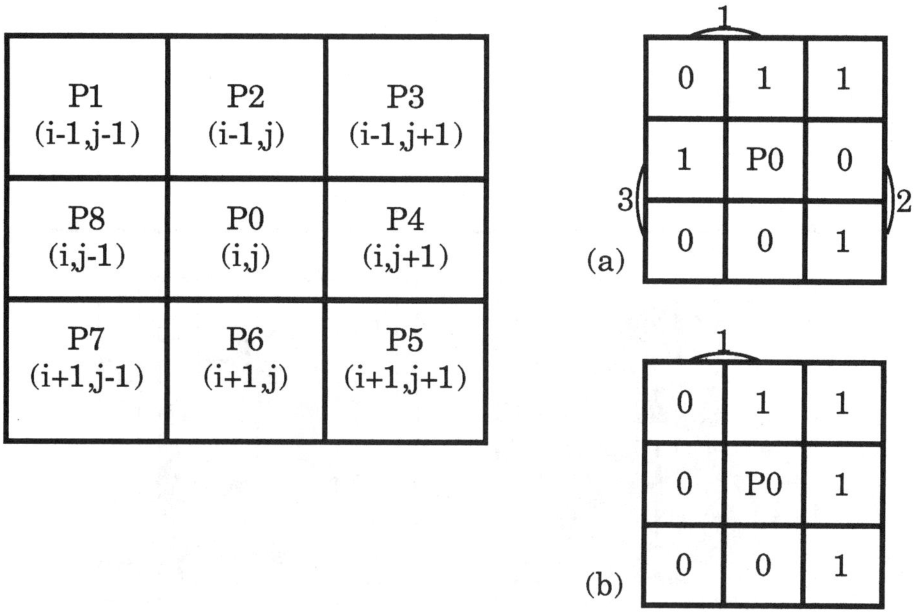

P1 (i-1,j-1)	P2 (i-1,j)	P3 (i-1,j+1)
P8 (i,j-1)	P0 (i,j)	P4 (i,j+1)
P7 (i+1,j-1)	P6 (i+1,j)	P5 (i+1,j+1)

(a)

0	1	1
1	P0	0
0	0	1

(b)

0	1	1
0	P0	1
0	0	1

In (a) A(P0) = 3, B(P0) = 4, P2*P4*P6 = 0, P4*P6*P8 = 0.
    P0 cannot be deleted because A(P0) <> 1.
In (b) A(P0) = 1, B(P0) = 4, P2*P4*P6 = 0, P4*P6*P8 = 0.
    P0 can be deleted.

Figure 5.13: The neighbourhood measurements used in Zhang and Suen's method.

(i) the number of zero pixels in a neighbourhood ($B(p_1)$) is computed, and (ii) the number of 01 pixel pairs as one circles around the neighbourhood ($A(P_1)$) (see Figure 5.13(b)) is computed. Along with these two computations, in the first subiteration, one also tests whether $P_2 \cap P_4 \cap P_6 = 0$ and $P_4 \cap P_6 \cap P_8 = 0$. In the second subiteration, these two tests are replaced by $P_2 \cap P_4 \cap P_8 = 0$ and $P_2 \cap P_6 \cap P_8 = 0$. If all 4 tests in each subiteration are satisfied, then the central pixel P_1 is deleted. An example of the algorithm in shown in Figure 5.14.

All the thinning algorithms we have described are data parallel. The Arcelli method is particularly suitable (with some modification proposed by Hilditch [Hilditch 83]) for bit-serial array processors because it uses neighbourhood template matching rather than counting. Therefore, Zhang and Suen's method is not likely to be as fast as Arcelli's method for an array processor such as CLIP4. Because the algorithms are data parallel, they can be implemented in MIMD multi-processor computers by distributing parts of the image (with

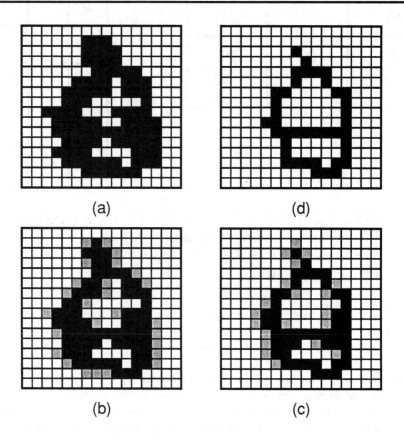

Figure 5.14: Example of thinning using Zhang and Suen's method.

one pixel overlap) and then carrying out one of these algorithms (Zhang and Suen's probably more suitable) in parallel. A problem with this approach is load balancing. The image is partitioned without knowledge of the scene and therefore some processors may have *thicker* objects to thin while others may have no useful data.

5.6 Feature and Shape Description

Many of the systems we describe in this section had their origin in the late 70s and a good survey of them is provided in [Binford 82]. Simple shape properties such as an object's perimeter length, area, eccentricity, number of corners and holes and their location and normalised central moments are used in many cases to uniquely label an object in a scene [Kitchen 83a]. However, if two objects are identical in all respects but their 'sense' (i.e. mirror images) or are partially occluded or touching, the above measures will not distinguish between them; a template-matching approach may be required. Any template-matching method considered must be capable of recognising translated objects at different orientations and indicate their relative orientations. As an extension, template matching schemes should also be invariant to scale changes and recognise partially occluded or touching objects.

There are numerous methods for object recognition involving digital images. However, using boundary or edge information is probably the most important. This statement is based partly on our ability to (1) recognise silhouette objects or line drawings and (2) recognise occluded objects even when only a partial boundary information is preserved and the loss of radial information. The importance of edge or boundary information in visual scene analysis is further indicated from psychological and psychophysical studies [Attneave 54, Lettvin 59, Hubel 62]. Of course, we are ignoring from our study another important source of visual information, namely *texture*. For example, both surface shape and orientation [Witkin 81] and spatial structure [Leu 85] can be recovered from texture information. A very useful introduction to texture analysis can be found in [Ballard 82].

5.6.1 Shape Description and Template Matching

To a large extent, object recognition involves some form of *template matching*, whether this be direct cross-correlation of grey-level images or classification based on an object property. For example, if the objects in an image are not occluded or touching, they can be classified using such simple properties as the area, perimeter, minimum bounding rectangle, semi-major and minor axis and/or the number, size and position of holes and corners [Kitchen 83a]. The ($area/perimeter^2$) measure has also been used as a 'shape' parameter [Rosenfeld 82, Barrow 71].

At this stage, a general discussion on the template matching method will be given. Given two functions $\mathcal{F}_1(x)$ and $\mathcal{F}_2(x)$, there are a number of methods for determining a match between them. The methods can be classified as being either a *similarity* measure or a *difference* measure. For the former, a high value and for the latter, a low value indicates a match.

1. Normalised cross-correlation (similarity measure)

$$\mathcal{C}(x) = \frac{\mathcal{P}(x)}{\mathcal{Q}(x)}$$

where

$$\mathcal{P}(x) = \sum_i \mathcal{F}_1(x+i)\mathcal{F}_2(x+i)$$

and

$$\mathcal{Q}(x) = \sqrt{\sum_i \mathcal{F}_1(x+i)^2 \mathcal{F}_2(x+i)^2}$$

2. Sum of absolute differences (difference measure)

$$\mathcal{A}(x) = \sum_i |\mathcal{F}_1(x+i) - \mathcal{F}_2(x+i)|$$

3. Sum of squared differences (difference measure)

$$\mathcal{S}(x) = \sum_i [\mathcal{F}_1(x+i) - \mathcal{F}_2(x+i)]^2$$

The computation of the normalised cross-correlation is the most expensive of the three. The sum of squared difference weights the values and therefore is more sensitive to some data points being widely separated. For this reason, the last method was chosen. The actual matching algorithm will be discussed in the following sections.

To distinguish between very similar or more complicated objects and to extract orientation information concerning them, sub-image template matching may be required. At its simplest, this method consists of moving a template over the search area in the image and measuring the similarity between the image area covered by the template and the template itself. The general method of grey-level template matching employs cross-correlation. Given a template **W** and an image **S**, there are several basic correlation techniques; these are discussed in [Hall 79]. The basic correlation technique can be defined [Wong 78] as

$$\mathbf{R}(x,y) = \frac{\sum\sum \mathbf{S}(i,j)\mathbf{W}(i-x,j-y)}{\sqrt{[\sum\sum \mathbf{S}^2(i,j)]}\sqrt{[\sum\sum \mathbf{W}^2(i-x,j-y)]}}$$

and

$$-1 \leq \mathbf{R}(x,y) \leq 1.$$

Note that for a given window or template **W**, $\sum\sum \mathbf{W}^2(i-x,j-y)$ is a constant.

However, there are several problems with using this sub-image template matching technique in detecting objects in *real* images. The template and the object may have different orientations. This will have to be corrected for, as well as any change in scale. Since the template window can be a large fraction of the size of the image, it can lead to a large amount of computation. Of course, the computation may be reduced by some form of hierarchical method. A second problem is that the method is relatively sensitive to noise [Wong 76a]. If noise is present, this could be in the form of small change in scale or deformation; the correlation function then produces a relatively broad peak, making it difficult to find the exact position of the correlation peak. Despite these problems, [Chien 74] have successfully used template matching to detect 'corners' on the lung boundary from chest radiographs.

However, the problems of matching for different scales and orientations in terms of the computation involved are more severe. What is required is a template matching scheme invariant to both scale and orientation. One such method is using *invariant moments* [Wong 78]. However, this method is still computationally intensive. To improve on this, a two-stage template matching method is proposed in [Goshtasby 85]. The first stage consists of a zeroth-order moment matching. This determines the likely regions for a second stage matching using second- and third-order moments.

The two-dimensional $(p+q)$th order moment of a digital image $\mathcal{F}$ is defined as

$$m_{pq} = \sum_x \sum_y x^p y^q \mathcal{F}(x,y) / \sum_x \sum_y \mathcal{F}(x,y)$$

where $\mathcal{F}(x,y)$ is the pixel value at location (x,y). The moments m_{pq} are made invariant by the following modification,

$$u_{pq} = \sum_x \sum_y (x - \mathbf{x})^p (y - \mathbf{y})^q \mathcal{F}(x,y) / \sum_x \sum_y \mathcal{F}(x,y)$$

where

$$\mathbf{x} = m_{10}/m_{00}$$

$$\mathbf{y} = m_{01}/m_{00}$$

u_{pq} are referred to as the $(p + q)$th order central moments. Using second- and third-moments, a set of seven invariant moments are calculated which are invariant to translation, rotation, and scale changes [Hu 62]. This method has been used for scene analysis in remotely sensed images [Sadjadi 78] and also in aircraft identification [Dudani 77]. A further point to be made is that the sub-image and invariant moment template matching scheme are SIMD [Flynn 72] parallel algorithms and can be efficiently implemented on SIMD architectures [Duff 78, Reddaway 79, Batcher 80]. Potter and Otto have developed an algorithm for parallel computation of centroid of many objects in the scene using a method [Potter 84] for isolating different regions on the CLIP processor array.

To further improve computational cost, we may consider matching only edge elements (*edgels*). Two such methods are the Hough transform [Hough 62] and the Fourier descriptors [Persoon 81]. Both of these methods will be considered in this chapter.

Starting with an arbitrary point on the boundary of an object, and tracking around the boundary, a sequence of complex numbers is obtained. The discrete

Fourier transform of these points is referred to as the Fourier descriptors. There
are several parameterisations of the Fourier descriptors; they are summarised in
[Persoon 81]. One of the advantage of Fourier descriptors is that the coefficients
are invariant to size, translation and rotation. Also, mirror images can be
distinguished by examining the phase angle. However, the coefficients cannot
be used to reconstruct the image.

For Fourier descriptors (FD) to work well, it requires *ideal input* (i.e. complete objects, not occluded). Although there are many boundary descriptors,
the usefulness of FD is described in [Zahn 72]. Given a closed curve γ, a point
on the curve can be described in the complex plane as $u(l) = x(l) + jy(l)$. The
FDs are defined as

$$a_n = \frac{1}{L} \int_0^L u(l) \exp\{j(2\pi/L)nl\}\, dl$$

and

$$u_l = \sum_{-\infty}^{\infty} a_n \exp\{jn(2\pi/L)l\}$$

Using the notations of Figure 5.15, with V_0 being the starting point, the FDs
are now

$$a_n = \frac{1}{L\left(\frac{n2\pi}{L}\right)^2} \sum_{k=1}^{m} (b_{k-1}) \exp\{-jn(2\pi/L)l_k\}$$

where

$$l_k = \sum_{i=1}^{k} |V_i - V_{i-1}|, \quad \text{for } k > 0 \text{ and } l_0 = 0$$

and

$$b_k = \frac{V_{k+1} - V_k}{|V_{k+1} - V_k|}, \quad \text{so } |b_k| = 1.$$

For further discussion on the interpretation and description of FDs, the reader
is referred to [Persoon 81, Zahn 72].

Persoon and Fu [Persoon 81] tested out their FDs on the Munson data set,
a collection of FORTRAN symbols (numerals, characters, and special symbols)
written by 49 persons [Zahn 72]. They observed an 11% error in classification.
Most of the miscalculations occur with the figure "8" because only the outer
boundary is used; this makes "1" and "8" very similar.

5.6.2   Using Boundary Descriptors to Recognise Objects

So far, we have considered some methods for object recognition based on global
feature measurements such as an objects area, size, shape measure, sub-image
template matching using cross-correlation or match-filtering and computation
of invariant moments. Furthermore, we have made several assumptions concerning our images: (a) the objects are rigid, (b) planar and (c) the object
plane is parallel to the image plane. The consequence of all these is that the
objects can only have uncertainties in their position, orientation and scale.
If the scale needs to be constrained, the perpendicular distance between the

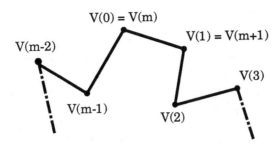

Figure 5.15: A polygon boundary notation.

object- and the image-plane can be fixed. Consequently, many of the object recognition systems described are restricted to recognising planar non-clipped, non-occluding objects (i.e. the objects needs to be complete and singular).

Although many of the features described above can be computed from the boundary information described by a chaincode [Freeman 61], such as an object's moments, Fourier descriptors, and an object's boundary length and area based on a discrete version of Green's theorem [Tang 83], we wish in this section to explore methods for object recognition both in non-occluding and occluding picture scenes using object boundary description methods.

In this section, we will, first, describe methods which are only applicable to non-occluding picture scenes and then look at methods for recognising objects in an occluded scene. For the later methods, instead of matching global features they attempt to match (what are commonly termed) *extended features*. Finally, we will briefly consider some methods for hypothesizing objects in complex image scenes by matching *local generic features* such as corners, holes and lines.

Along with a theoretical description, where appropriate we will describe some examples of the systems being employed in solving some real world problems, often in industrial inspection and quality control.

5.6.3 Recognising Single Objects

The systems we explore in this section are based on (modified) Freeman chain-coded data. Furthermore, many of the systems uses a radial model which requires the computation of the centroid of the objects — a global feature. By sub-sampling the data at a regular angle, the methods can be made invariant to scale.

The $r(\theta)$ Method

The centroidal profile [Freeman 78], or the $r(\theta)$ graph, is an orientation and scale-invariant shape descriptor based on the use of the Freeman chaincode [Freeman 61, Freeman 74]. However, the method is only applicable for closed curves. For the $r(\theta)$ method, as the contour of an object boundary is traced, its coordinates are recorded. By keeping cumulative values of its $x-$ and $y-$coordinates, the centroid of the object is calculated from its first moments. The centroid so calculated is insensitive to noise. The normalisation is carried out by computing the chaincode node furthest from the centroid using a chessboard distance measure, r_{max}; this is referred to as the vertical normalisation factor. All other chaincode nodes are then a ratio (r_i/r_{max}). Freeman [Freeman 78] set the abscissa arbitrarily to 100 points; this is referred to as the horizontal normalisation factor. Although the chaincode is normalised by taking 100 points at regular intervals, Freeman notes that accuracy may be better maintained by using Euclidean distance and angle measurements.

These improvements mentioned by Freeman are used by Hwang and Hall in their description of the $r(\theta)$ graph. They also note that the method is useful for data compression [Hwang 78]. Data compression of 10:1 is easily obtainable. Like Freeman, for graph normalisation they also compute r_{max}. This is then used as the reference axis for computing the angles. For example, if the centroid is (C_x, C_y) and the coordinate of the r_{max} point is (B_{mx}, B_{my}), then the gradient of a line drawn from the centroid to this point is

$$M_t = \frac{B_{my} - C_y}{B_{mx} - C_x}.$$

The gradient of a line drawn to any other point (B_x, B_y) is

$$M_b = \frac{B_y - C_y}{B_x - C_x}.$$

The angle between these two line is then

$$\tan\theta = \frac{M_t - M_b}{1 + M_t M_b}$$

given

$$(B_x - C_x)(B_{my} - C_y) - (B_y - C_y)(B_{mx} - C_x) > 0 \quad \text{for } \theta < \pi$$

$$(B_x - C_x)(B_{my} - C_y) - (B_y - C_y)(B_{mx} - C_x) < 0 \quad \text{for } \theta > \pi.$$

This centroidal profile method has been used by Yachida and Tsuji [Yachida 77] for recognising complex industrial parts and measure parameters for fabrication and location of holes in small industrial gasoline engines. They contend that the system designed for industrial inspection task (1) must work even in the presence of noise such as dirt and grease, (2) must classify complex parts, and (3) must be adaptable (i.e. it should be trainable to recognise new shapes).

The industrial images are segmented using a bottom-up processing scheme [Nagao 82]. The three major steps in the segmentation and recognition are preprocessing, feature extraction and classification.

To begin with, the object outline is extracted. From this one determines the size (S), thinness (T) and the shape. The thinness is defined as $T = k(S/L^2)$, where k is some scaling factor and L is the perimeter length. The measure of thinness is not sufficient to describe the shape; the shape is defined by the $r(\theta)$ graph. Furthermore, thinness or elongation is a particularly bad measure and better shape measures have been proposed elsewhere [Veillon 84].

The centroidal profile is computed for the chaincode. Given the centroid, the Euclidean distances are computed at intervals of three degrees.

$$R(\theta) = \sqrt{(x - \bar{x})^2 + (y - \bar{y})^2} \quad \text{for } \theta = \theta_i(\theta_i = 3, 6, \ldots, 360)$$

However, $R(\theta)$ for other than convex shapes is multivalued. Yachida and Tsuji's solution to this problem is to take the value for the furthest intersection. Note that unlike in [Freeman 78, Hwang 78], the profile distance has not been normalised.

To prevent matching of the input lists against all reference lists, some restrictions on the matching are imposed. For example, if a model M_i has an area S_i and thinness T_i, then the matching need only be carried out against those models for which the following conditions are true:

$$|S - S_i| < S_c \quad \text{and} \quad |T - T_i| < T_c,$$

where S_c and T_c are some threshold constants. The matching carried out is to find the minimum distance between the input object and a set of models.

$$D(\theta_s) = \sum_{k=1}^{120} |R_i(3k) - R'(3k + \theta_s)|.$$

The object $R'(\theta)$ and model $R_i(\theta)$ are similar if the distance function $D(\theta_s)$ is smaller than a given threshold. Yachida and Tsuji discuss further about generating and training model sets; also about locating other features once the object is recognised from its centroidal profile. The details of these can be found in [Yachida 77]. Their system was tested on objects consisting of 20 to 40 parts. The majority of the problems observed arose due to the failure to segment features of low contrast when considerable noise due to dirt and grease was present.

If one adopts the $r(\theta)$ scheme strictly, then as the contour is traced for non-convex objects, each angle will have one or more radii associated with it. Then a matching algorithm will have to operate in two dimensions, adding to computational complexity. Or for each angle, we could choose the radial distance to be the nearest boundary point. However, this scheme leads to several models for each object; one for every stable position of the object. Therefore, it may be better for matching purposes to ignore the angle information and match on a normalised string; now involving a one-dimensional match.

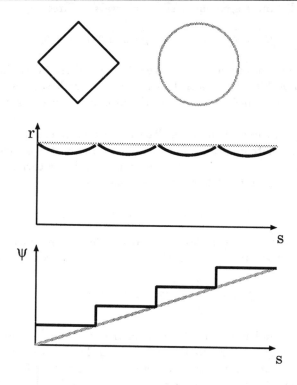

Figure 5.16: Boundary description using the $r(s)$ and $s(\psi)$ schemes for two simple shapes: (a) square and (b) circle.

If the string, i.e. the traced border of the object, is greater than 100 pixels, the radius, calculated with respect to the centroid, and the angle, with respect to the first scanned position, are calculated.

If the boundary signature, $r(s)$ for a model is known to be a^m, then matching coefficients for an object b are given by

$$c^m(j) = \sum_{i=0}^{p} \left(a_i^m - b_{i+j} \right)^2$$

This involves moving the signature string b with respect to the model string a^m and calculating the sum of the difference squared for the length of the string. If the matching coefficient $c^m(j)$ for a given $\{0 \le j < 100\}$ is found to be below a matching threshold, then a match between the object and the model has been found. The worst case complexity for matching two string of length p (but the objects were in different orientation) is $O(p^2)$. Furthermore, if there are m models and if we have to test against them all then the complexity is $O(mp^2)$. However, as we will show, use of some simple heuristics will greatly improve the complexity problem.

The models are generated by extracting the boundary of a given object with perimeter string bigger than 100 pixels. From this, the centroid of the object (assumed to have no holes) is computed. The perimeter string is rescanned to find a position which is the largest Euclidean distance, r_{max}, away from the centroid. Using this as the first position, we compute the normalised radial distance r_i/r_{max} at another 99 equally separated positions. This forms our normalised $r(s)$ graph. Furthermore, we also compute the area under the $r(s)$ graph. In our database, we store for each model: (i) the name of the model object, (ii) whether the r_{max} recorded is an unique position, (iii) the area under the $r(s)$ graph, and (iv) the 100 values of the $r(s)$ graph itself.

When it comes to matching our object data with the model data, the models are sorted using the area as a key. The object data is generated in the same fashion as for the model. Because, we know the area under the $r(s)$ graph for our unknown object, we need only match our object against a model base with similar $r(s)$ area constraints. This implies, that we need only match (at worst) against k models, where $k \le m$. Furthermore, for some of these models, we know that there is an unique position where a match could occur; we do not require to slide one graph over the other to obtain a best least-squares fit match. This indicates the best-case complexity is now $\Omega(kp)$ and our expected complexity is $\Theta(ap^2)$, where $a \le k$.

One other detail to be extracted is the orientation of the object. There are two ways in which we approach this problem depending on whether the matched model has a unique r_{max} value. If there is an unique r_{max}, then this is used for the orientation; otherwise, the orientation is determined by computing the semi-major axis for the object.

The objects the system currently has been trained to recognise are a subset of those suggested by Norton–Wayne and Saraga for benchmarking silhouette recognition systems [Norton-Wayne 84] and some plug-like objects (see

Table 5.1: The matching coefficients between the plug objects and the models.

match of *object - model*	*matching coefficients*
male - male	313 ± 15
female - female	323 ± 13
female - male	1225 ± 105
female - other models	$> 93 \times 10^3$
neck - neck	378 ± 111
noneck - noneck	550 ± 221
noneck - neck	4151 ± 277

Figure 5.17) which have very similar shapes. The latter objects allow us to determine the accuracy and the robustness of the method (Table 5.1).

Shape description from Freeman Chaincode

Although we have seen that the centroidal profile method is capable of distinguishing between similar objects and capable of recognising complex shapes in industrial images [Yachida 77], some have rejected its use as being too complex and time-consuming. Kitchen and Pugh [Kitchen 83a] contend that the image analysis "requirement is to verify the tentative identification of the component, and to identify its orientation rather than to perform 100% inspection by detailed comparison of input and reference outlines." Instead they suggest that feature extraction from chaincodes may be more useful since the method is orientation independent. Kitchen and Pugh uses Freeman chaincode to determine an object's (i) parameter, (ii) area, (iii) $x_{min}, x_{max}, y_{min}, y_{max}$, and (iv) centroid.

To obtain further information from the chaincode, Kitchen and Pugh propose the use of a *circles model*. The method consists of locating points at a certain radius from the centroid and computing angles. The angles are differenced from those near it to determine a list of invariant features. For object recognition, the input and the reference lists need matching. The number of intersections found for a given radius is used as a guide before any matching is carried out. To inspect between two or more similar shapes, it may be necessary to specify more than one circle. However, this still requires a far smaller matching set than that for the $r(\theta)$ scheme. The criteria used for choosing the circle radius are:

- the intersection points should uniquely define the orientation of the object,

- the number of intersection points should be small (typically < 8),

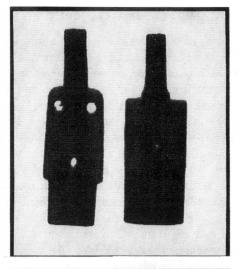

Figure 5.17: Example of some test shapes. These plug shapes are used to determine the accuracy and robustness of the radial profile method for object recognition.

- the radius should be large so that the angles may be computed accurately,

- the intersection points should not vanish or new ones appear for small change is radius, this prevents noise on the object boundary from adversely affecting the results.

5.6.4   Recognising Occluded Objects

In the last decade, there has been great interest in automating batch manufacturing processes and attempt to use image analysis methods in quality control. The initial work has usually involved (as we discussed in the last section) extraction of global features from silhouettes. In complex image scenes, objects often touch each other or overlap each other. However, the extraction of global features, apart from the problems associated with the methods which we have already discussed, requires that each object be isolated; this often requires hardwares such as shakers, bowl-feeders and conveyor.

We now wish to consider situations when, for one reason or another, it is not possible to isolate objects. In such situations, it is possible that object occlusion will occur. If objects are occluded, then information concerning global features is no longer available and therefore we need to consider extracting prominent or characteristic local features from which we can attempt to hypothesise the object. However, we should note, should any of the prominent features be missing, then the recognition task becomes more difficult.

However, there is a problem associated with matching many local features — that of combinatorial explosion. There are at least three ways to overcome this problem: (i) locate a few extended features instead of many local ones [Turney 85]; (ii) locate just one local feature and use this to restrict the search area [Bolles 83, Rummel 84]; and (iii) use parallel computation for graph searching, relaxation or histogram analysis.

We will only consider the two options for recognising partially occluded objects: (i) we can compute a modified $r(\theta)$ graph using an unique feature as the focus; and (ii) we can compute local tangents, ψ, at the boundary curve of objects. Plot these as an one-dimensional function $s(\psi)$ [Perkins 78]. From the $s(\psi)$ graph we will extract salient features [Turney 85]; match these salient features against a model database. However, these methods are only useful when specific boundary features exist.

The $r(\theta)$ Method

A scheme for recognising partially occluded objects using a radial profile, the $r(\theta)$ graph, has been presented by Berman *et al.* [Berman 85]. In the earlier example of the *radial profile*, we computed the $r(s)$ graph about the centroid of the object. However, in an occluded scene, computation of the centroid of objects is very difficult and often impossible. However, there is no reason for computing the radial profile about the centroid, except that the centroid is often easy to compute and is robust for general non-analytic shapes. However, if the shape contains an unique feature then this could be used as the *'centre'* about which the $r(\theta)$ graph could be calculated.

In the method described by Berman *et al.*, the radial profile is computed about a centre of a hole at 5° intervals. The hole is detected by searching for a region of given area and aspect ratio. Both the method for detecting holes and the sampling of radial profile at 5° intervals impose some severe limitations on this method and we will expand upon these below.

There are obvious problems of determining holes from a shape measure such as when, for example, the hole is only partially visible. This is a severe limitation when we are hoping to apply the method for occluded scenes. An improvement is to use an Hough transform to detect the holes. Note also that the $r(\theta)$ method can still be used even if a hole is not present in the object; we could use Hough transform to detect the 'centre' of an arc of a circle. Using this extension, we could recognise some shapes as shown in Figure 5.18. This further implies that using Hough transform to locate the 'centering' (or *salient*) features of an object, we could recognise more than one type of object in an image scene even when the objects are occluded provided sufficient information is present to locate the salient features.

The second problem arises from the 5° sampling. If the object to be recognised is concave or has multiple holes, then it is possible that some parts of the object will not be visible from the salient feature (see Figure 5.19). Furthermore, because of the different orientations of the object, the radial profile could very well be different. These problems could lead to the need for multiple models for the same object; one for each stable position of the object. These problems can be reduced if the sampling is increased. However, this increases the computation for a match to be found, although this might still be better if the number of models required is reduced.

Finally, we come to the choice of method for matching the radial profile of the object against model database. Berman *et al.*, uses a modified least squares fit technique.

$$deviation = \sum_{i=0}^{71} (template1[i] - template2[i])^2$$

For all i, such that $(0.5) \times template2[i] < template1[i] < (1.5) \times template2[i]$. However, because of occlusion and rotation, it might prove difficult to pick out the correct model because a few 'incorrect' values can bias the *deviation* measure. We expect a similarity measure would prove to be more reliable. We increase the similarity measure by one if and only if the radial profiles distance between object and model is smaller than a threshold. This does make the method very sensitive to changes in scale but then we also use circle Hough transform (CHT) to detect the salient features and CHT is also sensitive to changes in scale.

$$C(j) = \sum_{\substack{i=0 \\ \underbrace{\quad}_{i=i+3}}}^{357} \begin{cases} 1 & \text{if } (a_i - b_{i+j}) < thr \\ 0 & \text{otherwise} \end{cases}$$

Figure 5.18: Some shapes for recognition using the $r(\theta)$ method which do not have a hole but have unique circular arcs.

Figure 5.19: For a concave object some parts of the object are not visible from the salient feature.

If $C(j) > 72$, i.e. 60% of the 'boundary' is visible, then we assume the object is present and $j \times 3$ gives the relative rotation between the object and the model. Higher the $C(j)$ value, the more confidence we are of the recognition of the object.

It should be noted that the use of the $r(\theta)$ method in an occluded scene or for concave objects acts only as a means of recognition and location of the object. We cannot state whether the object is 'good' without further scrutiny. However, if the object is convex and in a non-occluding scene, then a high match confidence (maybe $\geq 95\%$) also does the scrutiny task.

An example of the use of the $r(\theta)$ method has been to recognise the shape shown in Figure 5.20(a) in the image scene shown in Figure 5.20(b). The image is $256 \times 256 \times 8$-bits. The salient feature being used is the 9-pixel radius hole. The radial profile about one such feature is shown in Figure 5.20(c); there is sufficient boundary visible for this object to be recognised. The scene has many possible salient features, more than the two objects present. All the salient features are first located as shown in Figure 5.21(a). Carrying out a match for each of the radial profile against the model leads to the recognition of the two objects present as shown in Figure 5.21(b).

The $s(\psi)$ Method

One of the problems associated with shape processing, object recognition, is that of describing a shape in a definitive manner. There are many approaches to solving this problem including Fourier descriptors, template matching, Hough transforms, moment calculations and matching centroidal profiles as we have already discussed. Another solution proposed by Attneave and Arnoult [Attneave 66] is to divide curves into segments and then use simple features to characterise each of the curves. It is important to note that for this method to work, the image segmentation should be correct. Having segmented the image into objects, the object boundary curves are encoded using an 8-direction Freeman chaincode [Freeman 61]. Using the encoded curves, a shape matching was proposed by Feder and Freeman [Feder 66] called the *chain correlation scheme*.

$$\phi_{ab}(j) = \frac{1}{n} \sum_{i=1}^{n} \cos\left[(a_i - b_{i+j}) \bmod 8 * \pi/4\right]$$

where a and b are the two chaincodes being matched. The problem associated with this form of correlation is that the chains are scale and orientation dependent. For matching, one of the chains will require to be scaled and rotated to align with the other. Furthermore, while the correlation function may be very efficient, it is also very sensitive to noise and small variations in scale.

To overcome these problems associated with scales, Freeman proposes a normalisation scheme based on the location of "critical points" [Freeman 78]. If the chord length between two critical points in object 1 is $\overline{A_1 B_1}$ and in object 2 its $\overline{A_2 B_2}$, then two features are normalised by scaling feature 2 by the ratio $\left(\overline{A_1 B_1} / \overline{A_2 B_2}\right)$. There are several methods for locating the critical points. One

Figure 5.20: The model shape (a) is to be located in image scene (b) using a radial profile, one such is shown in (c).

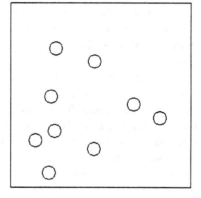

Figure 5.21: The salient holes located are shown in (a) and the two objects located is shown in (b).

method is to differentiate the chaincode and look for discontinuities [Freeman 78].

One of the earliest use of the $s(\psi)$ graph was described by Perkins for recognition of industrial parts [Perkins 78]. The method was restricted to: (i) non-textured objects (but the background could be textured), (ii) planar 2D models and (iii) the objects are at a known distance, i.e. the scale is fixed.

The method consist of extracting edge points using an Hueckel operator. This returns edge points, direction and magnitude, which are linked into chains using local criteria described in [McKee 75]. The linking uses knowledge of proximity, which is the most important, directional continuity and intensity continuity on both sides of the edge. If the result of linking at any point is a 3 or more branch node, this is pruned down to 2. This is achieved by, first removing all short branches; then all but the two branches with the best local continuity are removed.

From the continuous chain, a set of *concurves* is generated by fitting straight lines and curves to a proportion of the chain. The difficulty of this method is in grouping the chain into the two classes. This is achieved using the $s(\psi)$ graph and the curvature measure $(\Delta\psi/\Delta s)$ derived from it. The advantage of using concurves is in the tremendous reduction in data and the highly organised nature of the data. In the same fashion, both the models and the objects are broken down into a list of concurves. For object recognition, objects are compared to a model database for the following properties of the curves:

- descriptions of the curves, (i.e. circle, arc, ...);

- radius or total length of arcs;

- angle subtended by the arcs;

- number of straight line;

- number of arcs;

- bending energy (curvature function).

If closed concurves are detected within closed bodies, then the following properties are also tested:

- intensity direction;

- area of the holes;

- compactness $(area/perimeter^2)$;

- ratio of the minimum to maximum moment of inertia.

Once an object has been recognised in the scene, the transformation of the model, so that it overlies the object, is determined by locating the position and direction of one model point and the corresponding point in the object.

Use of the $s(\psi)$ graph in the analysis of objects in dynamic scenes is discussed in [Martin 79]. In such scenes objects often occlude each other. The object chain codes are generated using a method described in [McKee 75]. The $s(\psi)$ graph is decomposed into a set of straight lines. Dynamic objects are tracked on the basis of shape similarity calculated from the difference in the $s(\psi)$ graph between successive frames.

Another use of the $s(\psi)$ graph is in recognising partially occluded objects is described in [Turney 85]. Unlike the above two cases, in Turney *et al.*'s derivation the $s(\psi)$ graph is not simplified into *concurves* but operates directly on the graph. A model database of geometry of all the parts of the object is generated during the "learning phase" using a computer aided-design (CAD) system. This database consists of boundary descriptive (i.e. its $s(\psi)$ graph) templates generated by determining the stable portions of an object. From this, a set of subtemplates which are distinct to the object, called *salient* features, are determined.

The method described by Turney *et al.* is suitable for recognition of planar objects of known scale. However, if 3D object recognition is required, the method may be extended to use descriptions of *surface subregions* instead of *boundary segments*. Unfortunately, any system which is essentially based on boundary chaincoding suffers from quantisation problems. This is the change in arc length, s, (*length foreshortening*) we discussed earlier. A different solution is proposed by Turney *et al.*, instead of inserting extra copies of neighbouring chains, they propose fitting a third-order polynomial. Once a database of templates of salient features, τ, has been generated, object recognition is carried out by matching τ_i against an equal-length segment of the $s(\psi)$ graph of an object, β_j. However, before the matching can be carried out, one needs to correct the subtemplate orientation. The matching is carried out using a least-squares fit technique.

$$\gamma_{ij} = \sum_{p=1}^{h} \left[\psi_{\beta_j}(s_p) - \left(\psi_{\tau_i}(s_p) + \psi_{ij} \right) \right]^2$$

where h is the length of the τ_i chaincode. A matching coefficient is defined by

$$c_{ij} = \frac{1}{1 + \gamma_{ij}}.$$

c_{ij} will be a maximum if for a given ψ_{ij} a match between τ_i and β_j is found.

As for the $r(\theta)$, the $s(\psi)$ signature is generated as the boundary of an object is traced. If the string length is again greater than 100 pixels, then the local tangent, $\psi(0 \le \psi < 360)$ at each boundary point is calculated.

Now to the problem of matching the $s(\psi)$ signature of a model with those of objects. While the $r(\theta)$ is invariant to rotation since r is computed with respect to the centroid, this is not true for the $s(\psi)$ graph. Therefore, the matching algorithm has to cater for arbitrary rotation of the object with respect to the model as well as different scanning position.

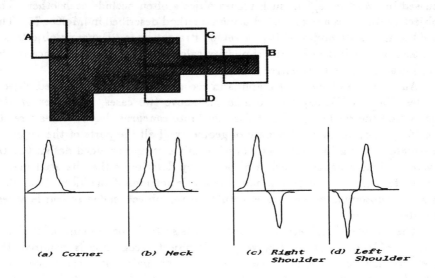

Figure 5.22: Symbolic representation of some simple salient features.

The matching algorithm is now minimising

$$c^m(\theta, j) = \sum_{i=0}^{p} \left(a_i - (b^m_{i+j} + \theta) \right)^2$$

for a given θ and j. There are four variables: the perimeter length of the object, P; there are m (≈ 16) model (salient features) of length p; and orientation θ, at 3 degree intervals there are approximately 2^7 different orientation we may have to check. Then the worst case complexity of matching a salient feature against the object is $O(Ppm\theta)$.

A method must be found to cut down on this large amount of computation. A solution to this is to look for a symbolic representation for the salient features. If the *symbol* exists then carry out an iconic match. The problem now is: how do we reduce a $s(\psi)$ graph into a series of symbols? A solution is to differentiate the $s(\psi)$ graph such that

$$\kappa(s) \longleftarrow \frac{\partial \psi(s)}{\partial s}$$

this gives us $|\kappa(s)| > 0$ where there are rapid changes in the boundary such as a corner (i.e. places where possible salient features may exist). Some simple examples of $s(\psi), \kappa(s)$ and symbolic representation are given in Figure 5.22.

During 'runtime', the boundary signature is converted to a series of symbols. If a symbol corresponding to a model salient feature exits then an iconic match is carried out. An advantage of this is that we know approximately the position and type of the salient feature; the real unknown still is the orientation of the feature. The matching complexity is now $O(\delta P \times p \times \delta m \times \theta)$; this should give rise to an improvement of $O(10^2 - 10^3)$ in speed of matching model and objects.

The advantages of the $r(s)$ method are: (i) the model representation is very compact; (ii) matching models to object data is fast, the worst case complexity is $O(mp^2)$ but the typical complexity is $\Theta(kp^2)$, where k is typically 3 or 4 and p is 100; (iii) the method is easily made scale-independent; and (iv) the computation of r is expected to be accurate provided the conditions stated below are satisfied. The disadvantages of the $r(s)$ method are: (i) the segmentation is critical, if there are 'fluffs' at the boundary of the object this affects the position of the centroid and the length of the object perimeter and hence the complete $r(s)$ graph since all r and s values are computed with respect to them; (ii) complete boundaries must be segmented, i.e. there can be no occlusion; (iii) the method only recognises the presence of a known object and the location of its centroid, if we need to know its orientation then more computation will have to be carried out; and (iv) there are problems with the segmentation and particularly the tracking algorithms at 'sharp' corner locations.

The advantages of the $s(\psi)$ method are: (i) the $s(\psi)$ values are computed locally, this means that only local and partial matching is required to recognise objects (provided a salient feature is visible) hence the method is particularly suitable when occlusion occurs or complete segmentation is not possible or in noisy images (however, note that we need to train the model database under 'ideal' conditions); and (ii) orientational information is readily available. The disadvantages of the $s(\psi)$ method are: (i) the database can be large (we will require to store the complete perimeter coordinates along with the boundary tangential directions and the tangential direction information for a number of salient features); (ii) the method is expected to be $O(10^2 - 10^3)$ slower than using the $r(\theta)$ method; and (iii) there could be approximately $10°$ error in determining the tangential direction if there is a one pixel error in segmentation.

Using local features

There are a number of object-recognition systems based on heuristic graph matching of local features. A system described by Bolles and Cain [Bolles 83] is called the *local-feature-focus* (LFF) method. This system works with two types of features, regions (i.e. holes) and corners. A hole is described by its colour, area and axis ratio, and a corner by the size of its angle. Bolles and Cain are attempting to recognise silhouettes of door hinges which have three types of primitives, holes and two types of corners.

The system generates runtime hypotheses which attempt to match an object cluster with a model cluster. To prevent a combinatorial explosion during matching, the method attempts to grow a cluster around a single feature. If this is not successful, then it chooses another feature. First it builds a cluster around a hole by searching a list of local features that fit the specification for features near a hole. If no holes are found then it uses a corner (A-type) to build a cluster around it.

The model and object feature clusters are matched using a graph-search technique, it locates the largest clusters of mutually consistent assignment. None in the graph are connected by an arc if the two assignments they represent are mutually consistent. The criteria for this are:

- Two object features are not referring to the same image feature.

- Two image features are not referring to the same object feature.

- Two image features must refer to object features which are part of the same object.

- The distance between two image features must be approximately equal to the distance between two object features.

- The relative orientation between two image features with respect to a line joining their centres must be approximately the same for the object.

The verification to the match is carried out by (1) looking for other object features which are consistent with the hypothesis, and (2) checking the boundary of the object. As matched features are found, these are added to a list of verified features; this improves the estimates of the position and the orientation of the object. If enough object features are found and its boundary verified, then the objects position and orientation is located.

The system is capable of recognising reflected objects because:

- It allows the model to have two orientations, that of the object and a mirrored version.

- The mirror-image orientation is allowed if this is 'mutually compatible'.

- It analyses three features. If the included angle differs in its sign, the system hypothesizes that the object has been reflected.

- If the above is true, all subsequent processing is carried out on this basis.

The LFF training system is divided into two parts: (1) model acquisition and (2) feature selection. As discussed earlier, the models can be constructed in one of two ways; either by 'teaching by showing' (the problem with this method is the large amount of statistics involved) or using a CAD system — Bolles and Cain uses the latter. The features are selected by:

- Identifying similar local features in different objects.

- Computing symmetries of objects by determining axis of reflection or rotation.

- Marking structurally equivalent features.

- Building a description of the objects.

- Selecting nearby features.

- Rank the focus features.

Details of these can be found in [Bolles 83].

Another system described by Rummel *et al.* [Rummel 82] is an extension of an work described in [Tropf 80]. In Tropf's system only corners were used, but Rummel *et al.*'s uses corners, straight lines and circles. These features are extracted in a manner similar to that described in [Perkins 78]. The straight lines are described by their start and end positions; the circles by their centre position and radii; circular arcs by their radius, centre position, angle and orientation; and corners are described by the intersection of two lines, their position, angle and orientation.

The local features are extracted using the following techniques: edge detection, line following, line segmentation using curvature measures and line classifications. For details of these methods see [Grasmuller 84]. The models are generated interactively using a CAD system. The models so generated consists of a *geometric model* and a *generative model*. The geometric model describes the relationship between local features.

Image scene analysis is based around a heuristic search. The system initially finds all the primitives features, then it chooses the best. It then hypothesizes where the next primitive in relation to itself should be located. If the hypothesis fails, then the system chooses another primitive and begins the search again. The search continues until all the primitive are found or are estimated and a similarity threshold with a predefined model is exceeded. Rummel and Beutel found that a depth-first search was more efficient than a breadth-first search. The matching is fairly fast; much of the computation is taken up in preprocessing and feature extraction. To speed this up, the many of the tasks are implemented in hardware called GSS — "Grey-scale Sensor System" [Rummel 86].

5.7 Lines and Splines Fitting

5.7.1 Linear Least-Squares Fit

If there is a distribution of pairs of data points (x_i, y_i) then a straight line $(y = m \cdot x + c)$ can be fitted to them — the straight line being specified by two parameters, the slope m and intercept c. Assume that the gradient of the line is small ($m \ll 1$). Then if (x_i, y_i) is one of the data points, the perpendicular distance between (x_i, y_i) and the best-fit line can be shown to approximately equal to the distance between (x_i, y_i) and (x_i, y_j) where (x_i, y_j) is a point on the best-fit line, Figure 5.23. Errors in m and c arise from errors in y_i since x_i and consequently $\sum x_i$ and $\sum x_i^2$ are assumed to have negligible errors.

If the equation of a line is $y = m \cdot x + c$ and x_i, y_i are observed pairs of values, $i = 1, 2, \ldots, n$, then the displacement of the ith observation from the line is $d_i = y_i - y$. The best estimate of y that can be obtained from an experimental value of $x (= x_i)$ is that $y = m \cdot x_i + c$ then

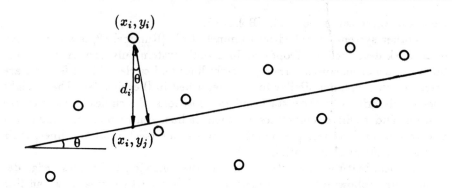

Figure 5.23: The least-squares minimisation of the vertical distance.

$$d_i = y_i - m \cdot x_i - c \tag{5.1}$$

$$\sum_{i=1}^{n} d_i^2 = \sum_{i=1}^{n} (y_i - m \cdot x_i - c)^2 = f(m, c) \tag{5.2}$$

The sum can be minimised by varying m and c. This requires

$$\frac{\partial f}{\partial m} = \frac{\partial f}{\partial c} = 0 \tag{5.3}$$

Solving these equations for m and c gives

$$m = \left[n \cdot \sum x_i y_i - \sum x_i \sum y_i \right] / \Delta \tag{5.4}$$

$$c = \left[\sum x_i^2 \sum y_i - \sum x_i \sum x_i y_i \right] / \Delta \tag{5.5}$$

where

$$\Delta = n \cdot \sum_{i=1}^{n} x_i^2 - \left(\sum_{i=1}^{n} x_i \right)^2 \tag{5.6}$$

On a serial computer, to compute the coefficients of the gradient and the intercept, m and c respectively, would require an $O(n)$ algorithm, where n is the number of data points. A serial implementation requires $2n$ multiplications and $5n$ additions. A parallel implementation requires each point in the array to be labelled with its corresponding Cartesian $x-$ and $y-$coordinates. This is achieved by generating $x-$ and $y-$ramp images using global propagation and labelling the data point image with the ramp images by ANDing point-wise the two images; every bit-plane of the ramp image is ANDed with the data point image. Thus ANDing the $x-$ramp with the data image labels each point in the data image with its unique Cartesian $x-$coordinates with reference to the left edge of the array, and similarly for the $y-$coordinate. Then the functions $\sum x_i$ and $\sum y_i$ are calculated by summing the resultant array. The functions

Algorithm 5.6: Linear least squares fit.

LINE_FIT(im)
begin
   xcoord $\leftarrow ANDS(RAMP(dirc[8], \lceil \log_2 X \rceil, \text{im});$
   ycoord $\leftarrow ANDS(RAMP(dirc[6], \lceil \log_2 Y \rceil, \text{im});$
   sum_x $\leftarrow VOLUME(\textbf{xcoord});$
   sum_y $\leftarrow VOLUME(\textbf{ycoord});$
   sum_x² $\leftarrow VOLUME(MULT(\textbf{xcoord}, \textbf{xcoord}));$
   sum_y² $\leftarrow VOLUME(MULT(\textbf{ycoord}, \textbf{ycoord}));$
end

$\sum x_i^2$ and $\sum x_i y_i$ requires the $x-$ramp to be squared and the $x-$ramp multiplied with the $y-$ramp respectively before ANDing with the data point image; the resultant array is summed in each case.

In summary, the parallel operations are given in Algorithm 5.6. The array is assumed to have a dimensions of $X \times Y$:

The array is assumed to be square, i.e. X=Y. Labelling each point with their unique coordinates requires $4(\lceil \log X \rceil + \lceil \log Y \rceil + 1)$ global propagation, and $2\lceil \log X \rceil$ point-wise operations. The functions $\sum x_i$ and $\sum y_i$ require $2\lceil \log X \rceil$ count operations, and $2\lceil \log X \rceil$-bit point-wise multiplications and $2\lceil \log X + \log Y \rceil$ count operations are required for calculating $\sum x_i^2$ and $\sum y_i^2$.

The algorithm is independent of the number of data points, but is data-position-dependent. For instance, if the data is confined to the bottom-left-quarter of the array, because CLIP is a bit-serial processor, the number of bits required to label the data is smaller than if there were data in the top-half of the array.

5.7.2 Least-Squares Fit by Minimising Moment of Inertia

If a collection of points, for which the best-fit line is to be determined, is regarded as forming a body, then the major axis, the best-fit line, is found by determining the axis about which the moment of inertia is a minimum [Meriam 51]. The axis about which the moment of inertia is a maximum is the minor axis. Furthermore, both the axes pass through the centroid of the body.

If the moments, M_{ij}, are defined by

$$M_{ij} = \sum (x_0 - x)^i (y_0 - y)^j$$

then the angle α with respect to the reference axes which minimises the mo-

Algorithm 5.7: Linear least squares fit by minimising moments.

$GRADIENT$(im)
begin
  xcoor $\leftarrow RAMP(dirc[8], \log_2(X - 1))$;
  ycoor $\leftarrow RAMP(dirc[8], \log_2(Y - 1))$;
  $Area \leftarrow VOLUME$(im);
  x_bar $\leftarrow VOLUME(ANDS(\text{xcoor}, \text{im}))/Area$;
  y_bar $\leftarrow VOLUME(ANDS(\text{ycoor}, \text{im}))/Area$;
  tmp_x $\leftarrow ANDS(SUB(\text{xcoor}, \text{x\_bar}), \text{im})$;
  tmp_y $\leftarrow ANDS(SUB(\text{ycoor}, \text{y\_bar}), \text{im})$;
  $sum\_x^2 \leftarrow VOLUME(MULT(\text{tmp\_x}, \text{tmp\_x}))$;
  $sum\_y^2 \leftarrow VOLUME(MULT(\text{tmp\_y}, \text{tmp\_y}))$;
  $sum\_xy \leftarrow VOLUME(MULT(\text{tmp\_x}, \text{tmp\_y}))$;
end

ments about the $x-$ and $y-$axis is given by

$$\alpha = (\frac{1}{2})\tan^{-1}(\frac{2M_{11}}{M_{20} - M_{02}}) + n(\frac{\pi}{2})$$

Although this calculation is computationally heavier, it has advantages over the linear least-squares method discussed previously. In §5.7.1 the gradients were assumed to be small ($m \ll 1$). For larger gradients, the approximations are no longer true, and it is the perpendicular distances rather than the vertical which needs to be minimised.

The computations that are carried out in parallel are given in Algorithm 5.7.

Much of the algorithm is taken up in determining the centroid of the body, transforming the origin from the bottom-left corner of the array to the centroid, and labelling each point with its unique coordinates. The last three steps are spent in calculating the moments. This requires $3\lceil \log X \rceil$ multiplications and 3 summations of the resulting images. As for the linear least squares fit algorithm, the GRADIENT algorithm is independent of the number of data points but is data-position-dependent as for the previous reasons.

5.8   Joining Points by a Straight Line

Two methods for joining points by a straight line will be considered here. They are the expand region and skeletonise method used by Pass [Pass 81] and an improvement over the analytical method [Otto 84].

Pass's method consists of iteratively expanding [Goetcherian 80] the points to generate octagonal regions until two of the regions join, then the image is

Algorithm 5.8: Joining points to form straight line.

```
loopcount ← max;
result ← 0;

while loopcount ≠ 0
      begin
          im ← EXPAND(im);
          loopcount ← loopcount − 1;
          tmp ← SKELETONISE(im);
          if any skeleton > 2 then
                                 result ← ADD(result, tmp);
                                 remove corresponding parent region from im;
                                 if im isempty then
                                                  STOP; fi

          fi
      end
```

skeletonised [Arcelli 75] to join the points by a straight line. This process is best described by the algorithm given below. The maximum number of iterations allowed is *max*, the points to be joined are in image **im** and the resultant lines are in the image **result**.

With this method, a significant problem arises. Consider the problem in Figure 5.24. If the distance separating two points to be joined is greater than the distance of any of the points from the edge of the array, the symmetry required for successful joining of the two points by a straight line no longer exists with the subsequent result shown in Figure 5.24. However, despite some of these problems, Pass [Pass 81] has used the algorithm successfully to join points to complete parameters of incompletely segmented objects. The second method to be discussed is more parallel [Otto 84].

Otto's method is simple and very elegant. The points A and B are to be joined together by a straight line (Figure 5.25). If these two points are enclosed by their bounding rectangle, then a point C inside the bounding rectangle is below the line AB if, and only if

$$\frac{y_1}{x_1} \leq \frac{y_2}{x_2}$$

and allowing for symmetry, iff

$$\left| \frac{y_1}{x_1} \right| \leq \left| \frac{y_2}{x_2} \right|$$

i.e. iff

$$f_c = y_2 x_1 - x_2 y_1 \geq 0$$

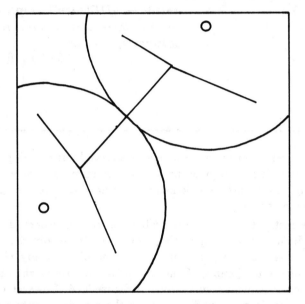

Figure 5.24: Problem of using Pass's algorithm when the data points are too close to the array edge.

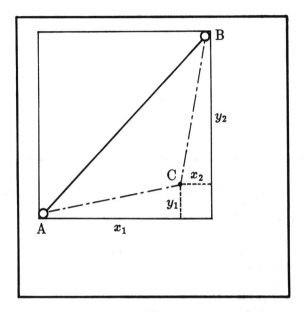

Figure 5.25: A bounding rectangle enclosing the points A,B to be joined together by a straight line.

This condition is satisfied by detecting the zero-crossings inside the bounding rectangle.

Otto's algorithm has a complexity $O(l)$, where l is the dimension of the array; the operation is independent of the separation of the two points to be joined together for a given CLIP4 array, unlike Pass' algorithm which has a complexity $O(d)$, where d is the separation distance of the two points since $O(\lceil d/2 \rceil)$ expand operations are required to join the two regions and skeletonisation also has a $O(d)$ complexity. However, the latter method does require considerably more memory while the former requires only a single-bit plane. For these reasons, the Pass's method is faster for joining points when their separation is small.

The author has attempted to extend Otto's method to join many points in parallel to complete parameters of a partially segmented feature. The algorithm JOIN_LINE does this. However, the algorithm has no intelligence. There is no set order or lists which are to be joined by straight line; the data is in the form of a segmented binary image. The partially segmented lines are joined to each other to form a complete parameter by use of a very simple rule: join the free end of one line to the free end of another line, and not to the other free end of its own line even if it is the nearest free end.

To achieve this goal, all the free ends need to be found along with the

Algorithm 5.9: Joining many points by straight-line segments.

$JOIN\_LINE$(line)
begin
   (1)$MASK\_END\_PTS$(line, mask, end_pts, *dir_list*);
   (2)tmask ← end_pts;
   (3)tmask ← E_{-1}^{n}(tmask);
   (4)tmask ← tmask · ¬mask;
   (5)tmask ← $BOUNDING\_RECTANGLE$(tmask, end_pts);
   (6)tmp_line ← $DRAW\_LINE$(end_pts, tmask);
   (7)line ← line ∪ tmp_line;
  where
        proc $MASK\_END\_PTS$(im, mask, end_pts, *dir*) ≡
           (1)mask ← $SINGLE\_POINT$(im);
           (2)end_pts ← mask@E_{048}(mask);
           (3)tmp ← $REMOVE\_SINGLE\_POINT$(im);
           (4)ttmp ← $CHEW$(tmp);
           (5)end_pts ← end_pts + (tmp@ttmp);
           (6)mask ← mask + (tmp@$CHEW$(ttmp));
           (7)mask ← $SPREAD$(mask, *dir*); .
  end

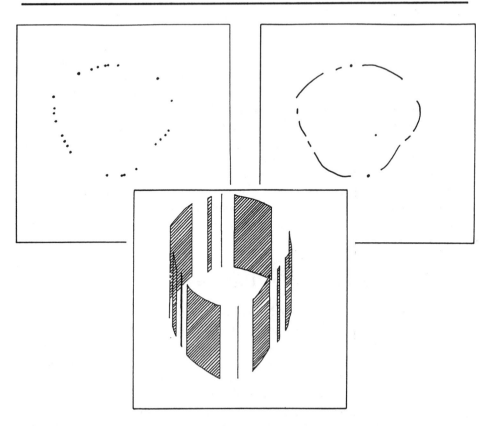

Figure 5.26: Joining points to complete a parameter.
(a) The data points consisting of lines and points as a result of incomplete segmentation. (b) The free ends, (end-pts). (c) The masking regions where no lines can be drawn (**mask**).

regions where no lines can occur so that a free end cannot join to its partnered free end. This is achieved by the algorithm MASK_END_PTS. This algorithm requires two inputs, **im** which contains amongst others the points to be joined together and *dir*, the direction of the masks. Two parameters are also returned by this routine, **end-pts**, which contains the free ends, and **mask**, which limits the regions where the new lines could be formed. This process is best described by the use of Figure 5.26.

Having defined an algorithm which given a set of points (Figure 5.26(a)) and a direction (dirstr("026") or ↕) for generating the masking regions, two new data sets are generated as described above, the free ends (**end-pts**, Figure 5.26(b)) and the masking region (**mask**, Figure 5.26(c)). This MASK_END_-PTS algorithm is embedded in the JOIN_LINE algorithm to join free ends by the rule given above to complete parameters.

The seven steps described above need to be repeated for the four directions ($\updownarrow$, $\nearrow$, $\leftrightarrow$, $\nwarrow$) or until there are no free ends left (i.e. the parameter is complete) or some distance rule is observed (i.e. the remaining free ends are too far apart). If points separated by n pixels is to be joined then $\lceil \frac{n}{2} \rceil$ 8-connected expands are required. From the given **end_pts** and the remaining **tmask** region, the bounding rectangle within which the DRAW_LINE algorithm (the function $f_c = y_2 x_1 - x_2 y_1 \geq 0$) is performed is created. Creating the bounding rectangles require four global propagations [Otto 84].

How should the spreading of the forbidden regions be achieved? On both CLIP4 and CLIP4S, this spreading must be computed iteratively by expanding the region bidirectionally. A considerably faster method would have been if CLIP machines clocked out global propagation after a given period determined by software, but this facility is not available. If global propagation is used to spread the data then the data will be spread bidirectionally across the length of the array and no guarantee can be given that the algorithm will work for all possible data configuration.

5.9   Spline Fitting in Parallel

Splines [Schoenberg 46] are named after the draughtman's device for drawing "fair" curves between specified points. The aim of this section is to give an overview of splines. The review is in no way comprehensive, but indicates some of the techniques available for generating smooth lines and surfaces. No attempts to derive or prove any of the spline functions have been made. The aims are restricted to understanding and exploring spline properties and the implementation of splines in parallel on an SIMD architecture. The splines considered here are the well-known B-spline [Riesenfeld 73] and the Beta-spline [Barsky 81]. The Beta-spline has two very interesting features, the bias and the tension parameters; the tension parameter has been exploited here as an image filtering tool. For other spline types, the reader is referred to [Pavlidis 83, Pavlidis 85].

Before spline surfaces were formulated, planar polygons had been used to approximate surfaces. Using such methods to generate smooth shading of surfaces has not proved very satisfactory since Mach bands [Rogers 85] are apparent at the borders of adjacent polygons, and also these forms of image representation require very large amounts of storage space. The use of non-linear parametric polygons to represent segments and patches joined together to form piecewise curves and surfaces was introduced by Coons [Forrest 72] and Bezier [Bezier 74]. More recently parametric B-spline [Riesenfeld 73] representation of polynomials has allowed greater flexibility, control and efficiency.

5.9.1   Introduction to Spline Lines and Surfaces

If piecewise cubic polynomials $\mathbf{Q}_i(u)$ are used to define a curve such that

$$\mathbf{Q} = \sum_i \mathbf{V}_i B_i(u)$$

where $\mathbf{V}_i = (x_i, y_i)$ are the control vertices and $B_i(u)$ is the basis function, then for the curve to appear continuously smooth, the curve must have positional, first derivative and second derivative continuity at the joints, i.e.,

$$\begin{aligned}
\mathbf{Q}_{i-1}(u_i) &= \mathbf{Q}_i(u_i) \\
\mathbf{Q}_{i-1}^{(1)}(u_i) &= \mathbf{Q}_i^{(1)}(u_i) \\
\mathbf{Q}_{i-1}^{(2)}(u_i) &= \mathbf{Q}_i^{(2)}(u_i)
\end{aligned}$$

A curve can be described as a piecewise cubic polynomial

$$\mathbf{Q}(u) = (\mathbf{X}(u), \mathbf{Y}(u)),$$

where $u \in [u_0, u_m]$. The values $u_0 < u_1 < \ldots < u_m$ are the joints, called knots, between successive polynomial segments. A spline of order k (degree $k-1$) is C^{k-2} continuous. A B-spline curve is a parametric curve defined by

$$\mathbf{Q}(u) = \sum_{i=0}^{m} \mathbf{N}_{i,k}(u)\mathbf{V}_i \tag{5.7}$$

where the B-spline basis function is given by a recurrence formula [de Boor 78]

$$\mathbf{N}_{i,k}(u) = $$

$$\frac{u - u_i}{u_{i+k-1} - u_i}\mathbf{N}_{i,k-1}(u)$$

$$+\frac{u_{i+k} - u}{u_{i+k} - u_{i+1}}\mathbf{N}_{i+1,k-1}(u)$$

for $k > 1$ and $u_i \le u < u_{i+1}$ and

$$\mathbf{N}_{i,1}(u) = \begin{cases} 1 & \text{if } u_i \le u < u_{i+1} \\ 0 & \text{otherwise} \end{cases}$$

The basis functions for a uniform cubic B-spline are [Riesenfeld 73, Gordon 74, de Boor 78],

$$\mathbf{N}_{i,4} = \begin{cases} (1/6)u^3 & u_i \le u < u_{i+1} \\ (1/6)(1 + 3u + 3u^2 - 3u^3) & u_{i+1} \le u < u_{i+2} \\ (1/6)(4 - 6u^2 + 3u^3) & u_{i+2} \le u < u_{i+3} \\ (1/6)(1 - 3u + 3u^2 - u^3) & u_{i+3} \le u < u_{i+4} \end{cases}$$

Then the ith segment of the curve is given in the matrix form by

$$\mathbf{Q}_i(u) = \begin{bmatrix} u^3 & u^2 & u & 1 \end{bmatrix} [\mathbf{S}] \begin{bmatrix} \mathbf{V}_{i-1} \\ \mathbf{V}_i \\ \mathbf{V}_{i+1} \\ \mathbf{V}_{i+2} \end{bmatrix} \tag{5.8}$$

For a closed curve in which $i = 0(1)m$, then

$$\begin{aligned} \mathbf{V}_{-1} &= \mathbf{V}_m \\ \mathbf{V}_{m+1} &= \mathbf{V}_0 \\ \mathbf{V}_{m+2} &= \mathbf{V}_1 \end{aligned}$$

and $u \in [0,1]$. For an open curve with $m+1$ vertices $(\mathbf{V}_0, \mathbf{V}_1, \ldots, \mathbf{V}_m)$, the end positions can be found from equation (5.8)

$$\begin{aligned} \mathbf{Q}_0(0) &= (\tfrac{1}{6})(\mathbf{V}_{-1} + 4\mathbf{V}_0 + \mathbf{V}_1) \\ \mathbf{Q}_{m-1}(0) &= (\tfrac{1}{6})(\mathbf{V}_{m-1} + 4\mathbf{V}_m + \mathbf{V}_{m+1}) \end{aligned} \tag{5.9}$$

Furthermore, at the ends there needs to be zero curvature

$$\begin{aligned} \mathbf{Q}_0^{(2)}(0) &= \mathbf{V}_{-1} - 2\mathbf{V}_0 + \mathbf{V}_1 = 0 \\ \mathbf{Q}_{m-1}^{(2)}(0) &= \mathbf{V}_{m-1} - 2\mathbf{V}_m + \mathbf{V}_{m+1} = 0 \end{aligned} \tag{5.10}$$

Substituting equation (5.10) into equation (5.9) gives the end conditions

$$\begin{aligned} \mathbf{Q}_0(0) &= \mathbf{V}_0 \\ \mathbf{Q}_{m-1}(1) &= \mathbf{V}_m \end{aligned} \tag{5.11}$$

For the closed uniform cubic B-spline, the nodes (the $m+1$ selected points on the curve) are given by equation (5.8) with $u = 0$

$$\mathbf{Q}_i(0) = (1/6)(\mathbf{V}_i - 1 + 4\mathbf{V}_i + \mathbf{V}_i + 1) \tag{5.12}$$

Since the coefficients are constant, the system of linear equations needed to determine the controlling polygon is given by

$$\frac{1}{6} \begin{bmatrix} 4 & 1 & & & & 1 \\ 1 & 4 & 1 & & & \\ & & \cdot & & & \\ & & & \cdot & & \\ & & 1 & 4 & 1 & \\ 1 & & & & 1 & 4 \end{bmatrix} \begin{bmatrix} \mathbf{V}_0 \\ \mathbf{V}_1 \\ \cdot \\ \cdot \\ \cdot \\ \mathbf{V}_m \end{bmatrix} = \begin{bmatrix} \mathbf{Q}_0(0) \\ \mathbf{Q}_1(0) \\ \cdot \\ \cdot \\ \cdot \\ \mathbf{Q}_m(0) \end{bmatrix} \tag{5.13}$$

For an open curve with the end condition given in equation (5.11), the set of linear equation is slightly different

$$\frac{1}{6} \begin{bmatrix} 6 & 0 & & & & 1 \\ 1 & 4 & 1 & & & \\ & & \cdot & & & \\ & & & \cdot & & \\ & & 1 & 4 & 1 & \\ 1 & & & 0 & 6 & \end{bmatrix} \begin{bmatrix} \mathbf{V}_0 \\ \mathbf{V}_1 \\ \cdot \\ \cdot \\ \cdot \\ \mathbf{V}_m \end{bmatrix} = \begin{bmatrix} \mathbf{Q}_0(0) \\ \mathbf{Q}_1(0) \\ \cdot \\ \cdot \\ \mathbf{Q}_{m-1}(0) \\ \mathbf{Q}_{m-1}(1) \end{bmatrix} \tag{5.14}$$

where matrix $\mathbf{S}$ is given in equation (5.8) and $\mathbf{V}$ is a matrix of the control vertices.

Each B-spline curve segment is constrained to remain within the convex hull of only those vertices that define that segment, and a B-spline surface is constrained to lie within the union of the convex hull of the control graph. Since the B-spline basis function is non-zero only over k spans or $k \times k$ surfaces, the result is strong local control of curves or surfaces; if a vertex is altered then only k curve segments are affected. Although a cubic B-spline is continuous in its second derivative, a higher-order continuity is possible by generating a spline of order $k > 4$ (degree $k - 1$) which then has C^{k-2} continuity. Using too-high an order of spline gives rise to problems similar to those for high order polynomials to describe a curve; there are possibilities of oscillations between control vertices.

Although B-spline derivation requires C^2 continuity for the curve to appear smooth, Barsky has shown [Manning 74, Barsky 81, Barsky 83, Goodman 86] that only the tangent and the curvature (G^2) need to be continuous. For instance, if there are two curves $L(t)$ $t_0 \leq t \leq t_1$ and $R(u)$ $u_0 \leq u \leq u_1$, then the two curves meet with n^{th} order parametric continuity (C^n), if they satisfy the condition

$$\frac{d^k R}{du^k}\Big|_{u=u_0} = \frac{d^k L}{dt^k}\Big|_{t=t_1} \quad k = 0, 1, ..., n$$

But if there is a linear change of parameter $u = \beta \nu$; $\beta > 0$ then the shape of the curve R(u) will not be changed, but

$$\frac{d^k R}{d\nu^k}\Big|_{\nu=\nu_0} = \beta^k \frac{d^k R}{du^k}\Big|_{u=u_0}$$
$$= \beta^k \frac{d^k L}{dt^k}\Big|_{t=t_1} \neq \frac{d^k L}{dt^k}\Big|_{t=t_1}$$

Thus there is no longer C^n order parametric continuity between the curves $R(\beta \nu)$ and L(t) although the curves are geometrically equivalent. Thus Barsky introduced the concept of geometric continuity [Barsky 81], such that the curve meet with n^{th} order geometric continuity (G^n) if $\beta > 0$ such that

$$\frac{d^k R}{du^k}\Big|_{u=u_0} = \beta^k \frac{d^k L}{dt^k}\Big|_{t=t_1} \quad k = 0, 1, ..., n$$

For cubic polynomials (G^2 *continuity*) and normalisation

$$\sum_{r=-2}^{1} B_{i+r}(u_i) = b_1(0) + b_0(0) + b_{-1}(0) + b_{-2}(0) = 1$$

gives 16 equations with 16 unknowns. Solving by symbolic manipulation gives the following basis function

$$
\begin{aligned}
b_1(u) &= \tfrac{1}{6}[2u^3] \\
b_0(u) &= \tfrac{1}{6}[2 + (6\beta_1)u + (3\beta_2 + 6\beta_1^2)u^2 \\
&\quad -(2\beta_2 + 2\beta_1^2 + 2beta_1 + 2)u^3] \\
b_{-1}(u) &= \tfrac{1}{6}[(\beta_2 + 4\beta_1^2 + 4\beta_1) + (6\beta_1^3 - 6\beta_1)u \\
&\quad -(3\beta_2 + 6\beta_1^3 + 6\beta_1^2)u^2 \\
&\quad +(2\beta_2 + 2\beta_1^3 + 2\beta_1^2 + 2\beta_1)u^3] \\
b_{-2}(u) &= \tfrac{1}{6}[(2\beta_1^3) - (6\beta_1^3)u + (6\beta_1^3)u^3 - (2\beta_1^3)u^3]
\end{aligned}
$$

and
$$\delta = \beta_2 + 2\beta_1^3 + 4\beta_1^2 + 4\beta_1 + 2 \neq 0$$

The basis function for some β_1 and β_2 are given in Figure 5.27. When $\beta_1 = 1$ and $\beta_2 = 0$, the basis function is same as for B-spline basis function (Figure 5.27(a)), the result is a unbiased, untensioned curve. Tension to the curve is increased by increasing β_2 (Figure 5.27(b)), so that the curve approaches the control polygon symmetrically.

5.9.2   Parallel Implementation of Splines on CLIP

Curve and surface representation in splines is an interpolation of discrete data by means of a "weighting function" or interpolants [Gordon 69, Rogers 76].

How are splines to be implemented in parallel on an SIMD architecture? A significant problem on an SIMD array is that there are no graph relations (except near-neighbour) of data in the array and data cannot be addressed but rather has to be shifted around the array, a very time-consuming process, as was discovered for geometrical transformations. Any method chosen to implement splines on an SIMD array must fit the structure of the array and use the strength of the array connectivity to move data.

In B-spline, the basis function coefficients are constants and if the vertices are equally spaced, the segments consisting of piecewise polynomials can be evaluated using difference equations [Gordon 69]. A more promising method for spline implementation on SIMD arrays is by recursive subdivision methods [Catmull 78, Doo 78, Lane 80, Cohen 80].

If one considers the formulation for a spline surface

$$Q_{k,l}(u,v) = \sum_{i=0}^{m}\sum_{j=0}^{n} N_{i,k}(u)N_{j,l}(v)V_{i,j}$$

and notes that $N_{i,k}(u)N_{j,l}(v)$ can be precalculated in a discrete form over a finite grid space, then the problem can be solved as series of convolutions. The convolutions, of course, can be performed in parallel on SIMD array computers.

Furthermore, the knot points in a spline can be calculated as an average weighting function of the three neighbouring vertices with the masks [1 4 1] for a cubic B-spline. Then the surface knots can be computed by convolving the separable masks

$$[1\ 4\ 1] \begin{bmatrix} 1 \\ 4 \\ 1 \end{bmatrix}$$

with the control graph. First, implementing this method for surface splines will be considered and then the concept will be extended and modified for generating line splines. The problem of implementing line spline fitting has arisen because the vertices are not being addressed in parallel; in fact the control vertices cannot be addressed (in parallel or otherwise) because the CLIP4 chips are simple Boolean processors with limited memory. Furthermore addressing data of often unknown and varying geometry on a grid structure would prove

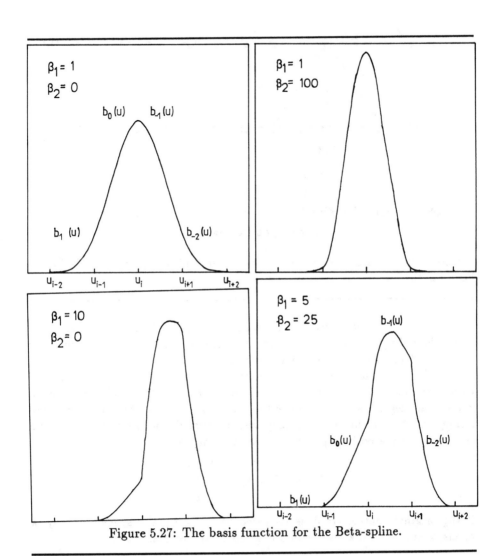

Figure 5.27: The basis function for the Beta-spline.

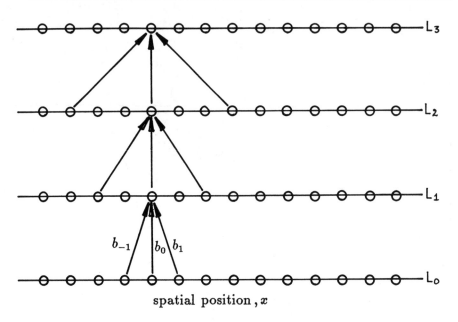

spatial position , x

Figure 5.28: Recursion using a modified hierarchical discrete correlation to generate the knot position.

to be both computationally and conceptually very difficult; thus it is easier to consider the points being joined in the first approximation by straight lines forming ridges on a two-dimensional surface.

Problems associated with the inability to address data is not only confined to line splines but to surface splines as well. However, the solution is both simple and elegant; start by joining the control graphs with first order interpolations. Each point in the array may then be regarded as a control graph and the knots calculated by separable convolution masks

$$b_i b_j \equiv [b_{-1} \ b_0 \ b_1] \begin{bmatrix} b_{-1} \\ b_0 \\ b_1 \end{bmatrix}$$

If the original control graph is regarded as being equally spaced by, say, k pixels, then the final knot position is given by the recursion

$$g_0(x,y) = f(x,y)$$
$$g_k(x,y) = \sum_{j=-1}^{1} \sum_{i=-1}^{1} b_i b_j g_{k-1}(x+ik, y+jk), \qquad k \geq 1$$

where $g_0(x,y)$ is the original control graph and $g_k(x,y)$ are the knot position (Figure 5.28).

This form of convolution can be regarded as a linear multiple of a first-order hierarchical discrete correlation [Burt 81]. In one dimension, the odd hierarchical discrete correlation (Odd HDC) for a function f(x) is defined as a set of correlation functions $g_k(x)$ for a discrete weighting function $w(x)$ by

$$(\text{Odd} \quad \text{HDC}) \quad \begin{aligned} g_0(x) &= f(x), \\ g_k(x) &= \sum_{i=-m}^{m} w(i)g_{k-1}(x + ir^{k-1}) \quad , k \geq 1 \end{aligned}$$

The sample distance grows geometrically by a factor r from level to level, thus r is the order of the HDC; $k = 2m+1$ is the width of the generating kernel. The hierarchical discrete correlation function has been used to spline image mosaics to give smooth continuous output without seams [Burt 83]. Burt's method for splining images is very different from what is being proposed here. In the multi-resolutional spline [Burt 83], the images to be splined are first decomposed into a set of band-pass filtered component images by the use of the HDC. Next, the component images in each spatial frequency band are assembled into a band-passed mosaic and components are joined using a weighted averaging technique only within the transitional zone proportional to the wavelength of the band. Finally, the band-passed mosaic images are added to give the splined image.

Returning to the problem of line splining, the control vertices are interpolated between by use of straight-line segments. Convolutions are not possible to sufficient precision with binary images using the given masks; therefore the line is made to have typically 10-bit precision by adding zero planes at the bottom of the bit-plane stack. The convolution is now conducted as for splining surfaces. After each convolution, the local maximum along the curve is the splined curve. This process is repeated recursively until the desired vertex spacing is obtained. Obtaining the local maxima using global propagations is difficult. This is because of the truncation errors and different global propagation paths picking up different maxima, leading to clumping at the corners. Furthermore, the curves are not continuous at all places because of the global propagation already having found a maximum in its path. Any local means of tracing maxima by neighbourhood connectivity is essentially serial and very slow on CLIP4 or CLIP4S. Fortunately, there is a very simple solution. Provided one chooses a basis function such as Barsky's Beta2-spline [Barsky 85] such that the spline has tension but no biasing, the convolution on a line is a uniform spreading function, the maximum of which may be found to a good approximation by choosing the medial axis [Arcelli 75] having suitably thresholded the spread image. Some examples of spline curves by the above method are given in Figure 5.29.

Filtering an image often involves operations on the image such that a smooth continuous curve is fitted to a discrete set of points; this is an interpolation process. The B-spline and the Beta-spline have these properties of second order continuity. If the noisy image is regarded as being the control graph, the spline surface fitted to it passes through the convex hull reducing the high-frequency information; much of the impulse noise, i.e., salt-and-pepper noise, is inherent in the high-frequency domain of the image. If the high-frequency component of the image must be maintained in certain regions of the image then the image can be filtered using the Beta-spline basis function with tension being set to greater than zero in these regions. The effect of this is to reduce

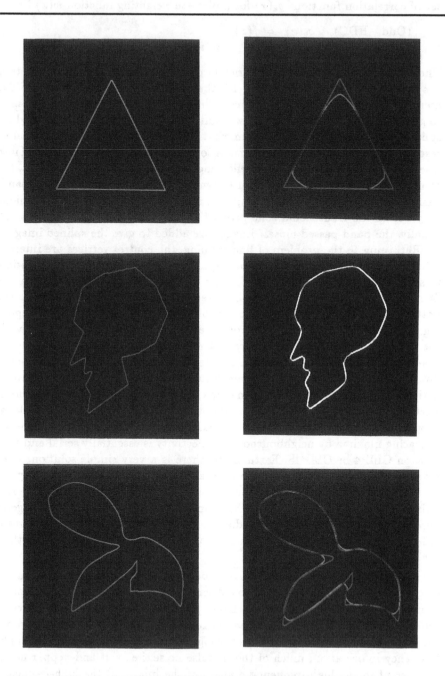

Figure 5.29: Line splines.

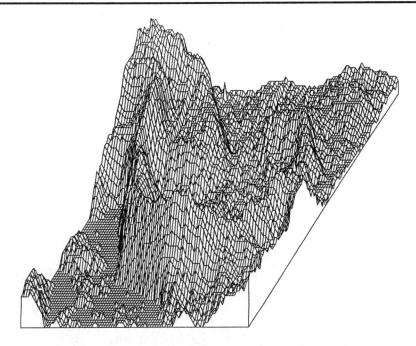

Figure 5.30: Topographical representation of a noisy sella image.

smoothing and maintain the original shape in these regions. The action of filtering using the Beta-spline basis function is shown in Figures 5.30–5.33 and 5.34–5.37 where comparisons of the "Beta-spline filter" with the more common median and weighted average filters are made.

5.10 Convex Hull: An Introduction

The convex hull of a set $\mathbf{S} = p_1, p_2, \ldots, p_n$ of n points has geometrical properties which have proved useful in pattern recognition [Toussaint 82, Nevatia 82]. An $O(n \log_2 n)$ algorithm which is of more interest here because of its intrinsic parallelism is due to Akl and Toussaint [Akl 78, Bhattacharya 83]. Furthermore, the Akl–Toussaint algorithm incorporates many of the fundamental ideas behind convex hull generation and many other convex hull algorithms borrow heavily from the Akl–Toussaint algorithm. The Akl–Toussaint algorithm has almost become a benchmark against which other convex hull algorithms are judged. A brief description of this algorithm is given below.

1. Determine the extreme points (Figure 5.38) in the $x-$ and $y-$directions, E_i ($i = 1,2,3,4$).

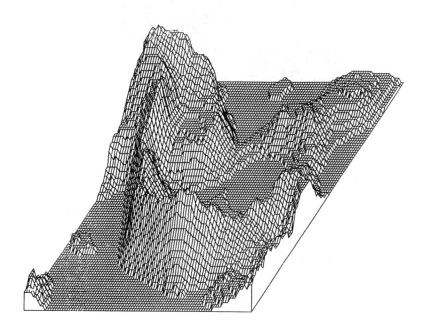

Figure 5.31: Result of applying a Beta-spline filter.

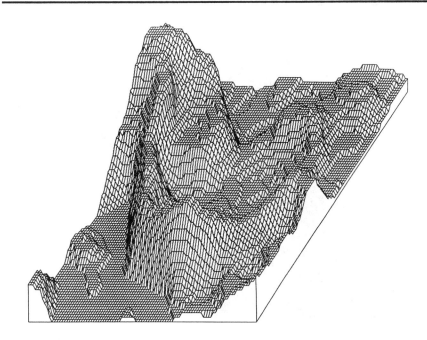

Figure 5.32: Result of applying a 7×7 median filter to the sella image in Figure 5.30.

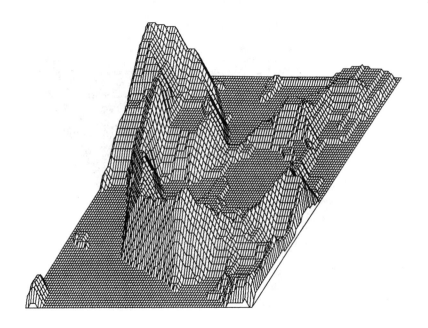

Figure 5.33: Result of applying a 7 × 7 weighted average filter to the sella image.

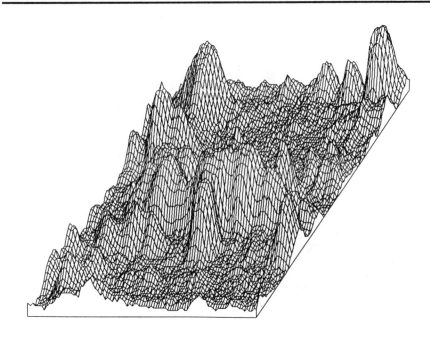

Figure 5.34: Topographical representation of a noisy electrophoresis gel image.

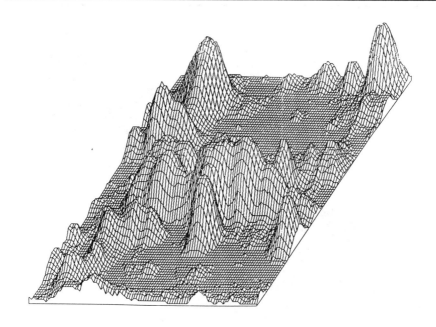

Figure 5.35: Result of applying a Beta-spline filter $(\beta 1 = 1, \beta 2 = 25)$ to image in Figure 5.34.

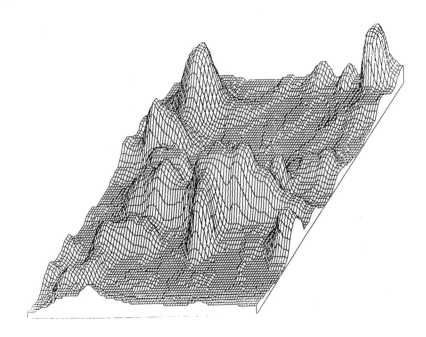

Figure 5.36: Result of applying a 7×7 median filter to the electropheresis gel image in Figure 5.34.

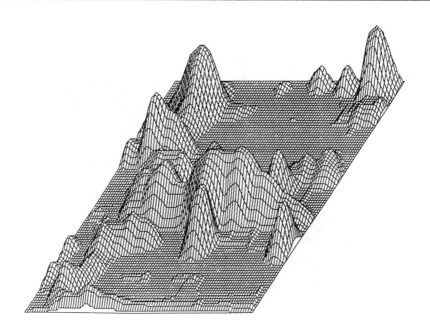

Figure 5.37: Result of applying a 7 × 7 weighted average filter to the image in Figure 5.34.

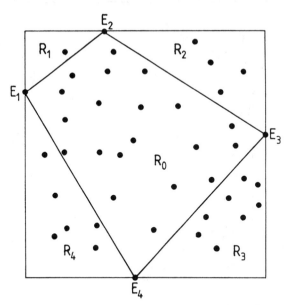

Figure 5.38: The extreme points in the $x-$ and $y-$directions.

2. Discard points in R_0. Determine extreme points E_i' ($i = 1,2,3,4$) in the $x + y$ and $y - x$ direction (Figure 5.39).

3. Identify points in R_i lying inside the triangle $E_i E_i' E_{i+1}$ and discard them.

4. The remaining points are sorted [Knuth 73, Horowitz 78] by their x-coordinates in ascending order if the points are in R_1 or R_2 and in descending order if in R_3 or R_4. The monotoned ordered points when joined form a polygon (Figure 5.40).

5. The convex hull of the monotone polygon is determined using a back-tracking algorithm [Sklansky 72].

An algorithm with some similarities to the Akl–Toussaint algorithm has been reported by [Sklansky 82] to find the convex hull of simple (i.e. not self-intersecting) polygons with $O(n)$ complexity.

The Akl–Toussaint algorithm can be implemented in parallel on SIMD computers such as CLIP4 and CLIP4S by the following method:

1. The extreme points can be extracted using iterative local neighbourhood operations. For example, the extreme points in the y direction can be extracted by generating a line at the top and bottom of the array. These lines are expanded in the N–S direction. If one of the expanded lines collides with a point, the point is noted as being an extreme point and

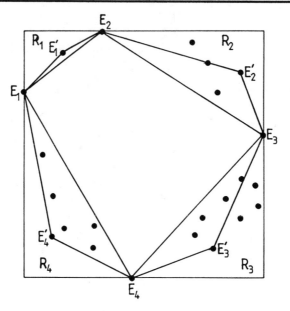

Figure 5.39: The extreme points in the $x + y$ and $y - x$ directions.

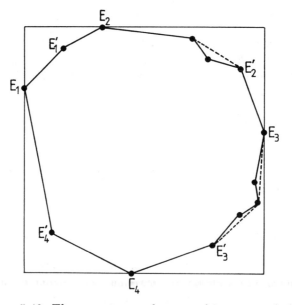

Figure 5.40: The monotone polygon and its convex hull.

the collision line is removed. The process is continued until the other extreme point is found. This process is repeated in the E–W direction to detect the extreme points in the $x-$direction.

2. To detect the extreme points in the $x + y$ and $x - y$ direction, a point is generated at each of the four corners of the array. The process as above is carried out, but now the points are expanded by iterative 4- and 8-connected expands, generating octagonal masks from the corners.

3. Join up the points to form an octagonal shape. Remove all the points from inside the shape by filling in the shape. Then negate the filled shape and boolean AND with the original data. This should leave behind the extreme points and any other points lying outside the hull.

4. Join up the remaining points cyclically to generate a monotone convex hull.

This algorithm will fail to determine some small concavities because the array is only 8-connected and therefore propagation paths are limited. In the following section, an algorithm is proposed to approximate the convex hull whilst recognising the connectivity limitations of SIMD arrays.

5.10.1 A Fast Approximation to a Convex Hull

Although the Akl–Toussaint algorithm is intrinsically parallel and its implementation on a SIMD array computer is relatively simple, a convex hull algorithm would be faster still if more use of global propagation could be made. A fast approximation to a convex hull is proposed using global propagation.

An algorithm to approximate the convex hull of a finite grid point set is by approximating it to a p-hull polygon [Klette 84]. This process has complexity of $O(pn)$, where p is number of sides of the polygon and n is the number of points in the data set. The CONVEX_HULL algorithm given below in Algorithm 5.10 has some similarities with the p-hull polygon [Klette 84]. The major differences are that CLIP4 and CLIP4S are eight-connected arrays and therefore generating p-hull polygons of $p > 8$ is difficult using only single global propagation operations. It is proposed here to generate an 8-hull polygon and a 16-hull polygon using 8-global propagations in each case and taking the median of the difference of the two hulls. The 8-hull polygon is referred here as being *over-determined* since the polygon encompasses the true convex hull and the 16-hull polygon is said to be *under-determined* because it has concavities and lies within the true convex hull.

The GENERATE_HULL function determines the **hull** of a given set of points (**pts**) by using global propagation. Initially, the whole array is regarded as being a possible convex hull and is set to one. In the GENERATE_HULL function, the image **pts** initiates propagation in the given directions. The signal propagated in the given directions is one if a point in the image **hull** is one and receives a propagation of one. The output at each pixel is one if the **hull**

Algorithm 5.10: Parallel fast approximation to a convex hull.

$CONVEX\_HULL$(pts)
<u>begin</u>
   /* Convex hull data is in image pts */
   hull1 ← 1;
   /* Find the over-determined octagonal hull */
   let \$temp = [8 rotations of the direction list]
   <u>for</u> \$*temp* <u>do</u>
       hull1 ← $GENERATE\_HULL$(hull1, pts, *dir_list*); <u>od</u>
   /* Find the under-determined hull */
   hull2 ← hull1;
   <u>for</u> \$*temp* <u>do</u>
       hull2 ← $GENERATE\_HULL$(hull2, pts, *dir_list*); <u>od</u>
   diff ← $DIFFERENCE$(hull1, hull2);
   hull2 ← OR(hull2, $THINR$(diff));
   hull2 ← OR(hull2, $CAVITY$(hull2));
   convex_hull ← $OUTER\_EDGE$(hull2);
<u>end</u>

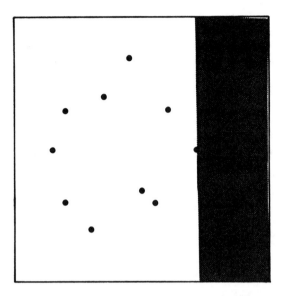

Figure 5.41: Result of the first-pass of the GENERATE_HULL algorithm. The data is spread in the direction the shaded area is where no propagation has occurred. On the border of the black-white is one of the extreme points.

is one at that pixel and receives a one propagation signal, otherwise the pixel remains, or turns to, zero. The result is shown in Figure 5.41. The shaded area is where the **hull** has received no propagation and has become zero.

If the GENERATE_HULL function is applied iteratively to the modified **hull** and the direction list rotated, an octagonal hull for the given set of points is generated (Figure 5.42). The inner angle BÂC at a point A is less than or equal to π radians. Since the array is eight connected, and the direction list chosen such that the result is an octagonal hull, the angle made by the octagonal hull at A will be greater than BÂC, and only equal when BÂC is π radians. Thus the octagonal hull is said to be over-determined since the area enclosed by an octagonal hull for a given set of points is always greater than or equal to that enclosed by a convex hull. Furthermore, a π radian propagation from the extreme data points ensures that there can never be any concavities in an octagonal hull so generated.

A better approximation to a convex hull is required. A better approximation would be to take the average of the over-determined octagonal hull and an under-determined hull. An under-determined hull can be generated by using the GENERATE_HULL function but now limiting the angle of propagation to $3\pi/4$ radians and using eight rotations of this angle in steps of $\pi/4$ radians. The result is shown in Figure 5.43.

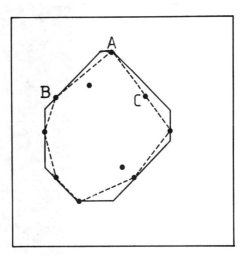

Figure 5.42: The solid line is the over-determined octagonal hull, the broken line is the proper convex hull.

A good approximation to a convex hull is then the mid-point between the over- and under-determined hulls. The medial line is obtained by calculating the difference between the two hulls by the exclusive OR of the two hulls and then skeletonising the result (THINR) [Arcelli 75, Hilditch 83]. The process of obtaining the medial line may create some concavities. Some of these concavities can be detected by global propagation and ORed with the skeletonised hull, **hull2**. The outer edge of the convex hull body is determined by a single global propagation operation. The propagation signal is sent from the edge of the array through a four-connected background. If a pixel in the array is one and receives a propagation then the output is one, i.e. it forms part of the outer edge. If the over-determined octagonal hull is of size Φ and the under-determined hull is Ψ, then the complexity of the algorithm given below is $O(\text{width}(\Phi @ \Psi))$, where @ is the EXOR function. We are ignoring the time for the 16 global propagation operations because they are a fixed overhead. Currently this takes 10.7 ms and for the new CLIP4A machine only 0.4 ms.

The aim has been to develop an algorithm for a fast approximation to a convex hull. The CONVEX_HULL algorithm has been run on figures of 17 alphabets and some of the results are shown in Figure 5.44. The figures consisted of 2110+330 points and run on CLIP4 in 19.7+2.9 ms, the variation in time is purely because of the skeletonisation process and therefore is a function of the concavity of the data. These times compare very favourably with the Akl–Toussaint algorithm run on an Amdahl V7 computer which took typically

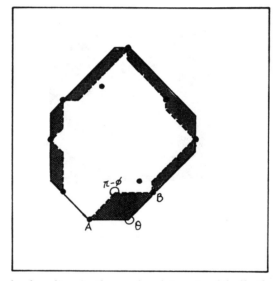

Figure 5.43: The broken line is the under-determined hull, the shaded area is the difference between the under-determined and the over-determined hull. Skeletonising this region gives the convex hull.

Figure 5.44: Convex hull of some figures.

100 ms [Bhattacharya 83].

5.10.2   Finding more than one Convex Hull in Parallel

By a simple modification of the algorithm, more than one convex hull can
be computed in parallel. For the CONVEX_HULL algorithm, initially the
complete array was set to one and propagation was allowed in the whole array. However, by limiting the propagation paths to within rectangles bounding
sets of points forming convex hulls, each of the convex hulls lying within the
bounded rectangles [Otto 84] can be computed in parallel. If the set of points
constituting each convex hull is not a random collection of points but rather a
conglomerate such the letters given in the examples (Figure 5.45), their bounding rectangles can be computed in parallel by use of four global propagations
[Otto 84]. The method is one of detecting regions which are in shadow to
global propagation from certain directions. The method consists of starting
propagation from the edge of the array which has been set to one in the direc-

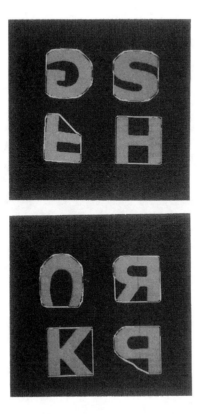

Figure 5.45: Calculating more than one convex hull in parallel.

tion   (and subsequently its $\pi/2$ radian rotations) through the background (zero points), and setting the result to one where the object was one or did not receive a propagation signal. Once the bounding rectangles have been determined, this forms the **hull1** image in the CONVEX_HULL algorithm. The rest of the algorithm proceeds as before, generating a convex hull in parallel in each of the bounding regions.

Chapter 6

Computational Vision

6.1 Introduction

In computational vision, the objects need to be described using geometric shapes [Marr 82]. However, the problem of extracting 3D information is ill-posed. Solutions are computationally expensive using regularization theories [Poggio 85]; also they do not deliver the right answer, only some smoothed approximation to it. However, with ill-posed problems, solutions often do not exist or they are not unique, or they do not depend continuously on the initial data. Examples of ill-posed problems involve stereo matching, structure from X (e.g. shading, motion), optical flow, edge detection and surface reconstruction. Some of these problems we will discuss in this chapter.

If we take two or more images, from the difference in the images, it is possible for the relative motion of the camera and the scene (or the motion of the objects or the relative displacement of the two cameras or the relative position of the objects) to be computed. This difference in the two images of the points in the 3D scene is called the *image disparity*. The difference we observe of a projected point in a 3D scene onto pair of perspective images is called *parallax*. The parallax is caused by changes in the position of the perspective centre and optical axis orientation. The computation of disparity is an important process in the human vision system. Along with the information of texture gradient and linear perspective, disparity is important for spatial perception. As we have motioned, these disparities arise because of binocular parallax, motion parallax, object motion or any combination of these.

Julesz has carried out experiment with *random dot stereograms* which show that visual process can fuse two images not containing usual structures [Julesz 71]. The stereograms, consisting of uncorrelated dots, is artificially created by shifting a part of the image either to the left or the right to form a second image which are then simultaneously recorded by two laterally separated cameras. Each image has no depth information, but our visual system can fuse then to give a perception of depth resulting from the shift of the parts of an image. The importance of this experiment is that it shows the pair of images contain information which are not present in any single image.

6.2   Camera and Image Geometry

Scaling, rotation and translation of objects in three-space is covered in most graphics textbooks [Watt 89, Foley 90]. A point ($\mathbf{v}$) transformed under translation ($\mathbf{T}$), scaling ($\mathbf{S}$), and rotation ($\mathbf{R}$) (which make up a Euclidean transformation) can be written in a matrix form as

$$\mathbf{v}^* = \mathbf{R}\left[\mathbf{S}\left(\mathbf{Tv}\right)\right]$$

or

$$\mathbf{v}^* = \mathbf{Av}$$

where

$$
\begin{aligned}
\mathbf{A} &= \mathbf{RST} \\
&= \begin{bmatrix} \cos\theta & \sin\theta & 0 & 0 \\ -\sin\theta & \cos\theta & 0 & 0 \\ 0 & 0 & 1 & 0 \\ 0 & 0 & 0 & 1 \end{bmatrix} \begin{bmatrix} S_x & 0 & 0 & 0 \\ 0 & S_y & 0 & 0 \\ 0 & 0 & S_z & 0 \\ 0 & 0 & 0 & 1 \end{bmatrix} \begin{bmatrix} 1 & 0 & 0 & t_x \\ 0 & 1 & 0 & t_y \\ 0 & 0 & 1 & t_z \\ 0 & 0 & 0 & 1 \end{bmatrix}
\end{aligned}
$$

In this example we have considered rotation about a single axis.

An image is a projection of a set of 3D points onto a 2D plane; this form of projection is referred to as *planar geometric projection* because the projection involves straight rays (*projectors*) and planes rather than curved surfaces. There are two basic types of projection: *perspective* and *parallel*. The difference is in the position of centre of projection: in parallel projection the centre of projection is at infinity (see Figure 6.1). With radiographs, parallel (or *orthographic*) projection is often assumed because the distance between X-ray source and the object is much greater than object and image distance; thus rays are assumed to be almost parallel.

We will confine our discussion to perspective projection. A point $\mathbf{P}(x_v, y_v, z_v)$ in the viewing coordinate system with a viewing plane normal to the z_v axis and a distance d from the centre of projection is projected on to a point $\mathbf{P}'$ with coordinates $(x_s, y_s, 0)$ in the view-plane coordinate system (see Figure 6.1).

The relationship between $\mathbf{P}$ and $\mathbf{P}'$ can be found using the similar triangle method

$$\frac{x_s}{f} = \frac{x_v}{z_v} \qquad \frac{y_s}{f} = \frac{y_v}{z_v}$$

This transformation can be expressed as a 4×1 matrix.

$$\begin{bmatrix} x_s & y_s & f & 1 \end{bmatrix} = \begin{bmatrix} fx_v/z_v & fy_v/z_v & f & 1 \end{bmatrix}$$

Thus any viewing transformation, $\mathbf{T}_{view}$, which maps a point in the world to a point in the image can be expressed as

$$\begin{bmatrix} x_v & y_v & z_v & 1 \end{bmatrix} = \begin{bmatrix} x_w & y_w & z_w & 1 \end{bmatrix} \mathbf{T}_{view}$$

Let us consider a camera placed at world coordinate (e, f, g), and the camera has its own coordinate system (Figure 6.2). The camera is allowed to tilt

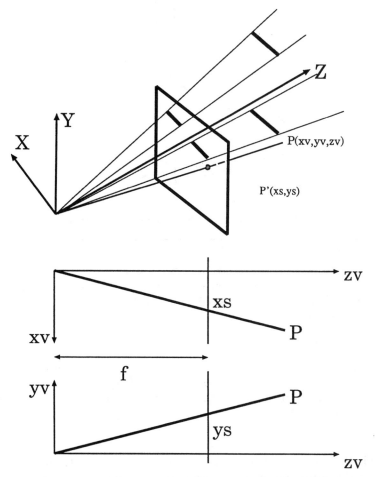

Figure 6.1: The geometry for perspective projection.

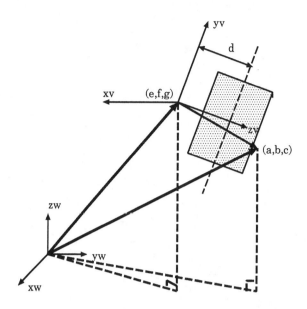

Figure 6.2: A general viewing camera setting.

through an angle α and pan through an angle θ. The geometrical arrangement as in Figure 6.2 can be achieved by: (i) moving the camera to the world coordinate origin, (ii) pan of the x-axis, (iii) tilt of the z-axis. This transformation (assuming no scaling) is

$$
\begin{aligned}
\mathbf{T}_{view} &= \mathbf{R}_\alpha \mathbf{R}_\theta \mathbf{T} \\
&= \begin{bmatrix} \cos\theta & \sin\theta & 0 & 0 \\ -\sin\theta\cos\alpha & \cos\theta\cos\alpha & \sin\alpha & 0 \\ \sin\theta\sin\alpha & -\cos\theta\sin\alpha & \cos\alpha & 0 \\ 0 & 0 & 0 & 1 \end{bmatrix} \begin{bmatrix} 1 & 0 & 0 & 0 \\ 0 & 1 & 0 & 0 \\ 0 & 0 & 1 & 0 \\ -t_x & -t_y & -t_z & 1 \end{bmatrix}
\end{aligned}
$$

then $\mathbf{v}^* = \mathbf{v}\mathbf{T}_{view}$.

This is the typical case in computer graphics where we know the object's position in the world coordinate and we project them onto the image coordinate system. However, in computer vision, we have the inverse problem. We know the objects position in the image and we wish to determine its position in the world. We need an *inverse transformation mapping* from the image coordinate system to the world coordinate system

$$
\mathbf{v} = \mathbf{T}_{view}^{-1}\mathbf{v}^*
$$

Assuming that there is no rotation and translation, $\mathbf{T}_{view}^{-1}$ is easily determined to be

$$\mathbf{T}_{view}^{-1} = \begin{bmatrix} 1 & 0 & 0 & 0 \\ 0 & 1 & 0 & 0 \\ 0 & 0 & 1 & \frac{1}{f} \\ 0 & 0 & 0 & 1 \end{bmatrix}$$

where f is the focal length of the lens. A point $[x_0, y_0, 0]^T$ in the image is mapped to world coordinate $[X, Y, Z]^T$. This is the simple perspective transformation. We find that $z = 0$ for any 3D point; this has arisen because mapping from the scene to the image is a many-to-one transformation. A point (x_0, y_0) in the image corresponds to a line passing through $[x_0, y_0, 0]^T$ and $[0, 0, d]^T$. The equation of this line (from Figure 6.1) is

$$x = \frac{x_0}{d} Z \qquad y = \frac{y_0}{d} Z$$

Therefore, we can only determine its position if one component of the world coordinate for the object is known. In many instances we assume that the objects are some height off the ground plane.

There are circumstances when we need to determine the camera parameters by using the information available in the image alone; this is referred to as the *camera calibration* problem. Refer back to

$$\mathbf{c} = \mathbf{T}_{view} \mathbf{w} \tag{6.1}$$

where

$$\mathbf{T}_{view} = \begin{bmatrix} t_{11} & t_{12} & t_{13} & t_{14} \\ t_{21} & t_{22} & t_{23} & t_{24} \\ t_{31} & t_{32} & t_{33} & t_{34} \\ t_{41} & t_{42} & t_{43} & t_{44} \end{bmatrix} \qquad \mathbf{c} = \begin{bmatrix} c_1 \\ c_2 \\ c_3 \\ c_4 \end{bmatrix} \qquad \mathbf{w} = \begin{bmatrix} w_1 \\ w_2 \\ w_3 \\ w_4 \end{bmatrix}$$

and

$$x = \frac{c_1}{c_4} \qquad y = \frac{c_2}{c_4} \tag{6.2}$$

Substituting equation (6.1) in equation (6.2) gives

$$\begin{aligned} xc_4 &= t_{11}w_1 + t_{12}W_2 + t_{13}w_3 + t_{14} \\ yc_4 &= t_{21}w_1 + t_{22}W_2 + t_{23}w_3 + t_{24} \\ c_4 &= t_{41}w_1 + t_{42}W_2 + t_{43}w_3 + t_{44} \end{aligned}$$

Substituting for c_4 gives two equations:

$$\begin{aligned} t_{11}w_1 + t_{12}w_2 + t_{13}w_3 - t_{41}xw_1 - t_{42}xw_2 - t_{43}xw_3 - t_{44}x + t_{14} &= 0 \\ t_{11}w_1 + t_{12}w_2 + t_{13}w_3 - t_{41}yw_1 - t_{42}yw_2 - t_{43}yw_3 - t_{44}y + t_{24} &= 0 \end{aligned}$$

These two equations can be solved if we can find six or more points in the image and we know the position of the corresponding object points in the world coordinate system. Then the twelve or more equations can be solved to determine the twelve unknown parameters in $\mathbf{T}_{view}$ using a numerical method. If the camera calibration is unknown then six points are need. Solutions fails to be unique if the six points and the optical centres of the camera are on a space curve of degree three. If greater constraints can be applied then we would require fewer points.

6.3   Computational Stereo

Computational stereo attempts to solve the problem of recovering three-dimensional information from two images of the same scene taken by cameras in different positions. The distance to the points in the field of view are estimated using *triangulation.* An image is formed by light coming from the object surface and passing through the lens centre to fall on sensor surface. A point on the image and the lens centre uniquely determines a line along which the ray would have travelled from the object surface. Therefore, the surface must lie along this line, but from a single image, the depth is unknown. Hence the need for stereo. The position on the surface can be uniquely determined if we find the corresponding line for the other images; the surface must lie where the two lines intersect and the depth can be determined by triangulation. An advantage of using stereo to determine the surface depth is that the method is passive. Methods for stereo matching to determine depth information has been reviewed in [Case 81, Konecny 81]. Stereo has been applied in passive visual navigation such as autonomous vehicle guidance [Hannah 80, Moravec 79, Moravec 81, Gennery 80], in industrial automation, and the interpretation of aerial images such as in the generation of maps. The later is usually uses motion stereo. The aeroplane contains a single camera but it flies in a known path.

A point in space and the lines through the lens centre from the corresponding projected image points determine a plane called the *epipolar plane,* see Figure 6.3. The intersection of the epipolar plane with the image plane is along a line called the *epipolar line.* The importance of determining the epipolar line is that points on this line in one image corresponds to points in a corresponding epipolar line in the second image. This construction confines the possible area of search for corresponding feature points along the epipolar lines thereby reducing the computation required for the feature matching process. If the cameras are set up such that there is only a horizontal displacement between them, then the disparity is only along the scan lines, the *epipolar lines.* The images are then said to be *in correspondence.* This constraint in camera position has been used by many [Grimson 80, Marr 77a]; however, in reality, the images are rarely *in correspondence* and one has to have relative camera model [Gennery 79]. It is then vital to recover good camera geometry, this problem has been considered in [Fischler 81].

There are five stages in the process of recovering depth from stereo images:

1. Camera calibration, this involves the determination of the stereo baseline, the focal length of each camera, and the computation of the epipolar lines.

2. Finding features such as corners, lines and curves.

3. Matching features to generate the disparity map. We find a feature point in one image. This defines an epipolar line in the image. We search along the corresponding epipolar line in the second image to find the matching point. The disparity map contains the separation information of the matching points in the two images.

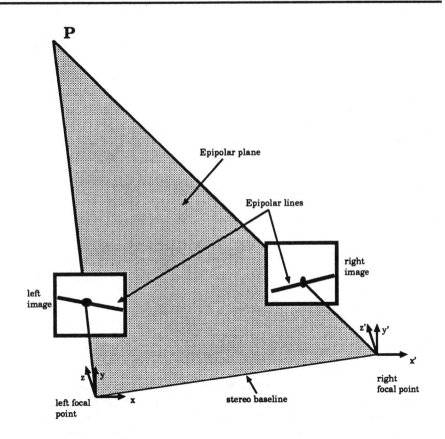

Figure 6.3: The epipolar geometry used in stereo vision.

4. Determining the depth at á finite set of points knowing the disparity and the camera parameters.

5. Interpolating the depth to obtain a surface.

Two important parts to computational stereo are feature detection and matching. If the camera geometry is unknown, then it is easier to use point-like features because the epipolar lines are unknown and also with larger features there is the problem of perspective distortion [Barnard 80]. If epipolar transformations are known, then edge features (particularly vertical lines [Crowley 90]) are used [Arnold 78, Baker 80, Grimson 79]. One of the earliest used operators for feature detection was Moravec's interest operator [Moravec 81]. Edge-based feature detection has been described in [Medioni 85] using the Nevatia–Babu algorithm [Nevatia 80]. Medioni and Nevatia have extended on the edge-based method by checking for edge connectivity. They describe edge lines by the coordinate of their end points, their orientation and edge strength. On a similar track, Lloyd *et al.* describe a parallel method of line matching extending over many epipolar lines by a relaxation labelling method [Lloyd 87a]. Others have used zero-crossing edges with sign information as features. Nishihara describes hardware for computing zero-crossing with sign changes and the matching is carried out in a hierarchical manner at different resolution [Nishihara 84]. The problem with many of these approaches is that the objects under observation are often man-made and have plain surfaces. This leads to the determination of sparse stereo maps. For dense stereo maps one needs texture. Nishihara introduces texture using *unstructured light*, this is a random-textured pattern from a projector situated between the two cameras [Nishihara 84]. The problem of using *structured light* is it produces surfaces with a repeated pattern and this leads to local ambiguity in matching.

Matching of features usually consists of computing a difference between the two features set and then using a search strategy to find a minimum between them. A question to answer is whether the matching should be area-based or feature-based. The limitation of using an area-based method is that the surface needs to have texture to find a good match; many man-made objects have no discernible texture. The method is also sensitive to changes in intensity, contrast and illumination which do occur with two separate cameras. There are also problems with surface discontinuities, these have to be first found and surface patches generated within which there are no discontinuities. Also the surface patches tend to be very different between the two images if the depth changes rapidly. However, while feature-based matching is more accurate, since features can be computed to sub-pixel accuracy, faster because the correlation areas are smaller, and less sensitive to changes in camera position, the method essentially leads to producing only a sparse depth map. The surface then has to be constructed by interpolation. In using the feature-based method, Marr and Poggio added two further constraints to give better match [Marr 77b]: (1) each feature gives rise to an *unique* match, thus only one disparity value; and (2) there is a *continuity* in surface, thus disparity changes smoothly except at edges.

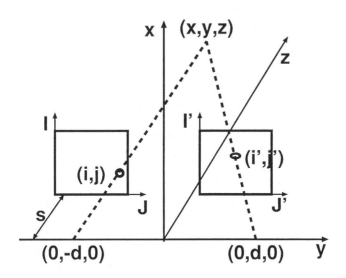

Figure 6.4: A simple stereo setup.

If we have two cameras with the setup as shown in Figure 6.4 with focal length s and focal point at $[0, \pm d, 0]^T$ and the line of sight parallel to the z-axis, then a point at $[x, y, z]^T$ will appear at (i, j) in the left image and at (i', j') in the right image. The epipolar line is along the y-axis (horizontal), then $i = i'$, therefore the correspondence search is confined to be along the scan line. The depth is calculated from:

$$y_1 = \frac{j}{s} Z_1$$

$$y_2 = \frac{j'}{s} Z_2$$

Now $y_2 = y_1 + 2d$ and $Z_1 = Z_2 = Z$. Then

$$Z = \frac{2sd}{(j - j')}$$

A stereo vision which generated from the work done by Moravec is FIDO [Thorpe 84]. The 512 × 512 left and right stereo pair images are reduced in resolution using a image pyramid generation scheme to generate images of size 256 × 256, 128 × 128, upto 8 × 8. It locates features in one of the images using an *interest operator*. Then for each feature located, a search is made for a corresponding feature in the other image within a search window using a correlation technique. Once the corresponding features have been found, the disparity computation can be carried out. This algorithm has been implemented on the

Warp systolic array computer [Clune 87]. The parallelism is exploited in various
ways: the two sets of *pyramid* images are generated in parallel in two clusters
using a systolic convolution type of approach. The feature points are found
by partitioning the images (with appropriate overlaps) amongst the available
processors. The feature correlation between image pairs is carried out in the
appropriate pyramid layer using a systolic approach.

Strong has taken the approach of implementing the Marr–Poggio algorithm
on the MPP [Strong 91]. As in the FIDO algorithm, a pyramid image is used
once again. In this approach, grey-level area correlation is carried out to find
regions where the stereo pairs matches locally and to create a disparity map.

The problem of matching grey-level patches is well known because the intensity of the images in the stereo pair may be different and because of noise.
As previously mentioned, feature-based methods are more robust. Matching
features in parallel using a CLIP4 SIMD processor array has been used in the
analysis of electrophoresis gels [Potter 84] to align a test gel against a reference gel. This approach could be used for matching stereo pairs. Unmatched
features in one image can be "slid over" the other image until all the matches
are found. To each feature a vector can be assigned such that, as they are
moved, the disparity is computed from its movement with respect to their original position. If we compute the epipolar geometry for each of the images,
then the two images can be rotated until the epipolar lines are aligned (say to
the horizontal). Then to match features, the test image need only be moved
"horizontal" to find matches and determine disparity. This would give rise to
a sparse disparity map. To determine dense disparity locally to these points
a relaxation approach (see §3.3.3) can be applied. Then the depth z at each
point can be computed in parallel using

$$z = \frac{2\lambda d}{disparity}$$

6.4   Motion Detection and Optical Flow

Optical flow is the measure of the apparent motion of image brightness. Both
optical flow and stereopsis give same type of information: displacement vector
fields, assuming that the changes in the image intensity is only due to motion
of objects.

A straightforward method for measuring velocity of image points in a sequence of image frames is to find feature points in successive images and match
them to determine the disparity map. Such a method for determining image velocity has been presented in [Moravec 79, Thompson 80, Barnard 80, Dreschler
81]. If one uses a long sequence of image frames, then the detected features
could be "superimposed" onto a single frame and the "flow field" could be determined using a Hough transform technique rather than explicitly matching
the features across the frames and solving the "correspondence" problem.

However, there are two general problems with such feature (*token*) matching
schemes. The ease or success of finding good correspondence between features

depends on the level at which the correspondence is sought. If the tokens are *edgels, points, blobs, or line segments* then this has the advantage that no extra processing is required to build higher primitives because the tokens are easier to find. This is important when real-time performance is sought. Furthermore, this method is more general in that no special objects need to be recognised. However, finding good correspondence is difficult. For such primitive token matching, it is usually assumed that the motion is small, this is achieved by using appropriate frame-rate. Then matching can often be carried on the basis of proximity alone [Potter 77]. Other methods for matching features have involved matching regions using cross-correlation techniques [Nagel 78], matching shape descriptors [Chow 77]. Other matching techniques can be found in [Tsuji 79, Jacobus 80].

When using structures or object description, finding correspondence is significantly easier since objects tend to be unique; however, as mentioned above, this is generally achieved at great computational cost. Also, the token-matching approach gives rise to a sparse flow map; if the map is not too sparse then a diffusion process can be applied to estimate the optical flow in other parts of the image. Nagel determined flow field by matching corners which were detected using the Moravec operator [Nagel 81].

There is another approach separate to token matching. If we assume that objects are moving over a background, then there is a relationship between the spatial and temporal changes in the intensity. We can express the intensity at a point in image as $I(x, y, t)$. Motion can be measured by change detection from subtraction of two successive frames [Jain 79a, Jain 79b] or by grey-level cross-correlation of successive frames [Smith 72, Leese 70, Lillestrand 72]. However, these methods suffer from the problem that we need to assume that the image as a whole moves between frames.

If we assume that a time-varying image is locally continuous and differentiable then it can be described by a function which can be expanded using the Taylor series.

$$I(x + \mathrm{d}x, y + \mathrm{d}y, t + \mathrm{d}t) = I(x, y, z) + I_x \mathrm{d}x + I_y \mathrm{d}y + I_t \mathrm{d}t + \mathcal{O}^2 \qquad (6.3)$$

where $I_x = \frac{\partial I}{\partial x}, I_y = \frac{\partial I}{\partial y}$, and $I_t = \frac{\partial I}{\partial t}$. If we assume that within a time interval $[t, t + \mathrm{d}t]$, objects undergo pure translation over a short distance $(\mathrm{d}x, \mathrm{d}y)$, and the surfaces are Lambertian then

$$I(x + \mathrm{d}x, y + \mathrm{d}y, t + \mathrm{d}t) = I(x, y, t) \qquad (6.4)$$

Combining equations (6.3) and (6.4), we have

$$I_x \mathrm{d}x + I_y \mathrm{d}y = -I_t \mathrm{d}t$$

or

$$I_x \frac{\mathrm{d}x}{\mathrm{d}t} + I_y \frac{\mathrm{d}y}{\mathrm{d}t} = \nabla \mathbf{I} \cdot \mathbf{v} = -I_t$$

This is called the *motion constraint equation*. From this we see that the component of flow is in the direction of brightness gradient (I_x, I_y) and has a magnitude $\frac{-I_t}{\sqrt{I_x^2 + I_y^2}}$. Note that I_x, I_y and I_t are all measurable; but there are difficulties

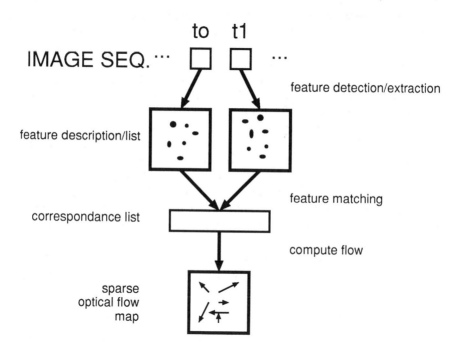

Figure 6.5: Feature-based method for determining velocity field.

when the grey-level gradients are low. However, we have a single equation but two unknown velocity components. To solve this equation and recover the full velocity, we need to make further assumptions. Some have assumed that the velocity is constant within a region [Fennema 79]. However, this assumption is only valid when the object translations occur parallel to the image plane and even then only near the origin. A more general assumption is that the objects are rigid, and the velocity varies smoothly between neighbouring points [Horn 81]. This condition can be expressed as the requirement to minimise

$$E_2 = \sqrt{(\bar{v_x} - v_x)^2 + (\bar{v_y} - v_y)^2}.$$

Also, because of noise, the flow equation can also be expressed as requiring to determine a minimum

$$E_1 = v_x \times I_x + v_y \times I_y + I_t$$

An expression for the velocity can now be derived by minimising the total error

$$E^2 = \alpha^2 E_1^2 + E_2^2$$

This we can rewrite as

$$v_x = \bar{v_x} - I_x \times (I_x \times \bar{v_x} + I_y \times \overline{v_y + I_t})/(\alpha^2 + I_x^2 + I_y^2)$$

$$v_y = \bar{v_y} - I_y \times (I_x \times \bar{v_x} + I_y \times \overline{v_y + I_t})/(\alpha^2 + I_x^2 + I_y^2)$$

If $\bar{v_x}$ and $\bar{v_y}$, the local average velocity, are known then we can solve for v_x and v_y.

Horn and Schunck present a method for computing the velocity vectors using an iterative solution to solve a differential equation which is obtained using the calculus of variations[Horn 81].

$$v_x^{n+1} = \bar{v_x}^n - I_x \left[\frac{I_x \bar{v_x}^n + I_y \bar{v_y}^n + I_t}{(\alpha^2 + I_x^2 + I_y^2)} \right]$$

$$v_y^{n+1} = \bar{v_y}^n - I_y \left[\frac{I_x \bar{v_x}^n + I_y \bar{v_y}^n + I_t}{(\alpha^2 + I_x^2 + I_y^2)} \right]$$

Initially (v_x^0, v_y^0) is set zero everywhere.

Computing the flow at each point using such an iterative method is generally slow. However, we observe that the computations required are local and need to be computed at all points and there the algorithm is potentially massively data parallel and maps well onto the SIMD paradigm of computation on SIMD array computers. Also, Wang has described an implementation of the modified Hildrith algorithm [Gong 90] on a network of transputers [Wang 90]. Hildreth computes the optical flow using an edge-based method [Hilditch 84]. The flow is first computed normal to the zero-crossing contours using the motion constraint equation. Then the full-flow is estimated by regularization around the closed contour. This gives arise to the Euler–Lagrange equation which can be solved using a conjugate gradient decent algorithm. However, such a solution is essentially sequential. This method is improved on by Gong and Brady [Gong 90] who computes the tangential component along with the normal component of flow. Unlike the normal component, the tangential component is computed only at locations near corners where the image Hessian is well constrained [Gong 89]. These local flow estimates are then propagated along the contours using a wave/diffusion process [Scott 88]. Wang's implementation then consists of: (1) edge segmentation using the Canny edge operator [Canny 86] and generating closed contours; (2) detecting corners by looking for sharp curvatures after B-spline fitting to the curves; (3) flow is computed at corners by matching using cross-correlation using 5×5 windows and 8×8 search area; and (4) the local flow is propagated along the contour using the wave/diffusion algorithm. The first three stages of the computation are carried out using a pipelined architecture; the wave/diffusion algorithm which requires dynamic sheduling is implemented using a processor farm. However, we should note that optical flow involves equation with derivatives. These computations are noise-sensitive and hence unreliable when dealing with real images where the noise content is often not known.

As a summary, we should note that both feature matching and variations in intensity method plays import part in the visual motion detection process and they are complementary to each other. The intensity-based methods act as the "attentive vision", it computes the local flow field and also acts as the focus of attention using visual cues. The token-matching schemes play their role in recovery of structure from motion and in the long-term accurate tracking of objects. The measurement of the optical flow allows us to recover the following information from a two-dimensional projection of a moving scene: we can separate the moving objects from their surroundings based on their relative velocity; we can recover the three-dimensional shape of the object; and we can use motion information to carry out recognition of objects. All these three properties are important in biological visual systems. Motion-based segmentation work has been carried out by [Potter 75, Potter 77, Nagel 78, Thompson 80]. However, this task is very difficult because in using intensity-based methods assumptions of continuity of velocity are made and these assumption are violated at boundaries; hence there is a paradox in using these methods. Walloch and O'Connell were the first to carry out experiments which showed that humans are capable of recovering 3D structures from shadows of moving objects projected onto a screen [Wallach 53].

6.5   Shape from X

If we have a point $\mathcal{P}$ in space $\mathcal{S}$ such that $\mathcal{P}(x, y, z) \in \mathcal{S}$, then under perspective projection it forms an image $\mathcal{I}(u, v)$. If we know or can compute the normal vector $\vec{n}$ of surface $\mathcal{S}$ at $\mathcal{P}(x, y, z)$, then we know the surface of the object. If at every point (u, v) in the image $\mathcal{I}$ we know the normal to the surface, i.e. $(\partial z / \partial u, \partial z / \partial v)$, the surface gradient in the image, then we can compute the depth $z(u, v)$ up to a scale factor. In this section we will explore what information and methods we can employ to compute the surface normals and thereby compute depths.

6.5.1 Shape from Shading

Shading information plays an important role in our perception of surfaces. With changing illumination, the observed surface characteristics change. Under certain illumination condition, a surface might appear concave when it is actually convex. However, most of the time, the shape we observe from the presence of shading is a robust cue.

Consider a function $\mathcal{H}(u, v)$ describing a surface in three dimensions

$$z = \mathcal{H}(u, v)$$

This surface we observe as a light-intensity map, an image $\mathcal{I}(u, v)$, the value $\mathcal{I}(u, v)$ of which depends on its orientation, its surface property, the illumination and the position of the observer. A smoothly curved surface which has homogeneous reflective properties shows changes in its irradiance (or *shading*) which are proportional to the surface gradient. The *shape from shading* problem attempts to recover these properties of surface gradient, reflectance and illumination given only the observed image irradiance. This is then an inverse to the image formation process and is ill-posed. This problem cannot be solved without making assumptions or having additional knowledge. For *shape from shading* one needs to know or compute the *reflectance map*; this tells us the surface orientation as a function of its orientation [Horn 77b].

A typical viewing setup we might have is shown in Figure 6.6. A patch of the object surface at (x, y, z) is illuminated by a light source and the reflected light is observed in the image at a point (u, v). If for simplicity, we assume an orthographic projection and the incident illumination is parallel and constant, then this is equivalent to both the observer and the illumination source being far away from the surface. Under these conditions, the reflectance characteristics of the surface can be represented by a function $\mathcal{F}(\theta, \phi, \psi)$ which is dependent on the angle between the emergent and the incident rays, ψ, the angle between the emergent ray and the surface normal, θ, and the angle between the surface normal and the emergent ray, ϕ.

For convenience, let us define the partial derivatives of $\mathcal{H}(u, v)$ along u and v as

$$p(u, v) = \frac{\partial \mathcal{H}(u, v)}{\partial u}$$

$$q(u, v) = \frac{\partial \mathcal{H}(u, v)}{\partial v}$$

Because **p** and **q** are tangent vectors, the surface normal at (u, v) is perpendicular to a plane determined by the vectors $[1, 0, p]^T$ and $[0, 1, q]^T$, i.e along the direction of $\mathbf{p} \times \mathbf{q}$

$$\vec{n} = \frac{1}{\sqrt{1 + p^2 + q^2}} [-p, -q, 1]^T$$

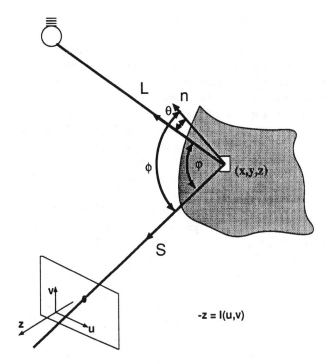

Figure 6.6: A typical viewing model for shape from shading.

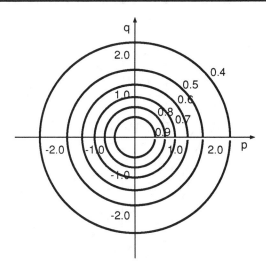

Figure 6.7: The reflectance map of a Lambertian surface for a light source close to the viewpoint.

The vector $\vec{L} = (p_L, q_L, 1)$ represents the direction of illumination and vector $\vec{V} = (0, 0, 1)$ is the viewer's line of sight under orthographic projection. For a Lambertian surface (one which is perfectly diffuse), the reflection is uniform for all viewing angles and varies as the cosine of the incident angle

$$\mathcal{F}(\theta, \phi, \psi) = \rho \cos \theta$$

where ρ is the reflectivity constant.

$$\begin{aligned} \mathcal{F}(\theta, \phi, \psi) &= \rho \cos \theta = \rho(\vec{L} \cdot \vec{n}) / \mid \vec{L} \mid\mid \vec{n} \mid \\ &= \rho(p p_L + q q_L + 1) / \sqrt{(p_L^2 + q_L^2 + 1)(p^2 + q^2 + 1)} \\ &= \mathcal{R}(p, q) \end{aligned}$$

The function $\mathcal{R}(p, q)$ is the reflectance map. The gradient space for a Lambertian surface of a single light source near the viewpoint is shown in Figure 6.7. When the light source is far away, the reflectance map is similar to that shown in Figure 6.8. The relation between shape and intensity is given by

$$\mathcal{I}(u, v) = \mathcal{R}(p, q).$$

Horn calls this the *image irradiance equation* [Horn 77a].

If the surface gradient is small, the reflectance map can be approximated by a series expansion

$$\mathcal{R}(p, q) = \mathcal{R}(0, 0) + p \mathcal{R}_p(0, 0) + q \mathcal{R}_q(0, 0)$$

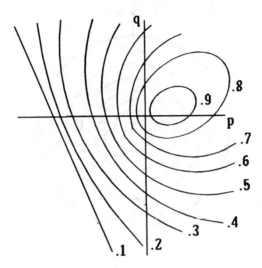

Figure 6.8: The reflectance map of a Lambertian surface for a light source far away from the viewpoint.

An early solution to the shape from shading problem was presented by Horn using a *characteristic-strip* expansion method [Horn75]. The method assumes that the surface is smooth (i.e continuous in its first and second derivatives everywhere). If at an image point (u, v) we know the height $\mathcal{H}(u, v)$ along the surface orientation p, q then we can compute a height profile along a curve called the *characteristic-strip*.

Let us assume that we are at a point (u_0, v_0) and we move a small distance $(\Delta u, \Delta v)$, then

$$(u, v) = (u_0, v_0) + (\Delta u, \Delta v)$$

Then the change in height we observe is

$$\Delta \mathcal{H}(u, v) = p(u_0, v_0)\Delta u + q(u_0, v_0)\Delta v$$

and also

$$\Delta p(u, v) = \frac{\partial p}{\partial u}\Delta u + \frac{\partial p}{\partial v}\Delta v$$

$$\Delta q(u, v) = \frac{\partial q}{\partial u}\Delta u + \frac{\partial q}{\partial v}\Delta v$$

Since $\mathcal{I}(u, v) = \mathcal{R}(p, q)$, then by differentiation

$$\mathcal{I}_u = \frac{\partial I}{\partial u} = \mathcal{R}_p(p, q)p_u + \mathcal{R}_q(p.q)q_u$$

$$\mathcal{I}_v = \frac{\partial I}{\partial v} = \mathcal{R}_p(p, q)p_v + \mathcal{R}_q(p.q)q_v$$

If for small distance Δs

$$(\Delta u, \Delta v) = (\mathcal{R}_p \Delta s, \mathcal{R}_q \Delta s)$$

Then

$$\Delta \mathcal{H} = (p \cdot \mathcal{R}_p + q \cdot \mathcal{R}_q) \Delta s$$
$$\Delta p = \mathcal{I}_u \cdot \Delta s$$
$$\Delta q = \mathcal{I}_v \cdot \Delta s$$

Another approach to solving *shape from shading* is to use the calculus of variation approach [Horn 86b, Ikeuchi 81]. The methods uses two general constraints: at any point there is only one surface orientation (*uniqueness*), and the orientation varies smoothly everywhere except at the boundaries (*continuity*). The first condition is determined by minimising $(\mathcal{I} - \mathcal{R})^2$ and the continuity is measured by computing the Laplacian of the gradient surface [Strat 79, Ikeuchi 81, Smith 82].

Because of the difficulties of computing gradients at occluding boundaries, Ikeuchi and Horn [Ikeuchi 81] changed from using (p, q) to (f, g) space. The relationship between these two spaces is given by

$$f = \frac{2p}{1 + \sqrt{1 + p^2 + q^2}}$$

and

$$g = \frac{2q}{1 + \sqrt{1 + p^2 + q^2}}$$

The goal remains to determine the best fitting surface by minimising the error at every point.

$$E(u, v) = (\mathcal{I}(u, v) - \mathcal{R}(u, v))^2 + \lambda \left(f_u^2 + f_v^2 + g_u^2 + g_v^2 \right)$$

where λ is the Lagrange multiplier. Differentiating this equation with respect to f and g produces the following two Euler equations

$$(\mathcal{I} - \mathcal{R})\mathcal{R}_f + \lambda \nabla^2 f = 0$$
$$(\mathcal{I} - \mathcal{R})\mathcal{R}_g + \lambda \nabla^2 g = 0$$

where ∇^2 is the Laplacian operator. These equations can be solve iteratively using finite-difference approximation techniques.

$$f = f_{av} + T(u, v, f, g) \frac{\partial \mathcal{R}(f, g)}{\partial f}$$

$$g = g_{av} + T(u, v, f, g) \frac{\partial \mathcal{R}(f, g)}{\partial g}$$

where $T(u, v, f, g) = \frac{\epsilon^2}{4\lambda}(E - \mathcal{R}(f, g))$, ϵ is the grid spacing, and f_{av} and g_{av} are the local averages, i.e. $f_{av} = f(u+1, v) + f(u-1, v) + f(u, v-1) + f(u, v+1)$. An iterative solution using a relaxation method (see §3.3.3) is

$$f^{i+1} - f_{av}^i + \frac{\epsilon^2}{4\lambda}(E - \mathcal{R}(f^i, g^i))\mathcal{R}_f(f^i, g^i)$$

$$g^{i+1} = g^i_{av} + \frac{\epsilon^2}{4\lambda}(E - \mathcal{R}(f^i, g^i))\mathcal{R}_g(f^i, g^i)$$

Once we have computed the surface gradients, the height from the gradient can be determined [Horn 86b, Horn 90]. Note, we cannot recover the absolute height but only relative height. The relative height given the surface gradients is

$$\delta z = p\delta u + q\delta v$$

As before, to find the best fitting surface we need to minimise the following function

$$\int\int_\Omega (z_u - p)^2 + (z_v - q)^2 \, dudv$$

The Euler equation of this integral is

$$\nabla^2 z = p_u + q_v$$

Using the iterative solution discussed earlier gives

$$z^{i+1} = z^i_{av} - \frac{\epsilon}{4}(h + t)$$

where z^i_{av} is the local average, and

$$h = \frac{1}{2}(p(u+1, v) - p(u-1, v))$$

and

$$q = \frac{1}{2}(q(u, v+1) - q(u, v-1))$$

6.5.2   Shape from Motion

In computer vision, largely horizontal disparity is computed by stereo matching to recover depth, but others have shown that both horizontal and vertical disparities are important in depth recovery. Also, if the observer is in known motion, then depth can be recovered by matching temporal images in a similar manner to matching spatially separated image as in stereo, but now using monocular vision. For example, if the motion is pure translation, points in the image move outwards from the focus of expansion (FOE) with a velocity proportional to their distance from the FOE and inversely proportional to their depth [Buxton 83].

In §6.4 we described methods for obtaining the optical flow in the *image*. This needs to be related to the objects velocity and structure in the *scene* by *inverse projection* to three-dimension from two-dimensional images. If we assume a simple camera model such that the image axis (x, y) is parallel to the scene axis (X, Y) and the image coordinate system has its origin at $(0, 0, 1)$, then a point $\mathbf{R} = (X, Y, Z)$ is projected by perspective geometry to a point (x, y) in the image,

$$x = \frac{X}{Z} \qquad y = \frac{Y}{Z}$$

If we assume that the camera is in known motion where $\mathbf{V}$ and $\boldsymbol{\Omega}$ are its translational and rotational velocities respectively, then the instantaneous velocity of a point $\mathbf{R}$ is given by $\dot{\mathbf{R}} = (\mathbf{V} + \boldsymbol{\Omega} \times \mathbf{R})$

$$\dot{X} = -V_x - \Omega_y Z + \Omega_z Y$$
$$\dot{Y} = -V_y - \Omega_z X + \Omega_x Z$$
$$\dot{Z} = -V_z - \Omega_x Y + \Omega_y X$$

From this the instantaneous image velocity $(u, v) = (\dot{x}, \dot{y})$ can be obtained [Prazdny 79, Longuet-Higgins 80, Waxman 87, Subbarao 87]:

$$u = \left(x\frac{V_z}{Z} - \frac{V_x}{Z}\right) + \left(xy\Omega_x - (1 - x^2)\Omega_y + y\Omega_z\right)$$
$$v = \left(y\frac{V_z}{Z} - \frac{V_y}{Z}\right) + \left((1 - y^2)\Omega_x - xy\Omega_y + x\Omega_z\right)$$

If we further assume that both the optic flow and the object surfaces are smooth, then (u, v) and Z can be expressed as a Taylor series:

$$Z = Z_0 + Z_x X + Z_y Y + \frac{1}{2}(Z_{xx}X^2 + Z_{yy}Y^2) + Z_{xy}XY + O_3(X, Y)$$

Making the substitute for (x, y),

$$Z(x, y) = \frac{Z_0}{1 - (Z_x x + Z_y y + \frac{1}{2}(Z_{xx}x^2 + Z_{yy}y^2) + Z_{xy}xy + O_3(x, y))}$$

Thus we have a relationship between optical flow, three-dimensional motion and structure of the object.

$$u_0 = -V_x - \Omega_y$$
$$v_0 = -V_y + \Omega_x$$
$$u_x = -V_z + V_x Z_x$$
$$v_x = -\Omega_z + V_y Z_x$$
$$u_y = \Omega_z + V_x Z_y$$
$$v_y = V_z + V_y Z_y$$
$$u_{xx} = -2V_z Z_x + V_x Z_{xx} - 2\Omega_y$$
$$v_{xx} = V_y Z_{xx}$$
$$u_{yy} = V_x Z_{yy}$$
$$v_{yy} = -2V_z Z_y + V_y Z_{yy} + 2\Omega_x$$
$$u_{xy} = -V_z Z_x + V_y Z_{xy} - \Omega_y$$
$$v_{xy} = -V_z Z_y + V_x Z_{xy} + \Omega_x$$

It is to be noted that Z_0 cannot be determined, depth can only be recovered to a scale. Furthermore, all curvature information is related to the translational velocity of the object. Curvature information can only be recovered if there is translational motion parallel to the image plane.

As for optical flow, the method gives rise to massively parallel solutions. The velocities u, v and their derivatives can be computed at each point and the above set of equations can be solved locally to give Z (the depth) at each point in the image using an SIMD data-parallel computation technique.

Figure 6.9: Texture gradient gives depth information.

6.5.3   Shape from Texture

Texture is very important in obtaining information about surface orientation.
Gibson was the first to look at the problem of recovering shape from texture
[Gibson 50]. He stated that texture is made up of small texture elements
called *texels*. Texels are irregular and non-canonical; however, Gibson assumed
that texels are distributed uniformly in the world plane. One observes textured
surfaces as being non-uniform because there is a *texture gradient*. Gibson states
that our perception of shape arise from our observation of both the *uniformity*
and the *texture gradient*. Figure 6.9 is an example of the depth cue we obtain
from texture gradient. If we observe a regular pattern (texture) then this in
itself have little depth information. However when putting a gradient to this
texture field, it gives an impression of a texture plane stretching to infinity.
Any regular pattern on this gradient field appears as being standing out of the
surface.

If we have two surface patches S_1 and S_2 and we compute that there are k_1
and k_2 texels in each of the area respectively, then from the uniform density
assumption

$$\frac{k_1}{\text{area}(S_1)} = \frac{k_2}{\text{area}(S_2)}$$

Under paraperspective projection, Aloimonos shows that [Aloimonos 88a] the
above relationship becomes

$$\frac{k_1(1 - A_1 p - B_1 q)^3}{S_1 c^2 \sqrt{1 + p^2 + q^2}} = \frac{k_2(1 - A_2 p - B_2 q)^3}{S_2 c^2 \sqrt{1 + p^2 + q^2}} \tag{6.5}$$

where (A_i, B_i) are the centre of mass of the two surface patches and $Z = pX + qY + c$ is the equation of the world plane. Equation (6.5) represents a line in $p-q$ space. By choosing a number of different pairs of patches, a Hough transform type of function could be used to compute the surface orientation.

Witkin has argued against the Gibsonian approach [Witkin 81] because of his concern regarding the assumption about the uniform density of texture and the difficulty involved in detecting texels.

While the methods discussed above are data-parallel, they are not massively data-parallel because they do not operate at the pixel level. However, we have shown previously (§5.2.3) that analysis of texture and computing texels do involve massively parallel computation.

6.5.4   Shape from Fractal Geometry

Textured surfaces show depth cue since texture elements (*texels*) grow smaller or their spacing shrinks with distance. This effect is caused by projection foreshortening and perspective gradient. The surface orientation can be determined if the depth at three or more points can be determined. Assumptions made for recovering shape from texture are similar to those for shape from shading. It is assumed that surfaces are smooth and uniformly covered by *texels*. This is generally not true for natural object surfaces.

Many natural and physical phenomena can be described by fractal geometry models [Mandelbrot 82]. Use of fractals in image analysis has been described by [Peleg 84, Pentland 84]. They have showed that the fractal dimension of a particular textured surface is nearly scale-invariant. However, Chen *et al.* have defined a Holder constant [Chen 90] which is scale invariant and therefore can be used to predict distance scale factor. If $\mathcal{F}$ is some one-dimensional real-valued function and T is the sampling distance such that $T = t_i - t_{i-1}$, then

$$| \Delta \mathcal{F}_i |=| \mathcal{F}(t_i) - \mathcal{F}(t_{i-1}) |$$

Holder constant is defined as

$$\alpha_i = \frac{\log(| \Delta \mathcal{F}_i |)}{\log(T)}$$

In two dimensions, $T = (\Delta x, \Delta y)$ and $\Delta \mathcal{F}_i = \mathcal{F}(x + \Delta x, y + \Delta y) - \mathcal{F}(x, y)$. The important property of the Holder constant is that it changes systematically with scale.

Several methods have been proposed for computing the fractal dimension of a surface. Peleg *et al.* describes a "blanket covering" method [Peleg 84] (see Figure 6.10. This is an extension to a method by Mandelbrot for measuring the length of a coast line. For the case of line curve, Mandelbrot proposed drawing a strip around the coast line such that no part of the strip is more than a distance ϵ from the curve. Then the width of the strip is 2ϵ and the length of the coast line is $L(\epsilon) = $ area of strip/2ϵ. Thus, as the ruler, i.e. the strip width ϵ, becomes smaller, the length of the coast line gets longer. For a surface, Peleg says the surface are to be filled with a blanket such that all points on the blanket surface is at a distance ϵ from the surface; the blanket thickness is 2ϵ, there is both a top and bottom surface, $\mathcal{U}_\epsilon$ and $\mathcal{L}_\epsilon$ respectively.

Then the surface area = volume of blanket / 2ϵ. Thus initially we have both the upper and lower surface at the grey-level value at that pixel.

$$\mathcal{I}(i,j) = \mathcal{U}_0(i,j) = \mathcal{L}_0(i,j)$$

Then for for subsequent blanket thickness

$$\mathcal{U}_\epsilon = \max\left\{\mathcal{U}_{\epsilon-1}(i,j) + 1, \max_{|(m,n)-(i,j)|\leq 1} \mathcal{U}_{\epsilon-1}(m,n)\right\}$$

$$\mathcal{L}_\epsilon = \max\left\{\mathcal{L}_{\epsilon-1}(i,j) + 1, \min_{|(m,n)-(i,j)|\leq 1} \mathcal{L}_{\epsilon-1}(m,n)\right\}$$

Note that this is essentially a local-neighbourhood operations, very much similar to grey-scale mathematical morphology (see Chapter 4). Arduini *et al.* use Peleg's method for computing the fractal dimension but in their method the mask size is adaptively set based on the mask sub-window variance measure. The volume of the blanket is given by

$$\mathcal{V}_\epsilon = \sum_{i,j} (\mathcal{U}_\epsilon(i,j) - \mathcal{L}_\epsilon(i,j))$$

Peleg defines the surface area at radius ϵ to be

$$A(\epsilon) = \frac{\mathcal{V}_\epsilon - \mathcal{V}_{\epsilon-1}}{2}$$

Then following on with Mandelbrot's definition for a length of a line

$$L(\epsilon) = F\epsilon^{1-D}$$

Peleg defines surface area as

$$A(\epsilon) = F\epsilon^{2-D}$$

Plotting $A(\epsilon)$ against ϵ on a log-log scale give a straight line of slope $2 - D$. D is the fractal dimension of the line or surface. If we have high values of $A(\epsilon)$ for small ϵ, then the surface has high-frequency signals; and high values of $A(\epsilon)$ for large ϵ implies low-frequency signal.

Furthermore, since fractals are self-similar, texture of two surfaces can be compared based on the difference in $A(\epsilon)$ at various ϵ. For two textured surfaces α and β, the difference between them can be quantified by

$$D(\alpha,\beta) = \sum_\epsilon \gamma\left(A_\alpha(\epsilon) - A_\beta(\epsilon)\right)$$

where γ is a normalising factor for different ϵ.

Using an example, if we have two points in the scene (x_1, y_1, z_1) and (x_2, y_2, z_2) and their projection onto the image is at points (u_1, v_1) and (u_2, v_2) respectively, the ratio of the distance of the two points on the surface to the viewer is

$$\lambda = \frac{z_1}{z_2}$$

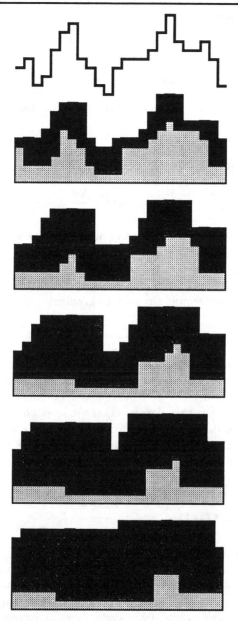

Figure 6.10: The "blanket" method for computing the fractal of a shape.

Then using perspective projection and normalising to the focal length of the camera $f = 1$, we have

$$u_1 = \frac{x_1}{z_1} \qquad v_1 = \frac{y_1}{z_1}$$

and

$$u_2 = \frac{x_2}{z_2} \qquad v_2 = \frac{y_2}{z_2}$$

Suppose we have a line L joining these two points in the scene

$$L = (\Delta x, \Delta y, \Delta z) = (x_2 - x_1, y_2 - y_1, z_2 - z_1)$$

The vanishing point of the line L is given by $(\Delta x/\Delta z, \Delta y/\Delta z)$, where

$$\frac{\Delta x}{\delta z} = \frac{u_2 - \lambda u_1}{1 - \lambda} \quad \text{and} \quad \frac{\Delta y}{\delta z} = \frac{v_2 - \lambda v_1}{1 - \lambda}$$

Thus we can compute the vanishing point of a single line without any knowledge of the three-dimensional coordinates of the points on a surface. Now, λ can be computed from the average Holder constant [Chen 90]. Since for n pair of points, n vanishing line can be computed, a linear least-squares fitting of these lines can be used to determine the surface gradient.

6.5.5   Shape from Contour

We are usually good at interpreting line drawings although such an image could have arisen from an infinite set of projections. This problem has been considered by Marr [Marr 77a] and Barrow and Tenenbaum [Barrow 81]. They distinguish the lines into two classes: (i) those lines constituting the external boundaries of the objects, the orientation of the surface is normal to line of sight and tangential to the line; and (ii) those lines which form the surface discontinuity boundaries, these lines form surface whose normal are orthogonal to the boundaries 3D tangent.

6.6   Active Vision

Vision is not passive. It is active. We adjust our pupils to control illumination level of the input signal, we move our gaze to find out more about our surrounding and we focus on particular objects to observe them in more details. Several systems are being designed which models these aspects of the biological vision systems using mobile stereo systems which have gaze control [Ballard 87, Poggio 87].

There are two avenues to active vision which have been explored: (i) there are the modelling and control strategies to be explored for best usage of vision [Bajcsy 88], and (ii) active vision provides solutions to *ill-posed* problems [Aloimonos 88a].

Aloimonos *et al.* have found that when the camera is in known motion then several of the ill-posed problem of the shape from X type becomes well conditioned and unique solutions can be found for them [Aloimonos 88b]. The motion of the camera leads to local transformations of the scene which are measurable and this provides the additional constraints which provides solution to some of these ill-posed problems.

Bajcsy *et al.* considers the problems of control of the vision system and flow of process and divides then into six levels [Bajcsy 88].

1. Control of camera: this involves the ability to have gross focus of the scene and control the aperture level to have appropriate contrast in the image; the ability to focus on a single object of interest and recover the range of the object from focus.

2. Control of low-level image processing algorithms: this involves the ability to find and set the similarity criterion for region merging; choosing appropriate threshold levels; segmenting the 2D image such that it has the maximum number of edges and both a maximum and a minimum number of regions.

3. Control of the generation of the $2\frac{1}{2}$D sketch images: this involves the detection of 3D boundaries and surfaces and produce depth maps and $2\frac{1}{2}$D sketch images.

4. Control of several views: this involves the integration of information concerning the scene from several viewpoints. This can be done through model matching, giving rise to a 3D description of the scene.

5. Control of the semantic interpretation: this involves determining the goodness of the model fitting and producing 3D object description.

This control structure is similar to that of Brooks [Brooks 87]. It should be noted that the execution of the control need not be sequential as presented, it would be invariable that several such control strategies would have to be executed in parallel each working on different aspects of the problem and co-operating with each other to find a description of the scene.

6.7 Concluding Remarks

This section summarises this and Chapters 4 and 5 which describes image processing and computer vision algorithms.

The important conclusion to be drawn from our discussion of the algorithms are: Early-vision processing such as filtering, segmentation, surface interpretation and description all involve highly data parallel, highly homogeneous local operations. These operations are most efficiently implemented on SIMD array processors by using local near-neighbourhood, global or relaxation algorithms. Many of the recognition tasks can also be carried out using data-parallel algorithms. Other forms of recognition algorithms require more task-oriented (MIMD) parallelism.

We have presented four parallel algorithms which attempt to join points with lines. The first two generate best-fit straight lines based (i) on linear least-squares (LSQ) method and (ii) minimising the moment (MM) of a collection of points. The second method determines the major axis. There are advantages and disadvantages to using either of the methods; their suitability is determined by the application. The major advantage of the linear least-squares method over the use of minimisation of moments is one of speed. The major disadvantage of the LSQ method is its accuracy for lines of steep gradient. Unlike for the minimisation of moments, the LSQ method minimises the vertical distance (d') between the best-fit line and the data points rather than the perpendicular distance (d). For a small gradient, $d' \approx d$. Then if the gradient of a line is small, the LSQ method may prove suitable.

For connected or otherwise set of n points, a parallel algorithm has been presented for a fast approximation to a convex hull. The method is based upon determination, using eight global propagations in each case, of an over-determined and an under-determined hull. The medial line of the boolean exclusive-or of the hulls is determined to be the convex hull. Within the limits of the 8-connectedness of an SIMD array, the method gives good results and is fast. The method has been tried out on 17 figures, each consisting of over 2000 points on the CLIP4 computer; running time for the algorithm has been determined to be 19.7 ± 2.9 ms, 10.7 ms of which is a fixed overhead of the global propagation operations. On the new CLIP4A computer, the times for the 16 global propagations have been reduced to 0.4 ms. The clock speed of the processor has also been increased from 1 MHz to 2.5 MHz. Therefore, on CLIP4A, the same algorithm is expected to run in better than 5 ms.

The convex hull method has been extended to calculate more than one convex hull in parallel for connected data sets. The global propagation previously occurred at all cells. To calculate multiple convex hulls in parallel, the global propagations need to be restricted to within the data sets. This has been achieved by generating bounding rectangles for each of the data sets and restricting the SPREAD function to within the bounding rectangles. The rest of the method is the same as before.

Finally, implementation of splines in parallel has been explored. Many of the standard spline-fitting techniques are unsuitable for the CLIP4 computer. The problem lies with SIMD array processors ability to access easily only local neighbourhood data; no addressing facility, except global, is available. To overcome this problem, splining was achieved using a hierarchical discrete correlation technique. This method presets the position of the control polygon or the control graph; then only the knot positions are calculated. The properties of the Beta-spline have been explored and the tension parameter has been found to be of use in image filtering. The result of the Beta-spline filter is compared with the median and the weighted-average filters.

Chapter 7

Applications

7.1 Introduction

In this chapter we discuss the application of image processing and computer vision techniques. In this chapter we will not discuss the technical details of the algorithms (these have largely been covered in Chapter 4, 5 and 6), rather we will discuss the methods employed in general and the reasons for automation of processes using IP and CV.

7.2 Remote Sensing and Aerial Imagery

The early manned space program showed the need for earth monitoring using low near equatorial and polar orbit meteorological satellites. Later NASA placed geostationary (GEOS) satellites in equatorial orbit. One of the first satellites used for earth observation was Landsat-1. This satellite was operational between July 1972 and January 1978. During its lifetime, NASA acquired some 272 thousand 4 spectral band images. Each image consists of about 28 million pixels; this is equivalent to the satellite sending for processing some 44 thousand pixels per second. These images need correction for visual processing such as contrast manipulation (through grey-level histogram modification and linear streatching) and spatial filtering.

For space shuttle flights very large format ($230mm \times 460$mm) cameras were developed which gave a ground resolution of $\sim 15m$ per line pair at an altitude of 300km. These images covered a ground area of 225×450km. With these cameras, different film-filter combinations could be used to get images in the different parts of the visible and near infra-red (NIR) spectrum. While the advantage of such a system is its relatively low cost, it had the major disadvantage of film load to the pay load of the shuttle, and the film had to be returned to earth for developing and processing. Landsat-4 (launched July 1982) and Landsat-5 (launched March 1984) have electro-optical sensors such as the multispectral scanners (MSS) and thermatic mapper (TM). These sensors can simultaneously take image in 7 bands ranging through the three visible band, NIR, short-wave IR, medium-wave IR and long-wave IR. The first six

bands have a ground resolution of 30m, and the 7th have a resolution of 120m.
By contrast, a 512×512 pixel image could cover an area of $280m \times 280m$.

Remotely sensed and aerial images require spatial registration so that large
image mosaics can be built from which temporal changes can be mapped and
compare images from different sensors. A significant problem is that the image
registration has to be conducted using only a few reliable features. A method
of registration is to choose a window with a feature and match it to another
image using a correlation technique.

There are a large number of applications involving remotely sensed images:
these include (i) agriculture – this could involve crop identification, forcasting
crop yield, detecting and predicting land erosion and desertification, carrying
out timber survey and detecting crop disease; (ii) geology – involving the ex-
ploitation of natural resources and the investigation of large scale features such
as plate tectonics; (iii) hydrology – to monitor the changes in glacier and snow
levels and managing water resources and measuring the mositure content in the
soil; (iv) oceanography – measuring the temperature and detecting pollution,
plankton levels and sedimentary deposition; (v) meteorology – measuring wind
velocity, atmospheric pollution levels and cloud cover, and tracking storms; and
(vi) carteography – generating topographic maps and map revision.

For many of the tasks described above to be carried out, images need to
be registered so that, for example, photographs taken at different times of the
year can be compared. However there are problems to image registration because there are geometric distortions in the images which occur due to optical
aberrations, sensor noise, and the changes in the attitude and altitude of the
sensors. For example, with the LANDSAT images, there is the movement of
the platform and the earth is rotating while images are being taken. These
distortions are constant over time and can be corrected. Kirchoff describes a
method for correcting distortion by registering some control points in the reference image with the same points in the distorted image [Kirchof 80]. If we use
four control points, then the coordinate transformation within a *quadrilateral*
can be modeled by a polynomial power series:

$$f(u, v) = T_c(f(x, y))$$

where

$$u = a_0 + a_1 x + a_2 x + a_2 y + a_3 xy + a_4 x^2 + a_5 y^2$$

and

$$v = b_0 + b_1 x + b_2 x + b_2 y + b_3 xy + b_4 x^2 + b_5 y^2$$

If we set the coefficients a_5 and b_5 to zero, then we can determine the remaining eight coefficients from the corresponding reference and the observed
quadrilaterals. The registration method used is a window based correlation.
This technique works well only if pure translation is involved. If there are
changes in velocity then we face the same problems as those in determining the
stereo disparity using grey-level region correlation.

The distortions involving *translation*, *rotation*, *perpective*, and *skewness*
(Figure 7.1) are due to the changes in the position and attitude of the sensor.

The real difficulty is in disambiguating these distortions with sensor distortions such as pin cushion, barrel, and optical abberations.

We have been discussing the need for registering images to detect changes in the land feature. There is a reverse problem with the GEOS images. In this case, the images are registered and we need to observe (for example) the cloud motion to detect the wind velocity. The requirement is that wind velocity must be determined to better than 3 knots. In this case we can use a technique such as optical flow or matching "cloud" features using a correlation technique. A problem is that clouds at different height move in different directions. Note, this problem is easily solved by a human manually using a loop sequence [Leese 70].

Nagao *et al.* have reported techniques for segmenting aerial photographs [Nagao 80]. The segmentation consists of edge-preserving smoothing of the images followed by segmentation on the basis of the continuity of the spectral property. From this the *cue regions* are extracted. A cue region could be a large homogeneous region, an elongated region, a shadow, a shadow-touching region, a vegetation region, a water region, or a high-texture contrast region. These cue regions are extracted independent of each other (therefore can be parallelised), hence a single region may appear in several of the cue regions. Once the cue regions have been extracted, analysis on them can be carried for secene interpretation.

A large homogeneous region found as a result of the segmentation processes could be agricultural fields, lakes, seas and grasslands. Most of these are flat regions. Other regions could have heights such as houses, buildings and trees. These regions are indicated by the presence of shadows. Shadow regions are present because aerial photographs are usually taken in good weather conditions. The presence of shadows allow us to discriminate between objects which are *flat* and those with *height*, indicating the three-dimensional nature of the objects. For example, houses are located first by finding a residential area, then individual houses are located within this region. The elongated regions (ie. those regions with length/width ratio greater than three) could be roads, rivers and railway tracks.

The spectral information can be used to distinguish within a cue region. For example vegetations have low content in the red to infra-red part of the spectrum. However, this property is also exhibited by a region which has a strong blue component. Therefore a vegetation region should have low amounts of blue and red components. In their segmentation process, Nagao merges small vegetation regions to a large adjacent vegetable region. A water region tends to be darker than its surrounding regions. However, there are problems of distinguishing it from a vegetation region when there are large amounts of weed in the water.

Woods and residential regions are characterised by their high-contrast texture. Texture regions are located on the basis of counting the number of edgels within a window region. Roads, which are elongated regions, are recognised on the basis that their width is constant and the length/width ratio along the skeleton is large.

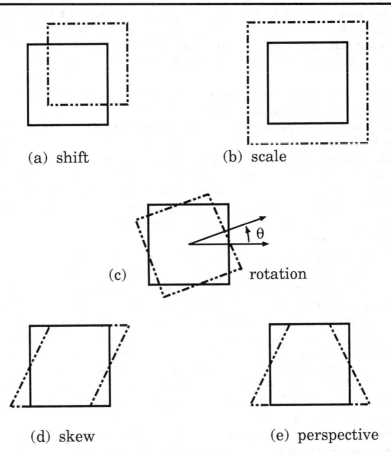

Figure 7.1: Some simple geometric transformations are shown.
For *shift* we have the transformation $u = a_0 + x, v = b_0 + y$, scaling has the transformation $u = a_1 x, v = b_1 y$, the rotation transformation is $u = a_1 x + a_2 y, v = b_1 x + b_2 y$, where $a_1 = b_2 = \cos\theta$ and $a_2 = -b_1 = \sin\theta$, the skew transformaation can be expressed as $u = x + a_2 y, v = y$ and the perpective transformation is $u = a_3 xy, v = y$.

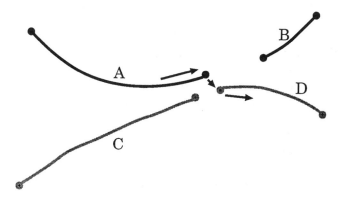

Figure 7.2: Alignment and proximity are important in determining how broken line segments should be joined together.
In the example shown, using proximity alone, line **A** would be joined to line **D**. However, when the alignment and continuity conditions are applied then line **C** is joined to **D** and line **A** is joined to line **B**.

The sytem which Nagao describes is a production-system using a large number of rules for both the segmentation and analysis of the regions. However, a limitation of the system is the weakness in its segmentation algorithm. For example, while it uses texture to locate woodland and residential areas, it does not use texture information to differentiate between different types of regions.

Common features extracted from aerial images include road extraction. This process is made difficult by the high road density present in many urban areas and the poor image quality. Furthermore, there are other features which have a similar appearance to roads such as field boundary and hedgerows. However, these later features vary with the season while roads are unchanged. Roads are also characterised by linear line segments which consists of antiparallel pair segments [Zhu 86]. The problem with this approach is if there are many fragmented edges. However, a Hough transform type of technique can be used to detect linear features. We can further help this process by applying some heuristic rules such as the road width is uniform, they tend to be long and straight and they have high contrast with respect to their surroundings. Because many edge fragmentations occur, Fischler *et al.* combine edge strength from multiple sources and choose the best edges using dynamic programming [Fischler 81]. Others have linked segments based on *proximity* and *alignment* [Vasudevan 88] (Figure 7.2).

It is diffult to extract the positions of buildings and roads; however once this is accomplished it can be used for solving the correspondance problem to either register an image to a map or an image to another image to obtain stereo information. The extraction of building structures is a difficult problem and few have attempted it [Liow 90]. The problems are: the buildings have shadow, they are low-contrast features, building are of variable shape, there are other structures (such as trees, roads, and shadows of other surrounding objects) from which buildings have to be distinguished, and perspective. Despite building shapes being variable, some have used generic shape model in segmentation and interpretation of urban scenes [Fau 87, Huertas 88]. These approaches require different algorithms for different building structures.

Because building structures have low contrast, region based methods are not suitable for their segmentation, and use of edge based methods lead to the production of many fragmented segments. Both Huertas and Liow have described using shadow information in detecting possible locations of buildings [Huertas 88, Liow 90]. The solution they have proposed is to combine a number of methods such as region- and edge-based segmentation and use of shadows which have high contrast. Therefore the shadow regions can be used as "seeds" in locating buildings. For example, edge detection is first carried out and those edge segments which have a neighbouring shadow regions are kept. These edges are then used to guide the region-based segmentation. If we know the direction of illumination, then the shadow region will have a "regular shape" perpendicular to this. Therefore, we can search, in parallel for all rays, for light to dark region transition. We can mark all such dark regions as possible shadow candidates. We can later verify this by checking their position and shape. The shadow regions should all be in the same direction (parallel to each other) and because buildings have "regular" shapes, the shadow regions should also have a regular shape and their boundaries be parallel to the building edge (Figure 7.3).

7.3   Vision Systems

Much of the pioneering work in image processing and computer vision has been carried out by Roberts [Roberts 65]. His scene consisted of polyhedral blocks such as a cube, a brick, a wedge and a hexagonal prism. The method consisted of edge detection using a derivative technique (Roberts-Cross Operator). These edges were grouped together into lines. Objects were recognised by scaling and rotating the models and matching with the lines found in the image. These early methods were data-driven (bottom-up) processing and involved little or no goal-driven (top-down) processing strategy. However, these techniques were the embryonic form of the later methods such as SCERPO [Lowe 87].

Since the early works of Roberts [Roberts 65] and Brice and Fennema [Brice 70], progress in image processing and computer vision has generally been slow compared to other branches of computer science. Vision is a very complex task and we are attempting to imitate human vision functions without really understanding much of the processes involved.

Image analysis involves *segmentation* by either detecting regions of change or detecting regions of uniformity in grey-level scene. Region-based segmen-

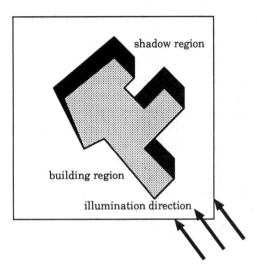

Figure 7.3: The detection of the shadow region helps to locate the building.

tation by *region growing*, where smaller regions are merged into larger ones if they have similar properties, was presented by Brice and Fennema [Brice 70] and Yakamovsky [Yakimovsky 73]. The opposite to this is to start with large regions and *split-and-merge* them to form uniform regions. This technique for region-based segmentation has been presented by Horowitz and Pavlidis [Horowitz 74]. Edge-based segmentation is usually easier because it is often easier to model. The discontinuities are detected by various forms of differential operators. These techniques are discussed in detail in Chapter 5. Two recent trends in image processing have been *rule based* segmentation [Nazif 84], and segmentation using texture and fractal dimension [Pentland 84].

Once regions and surfaces have been extracted, the scene understanding tasks need to be carried out. Scene understanding by making extensive use of heuristics was first used by Guzman for segmentation and classification of junctions in line drawings [Guzman 68]. The SEE program showed that complex reasoning about line drawings can be carried out using symbolic processing (cf. least-squares fitting of lines [Roberts 65]). The method does not provide a 3D interpretation of the scene but provides classification of the the junctions into eight classes. However, the segmentation has to be perfect for the method to work. Huffman and Clowes independently used a syntactic approach to interpretation of a polyhedral scene [Huffman 71, Clowes 71]. Using trihedral polyhedra, which has the property that three plane surfaces could meet at a vertex in only four ways forming convex and concave edges, scene interpretation can be carried out by describing the various ways the planes could meet to form junctions.

A 3D description of a scene can be object centered, however, a 2D description of a scene is viewer centered. A $2\frac{1}{2}D$ description of a scene can be generated by combining the attributes from both 3D and 2D descriptions of the scene. Important features of an object are its boundaries and the geometric measures obtained from the boundaries. These features can be global (such as the $r(\theta)$ and $r(s)$ graphs, centroid, area, and moments), local (such as line and arc segments, corners and holes), and relational (such as inter-feature distances). The local *structural features* can be combined with the relational features to form the *relational graph method* for object recognition. An example of this method is the *local feature focus* [Bolles 83]. The basic methods for feature detection and the use of relational graphs have been discussed in Chapter 5.

In geometric model based vision, given a set of observed features we need to determine which model gave rise to these features and recover the *pose* of the object, i.e. what is the relationship between the object and the viewing direction. The scene can be represented by a *world model* in which each object is signified by its position and orientation:

$$W = \{(\mathcal{O}_i, \mathcal{P}_i, \Theta_i)\}_{i=0}^{N_{obj}}$$

where $\mathcal{O}_i$ is an object, $\mathcal{P}_i$ is its position, and Θ_i is its orientation. If the scene is changing with time, then the *world model* needs to be time-varying and possibly needs to be observer centered.

Objects can be represented using either geometric- or nongeometric-models. In a geometric representation, often simple multi-view models are used when an object only has a small number of stable surfaces. Typical geometric models used are:

Wire-frame: The wire-frame models (Figure 7.4) consist of a list of 3D vertices and edge-list of vertex pairs. Its usage is common because it is simple and can be extended to solid models. The problem with using wire-frame models concerns the ambigious representation for determining the surface area and volume of objects and we also need techniques for hidden line removal.

Constructive solid geometry: The constructive solid geometry (CSG) models consist of a set of 3D volume primitives (such as blocks, cylinders, spheres and cones) and a set of boolean operations (such as union and difference). The model storage is via a binary tree which gives rise to compact representation (Figure 7.5). The CSG model does not suffer from the surface ambiguity problem when using wire frame models. However using this method it is difficult to obtain surface information and also difficult to represent complex natural surfaces.

A method similar to CSG is the use of *spatial occupancy* representation. This uses a set of *nonoverlapping* subregions to define an object. The spatial occupancy primitive could be a *voxel* (volume element) which is a small fixed size cube. While its use gives rise to simple algorithms, it is also expensive in its use of memory. An alternative commonly used is an *octree* representation [Meagher 82] which is a 3D version of the *quadtree* [Samet 84] (Figure 7.6).

Surface Boundary: The surface boundary consists of solid surfaces of the object. The surfaces could be built using triangular facets. Using this method, complex surfaces could be described by use of many small triangular facets. This is a common technique used in computer graphics [Foley 90].

Geometric models which are more commonly used in computer vision are: *generalised cone or sweep* [Soroka 83, Shafer 83] – this consists of a spine (the axis of the cone) which is a curve in three space about which a cross-section of a shape is swept.

7.4 Knowledge-based Systems for Vision

We describe a flexible system for recognising complex images involving partially occluded objects using planning and strategy (often heuristic in nature). The system uses a "blackboard" [Hayes-Roth 85a] to act as a short term memory and a means to share knowledge between different processes giving rise to data processing via a multiple instruction streams multiple data streams (MIMD) system. Amongst the different techniques being involved in the object segmentation are Hough transform to extract **lines, arcs** and **circles** (a separate and distinct small hole detector may be used). Other important generic features to be extracted are **corners** and **extended salient** features using the $s(\psi)$ graph. We will describe these segmentation technique, the data structure for the description of the objects and the use of the blackboard in the MIMD process.

Until recently, much of the work in Pattern Recognition and Image Processing (PRIP) has been associated with medium-level image processing; this is mostly image segmentation. Following segmentation, object recognition (with objects being completely present – no partial occlusion occurs) is achieved by measuring size, moments or template matching using geometrical or topographical *a priori* knowledge. Several such system exists; they have been reviewed in [Hussain 88c, Wallace 88] and discussed in §5.6. However, little attempt at interpreting complex image scenes has been made. Amongst the systems reported for recognising occluded objects are by Rummel [Rummel 84], Bolles [Bolles 83], and Turney [Turney 85]. The first two examples uses a *maximal clique* technique [Bolles 83, Ballard 82] involving detected corners and holes to hypothesize the presence of certain objects. The last example uses matching of salient features to recognise objects.

It is evident from what has been stated above that there are two types of processing going on (segmentation and recognition / interpretation). The segmentation is an *iconic* process which is best carried out procedurally and the image interpretation is a *symbolic* process which is probably best carried out

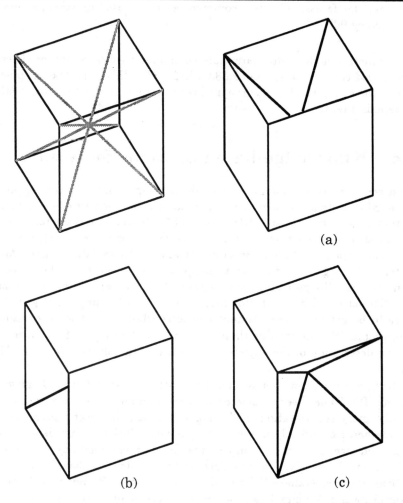

Figure 7.4: A wire-frame model of a cube.
Without hidden line removal shape the cube with the diagonal wire is ambigious; there are atleast three interpretations as shown.

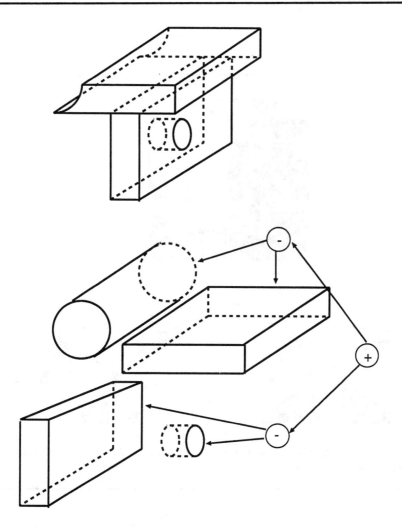

Figure 7.5: A CSG model and its primitive parts are shown.

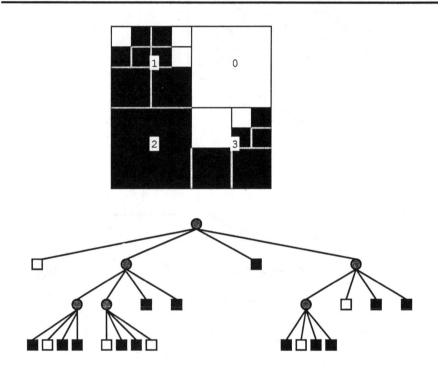

Figure 7.6: A quadtree for describing shapes in 2D. A 3D extension of a quadtree is the octree.

declaratively [Nagao 82]. However, in the examples given above this distinction is not made explicit.

Furthermore, from the iconic process we obtain symbols (such as a description of a corner, hole or a straight line) which we can store as a list of items on a *table* for subsequent symbolic processing. The symbols can be manipulated through predicate logic [Charniak 85, Rich 83]. It should be noted that we probably also require some conversion between boolean and probabilistic logic.

The *table* described above is often referred to as the *blackboard*. This was first used in association with the HEARSAY-II program [Erman 80] developed for speech recognition and interpretation. The HEARSAY-II system consists of *knowledge sources* (KSs) which are all independent modules. The blackboard is a shared memory structure to which all the KSs have access. When a KS is activated it uses the knowledge residing on the blackboard to create a new hypothesis and write the result back on the blackboard or it modifies existing knowledge. Furthermore, the KSs operate asynchronously; once activated they continue to operate until the task is finished. The KSs are *triggered* by *demons*, these are procedures when true activates the process. Which KS should be triggered is controlled by a master KS called the *scheduler*. The overall goal of the system is to use the blackboard to generate a single hypothesis.

A PRIP version of the HEARSAY-II program is called VISIONS [Hanson 78]. One of the interesting aspects of VISIONS is the concept of *expectation driven* reasoning; i.e. using context to drive the system. This system is suggested for use in interpreting outdoor scenes. Another blackboard system has been used in remote sensing [Nagao 82]. For recognising 3-D objects using geometrical reasoning is suggested via the ACRONYM system [Brooks 81].

Let us consider a parallel image processing system in which images are processed using data parallel methods to produce edge segmented images which contain the edge direction and magnitude information g_x, g_y and $\|g\|$ respectively. This is the *iconic* data. This iconic data can be shared amongst four independent *iconic-symbolic* processes. These four iconic-symbolic processes extract object primitives such as corners, lines, circles and extended features. The symbolic (or object description) of these is written on to the blackboard. A simplified blackboard system for these processes shown in Figure 7.7.

In the blackboard system described above we make use of only two image descriptions, the edge segments and the original image. This precludes any use of second image derivatives. This constraint affects the choice of our corner detector as we will explain later in this section. The line and circle detection are via the use of the Hough transform [Hough 62]. These three features can be considered as being generic. Other extended features, often termed **salient**, are detected using the $s(\psi)$ graph.

For our blackboard we extract two generic features using the Hough transform, lines and circles (or arcs). The circle detector we use is essentially that described by Kimme et al. [Kimme 75]. However, because of the rescaling needed to plot data in the parametric space, we do not use the line Hough transform described by Duda and Hart [Duda 72]. Rather we plot the result directly in to the image space. This is achieved by plotting the *foot of the*

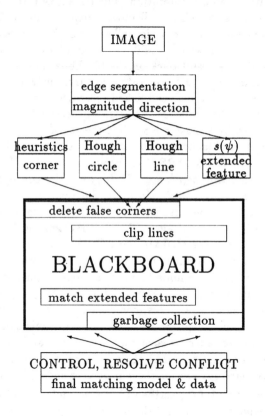

Figure 7.7: A schematic diagram of the blackboard processing system.

normal [Davies 86a]; this method has a problem which we will describe later.

Note, as a preprocess, we compute the x- and y-component of the edge gradient, g_x and g_y respectively, along with the edge gradient magnitude, $\|g\|$. Then to calculate the Hough transform for lines and circles respectively, we need to compute for each edge point (x, y), i.e. $\|g\| \geq thr$:

$$\nu = (xg_x + yg_y)/\|g\|^2$$
$$x_c = \nu g_x$$
$$y_c = \nu g_y$$
$$A_l[x_c, y_c] = A_l[x_c, y_c] + 1$$

and

$$x_c = x - r \times \left(\frac{g_x}{\|g\|}\right)$$
$$y_c = y - r \times \left(\frac{g_y}{\|g\|}\right)$$
$$A_c[x_c, y_c] = A_c[x_c, y_c] + 1$$

where r is the radius of the circle. The position of the lines and circles are held in the accumulators A_l and A_c respectively. Since these two Hough transformations are carried out in parallel, the time taken by these proceesses is determined by the longer process:

$$t = \max(t_l, t_c)$$

where t_l and t_c are the time taken by the line and circle transforms (i.e the algorithm given above plus the time taken to determine the peaks) respectively; rather than $t \approx t_l + t_c$ for the serial process.

The problem with the *foot-of-normal* method is that if any "peak" occurs near the origin, the line gradient cannot be determined accurately (the reciprocal law). If the peak is at the origin, then the gradient of the line is undefined. This is not as severe a problem as it may appear. We simply ignore any line passing through or close to the origin; we do not require the information about every line position and orientation to make hypothesis about the objects present in the image scene.

The other task at hand is to find a method for determining the peaks in the "parameter space" (which happens to be the image space in the Hough transformation derivations used). A typical method for determining the peaks from the general noise surrounding them is to store all the 'peaks' (their position and bin magnitude) which are above a threshold into a list of possible peak (i.e. *center of circles* or *foot-of-normal*) candidates. This list is sorted in order of the bin magnitude (highest to lowest). A large bin is a true 'peak'; any smaller peaks surrounding the 'true peak' are removed from the list. The remaining peaks in the list contains the coordinates of the centers of objects or the *foot-of-normal* of straight lines.

One of the problems associated with shape processing, object recognition, is that of describing shape in a definitive manner. There are many approaches to solving this problem including Fourier descriptors [Persoon 81], template matching [Chien 74], Hough transforms, moments calculations [Goshtasby 85] and matching centroidal profiles. Another solution proposed by Attneave and Arnoult [Attneave 66] is to divide curves into segments and then use simple features to characterise each of the curves. It is important to note that for this method to work, the image segmentation must be correct. Having segmented the image into objects, the object boundary curves are encoded using 8-direction Freeman chain code [Freeman 61], §5.6.

In the literature, several vision systems based upon matching of boundary profiles [Yachida 77, Perkins 78] have been discussed. These methods consist of generating a centroidal profile by computing the radial distance from the centroid at some predefined angular interval or computing the local tangents on the object boundary at a predefined number of points. Another object recognition method is to match the relative positions of object primitives (i.e. holes, corners and lines) with respect to a model which is interactively created [Bolles 83, Rummel 84]. Although, the object primitive matching schemes (often referred to as *Maximal Clique* methods) can recognise complex objects in an occluded scene, the implementations described above can only handle a single model at a time.

To allow for multiple models and very differing shapes a computer vision *toolbox* comprised of several object recognition methods is required. Within the *toolbox* facilities are provided for both automatic and semi-automatic methods for generating the object models. The vision system consists of two subsystems: (a) recognising objects in an occluded scene and, (b) recognising objects in a non-occluded scene. The latter method uses a modified version of the r-θ graph employed by Yachida [Yachida 77]. If the profile is generated by taking points at regular angular intervals then for non-convex objects there is a possibility of more than one boundary point for a given angle. To overcome this problem, the method has been modified so as not use angle 'scanning', but rather to scan spatially and pick n points at a regular interval on the boundary and compute the distance from the centroid. Furthermore, the graph is normalised using the method described in [Hall 79]. Therefore, object recognition through this method is invariant to translation, rotation and scaling. Using this method we've been able to distinguish between very similar objects such as electrical plugs.

For the former method we have developed a generic object recognition scheme based upon geometrical and model reasoning. The models consists of either circular objects, regular polygons (either with or without holes) and other more complex shapes which have either a hole or an arc of unique size as part of their shape. The object recognition occurs through geometrical reasoning about the object class and then instantiation of a matched model. For example, objects consisting only of circles are divided into five classes: DISK, RING1, RING2, RING3, and RING4 (see Figure 7.8).

To distinguish between these five classes, we need to carry out some geo-

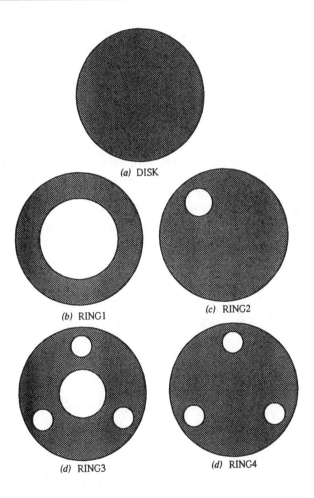

(a) DISK

(b) RING1   (c) RING2

(d) RING3   (d) RING4

Figure 7.8: Five classes of circular objects.

metrical reasoning. A brief description of the system is given below. We will
assume the objects are black and the background is white. We detect all the
circles in the image using the Hough transform [Kimme 75]. Then for any black
circle detected, if there are no white circles present within its boundary, then
the object can be of the *DISK* class. We carry out further computation to see
which black and white circles share the same center location. These circles are
stored in a special set called **ring**. If there is a black circle and it is not a
member of **ring** but has a single hole enclosed within its boundary then the
object is probably of class *RING2*. In a similar fashion: (i) if a black circle is
a member of **ring** and no other holes are enclosed within its boundary then
the object is of class *RING1*, (ii) if a black circle is a member of **ring** and has
other holes enclosed within its boundary, then if the 'spoke' radius and angle
are a 'constant', the object is of class *RING3*, and (iii) if a black circle which
is not a member of **ring**, has constant spoke radius and angle then the object
is of class *RING4*. Having determined which class the object may belong to
we need to carry out a search within the the determined model class for an
instantiation.

A cog is a circular structure with teeth at a constant radial distance and
the angle between two successive teeth is also fixed. The teeth have a shape
ranging from a parabola to half an ellipse; but to a first approximation it can
be represented by an arc of a circle. This makes recognition by the Hough
transform simple. Typical cog structures are shown in Figure 7.9.

Knowing the radius of the teeth, we can recognise and determine their
position using the Hough transform. For generating the model, we need to
know the radius of the cog *(R)* and the number of teeth. We can determine R
by a least squares method.

$$R^2 = \frac{1}{N}\{\sum\nolimits_{x^2} - 2\sum\nolimits_x \bar{x} + N\bar{x}^2 + \sum\nolimits_{y^2} - 2\sum\nolimits_y \bar{y} + N\bar{y}^2\}$$

where

$$\bar{x} = \frac{c_1 b_2 - c_2 b_1}{a_1 b_2 - a_2 b_1}$$

and

$$\bar{y} = \frac{a_1 c_2 - a_2 c_1}{a_1 b_2 - a_2 b_1}$$

and in the notation $\sum_x^2 = (\sum x)^2$, etc.:

$$a_1 = 2(\sum\nolimits_x^2 - n\sum\nolimits_{x^2}), \qquad b_1 = 2(\sum\nolimits_x \sum\nolimits_y - N\sum\nolimits_{xy})$$

$$a_2 = 2(\sum\nolimits_x \sum\nolimits_y - N\sum\nolimits_{xy}) = b_1, \qquad b_2 = 2(\sum\nolimits_y^2 - N\sum\nolimits_{y^2})$$

$$c_1 = (\sum\nolimits_{x^2}\sum\nolimits_x - N\sum\nolimits_{x^3} + \sum\nolimits_x\sum\nolimits_{y^2} - N\sum\nolimits_{xy^2}),$$

$$c_2 = (\sum\nolimits_{x^2}\sum\nolimits_y - N\sum\nolimits_{y^3} + \sum\nolimits_y\sum\nolimits_{y^2} - N\sum\nolimits_{x^2y})$$

By determining whether holes of known radius exist and finding their locations relative to the center, we can instantiate our model into one of five classes
of cogs (see Figure 7.9(a)–(e)).

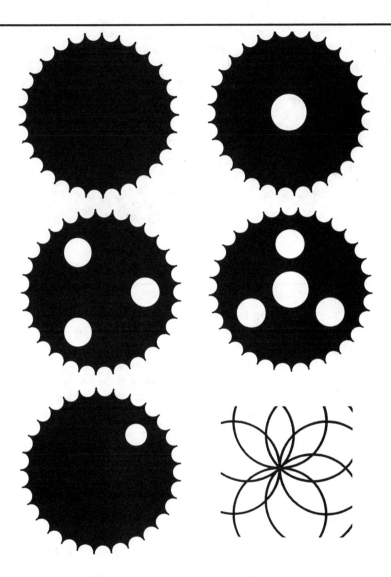

Figure 7.9: (a)-(e) The five classes of cogs. (f) The cogs are detected by finding the peak in the Hough space as a result of drawing a circle for each of the "teeth" circles found.

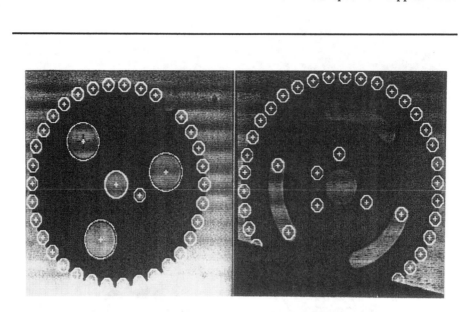

Figure 7.10: The instantiation of cog classes.

We begin the recognition task by extracting all concave and convex circles present in the scene and sorting these primitives both by their colour and radius. A cog is recognised by applying a second Hough transform. In the absence of convex circles, the presence of a large number of similar concave circles indicates the existence of a cog. For each of these teeth, we plot a circle of radius R. Then if a cog of radius R is present, these circles would all pass through the center (Figure 7.9(f)). The location of this peak gives the position of the cog.

Now we need to determine the **class** of the cog object. The classification rules can be simply expressed for each case in Figure 7.9 as:

(a) **cog & no holes** within structure $\Rightarrow$ **COG_DISK**

(b) **cog & center hole & no other holes** $\Rightarrow$ **COG RING1**

(c) **cog & no center hole & radial holes** $\Rightarrow$ **COG_RING2**

(d) **cog & centered hole & radial holes** $\Rightarrow$ **COG_RING3**

(e) **cog & one off–centered hole** $\Rightarrow$ **COG_RING4**

When the cog structure has been classified, an instantiation with a model in the determined class may occur. Figure 7.10 shows detection of cog structures, in (a) a COG_RING3 class of object is instantiated, but in (b) although a cog has been recognied it is not instantiated to any of the classes.

After this processing, there might still be circles which are unresolved. This could be due to incomplete segmentation leading to inconclusive reasoning, or else the objects do not consists only of circles. Further resolution of the scene is facilated by the use of the r-θ object recognition method.

Apart from detecting arcs of circles in the system we descibe, we also detect lines. To carry out further reasoning about object structures do need to determine which lines are parallel and which lines are perpendicular to each other (Figure 7.11). If we have two lines which are not parallel (i.e. the gradients are different) then the lines, expressed as

$$y_1 = m_1 x_1 + c_1$$

and

$$y_2 = m_2 x_2 + c_2$$

will intersect at a position, i.e. $(x_1, y_1) \equiv (x_2, y_2)$. Then,

$$m_1 x_1 + c_1 = m_2 x_2 + c_2$$

Solving gives,

$$x = \frac{c_2 - c_1}{m_1 - m_2}$$

$$y = \frac{m_1(c_2 - c_1)}{m_1 - m_2} + c_1$$

However, when extracting the lines using Hough transform, we are computing the *foot-of-normal* of the line. Thus the intercept, c, is never directly calculated. Therefore, we want to eliminate c from our calculations.

$$x = \frac{y_2 - y_1 + m_1 x_1 - m_2 x_2}{m_1 - m_2}$$

$$y = y_1 + m_1(x - x_1)$$

Given a circle and a line, there can be two positions, P_1 and P_2, where they can intercept.

The line is expressed as

$$y = mx + c \tag{7.1}$$

and the circle as

$$(y - b)^2 = r^2 - (x - a)^2 \tag{7.2}$$

where r is the radius of the circle and $a, b)$ are its center position. The intersection occurs where equations (7.1) and (7.2) are equal to each other. To solve, substitute equation (7.1) into equation (7.2) after expanding equation (7.2).

$$(mx + c)^2 - 2b(mx + c) + b^2$$
$$= r^2 - (x - a)^2$$

$$m^2 x^2 + c^2 + 2mxc - 2bmx - 2bc + b^2$$
$$= r^2 - x^2 + 2ax - a^2$$

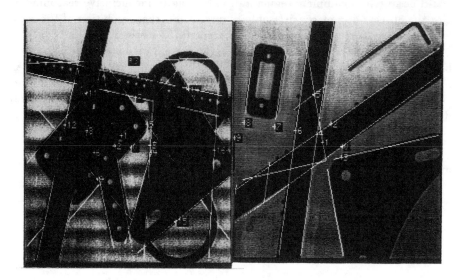

Figure 7.11: Many lines have been detected using the Hough transformation. We need to determine which lines are parallel or perpendicular to make inference about the shapes of objects.

Grouping gives,

$$x^2(m^2 + 1) + 2x(mc - mb - a)$$
$$= r^2 - (a^2 + b^2 + c^2) + 2bc$$

This can be written in the form

$$Ax^2 + 2Bx + K = 0$$

Solving for x gives,

$$x = \frac{-B \pm \sqrt{B^2 - AK}}{A}$$

where

$$
\begin{aligned}
A &= m^2 + 1 \\
B &= m(c - b) - a \\
K &= (a^2 + b^2 + c^2) - 2bc - r^2
\end{aligned}
$$

y is determined by substituting for x in equation (7.1).

We will consider the intersection of two circles of radius r_1 and r_2 respectively. The two circles will intersect if $r_1 + r_2 \leq d$, where d is the separating distance of the center of the two circles. To simplify the computation, we will temporarily transform our coordinate system to the center of one of the circles. Then the circles can be expressed as

$$y^2 = r_1^2 - x^2 \tag{7.3}$$

$$(y - b)^2 = r_2^2 - (x - a)^2 \tag{7.4}$$

where (a, b) is the relative coordinate of the center of the second circle.

First eliminate y from the equation; this we do by expanding equation (7.4) and subtracting equation (7.4) from equation (7.3), gives us

$$2by = -2ax + (a^2 + b^2) + (r_1^2 + r_2^2)$$

or

$$2yb = -2ax + K \tag{7.5}$$

Substituting for y in equations (7.3) and squaring both sides give

$$r_1^2 - x^2 = \frac{a^2}{b^2} x^2 - \frac{aK}{b^2} x + \frac{K^2}{4b^2}$$

or

$$\left(\frac{a^2}{b^2} + 1\right) x^2 - \frac{aK}{b^2} x + \left(\frac{K^2}{4b^2} - r_1^2\right) = 0$$

This is of the form

$$Ax^2 + Bx + C = 0$$

Solving gives

$$x = \frac{-B \pm \sqrt{B^2 - 4AC}}{2A}$$

$$y = -\frac{a}{b} x + \frac{K}{2b}$$

where

$$A = \frac{a^2}{b^2} + 1$$

$$B = \frac{aK}{b^2}$$

$$C = \frac{K^2}{4b^2} - r_1^2$$

$$K = (a^2 + b^2) + (r_1^2 + r_2^2)$$

However, note the problem with singularities if $b = 0$. This needs to be solved as a special case. Now equation (7.4) is modified to

$$y^2 = r_2^2 - (a - x)^2$$

Solving in a similar way gives,

$$x = \frac{(r_1^2 - r_2^2) + a^2}{2a}$$

$$y = \pm\sqrt{r_1^2 - x^2}$$

As soon as an iconic-symbolic KS (ISKS) finishes, a demon is triggered for symbolic processing of that data. We will give some of examples of this. The extended feature ISKS only extracts the $s(\psi)$ graph for each of the "unconnected" object-set. When this task finishes, a demon in the appropriate part of the blackboard is triggered to start a KS which matches known salient features in a database against the obtained $s(\psi)$ graph to hypothesise the presence of an object. The object's name and position is written back to the *symbolic blackboard*. Another example is with the line detection. The ISKS for the Hough line detector writes the *foot-of-normal* on the blackboard. Another symbolic KS is required to interpret and create new hypothesis. The new KS checks whether two lines are parallel, if so whether they belong to the same objects. This creates a new hypothesis or may confirm old ones. If two lines are not parallel, are they perpendicular to each other? (Where is the intersection?) This information is used to confirm corners detected by the corner detector ISKS.

Finally, when these secondary KSs are finished, the scheduler starts a final KS to resolve any contradictions that may still reside on the blackboard and generates a final hypothesis of the image scene.

Let us consider the requirements of a system for image processing:

- allow *backward chaining* (goal driven),

- allow *forward chaining* (data driven),

  1. provide *meta-level* control (means of selecting appropriate rules),

  2. provide means of grouping rules and context for efficiency,

  3. provide *conflict resolution* strategy,

- allow any data structures which may be required to best describe an object, i.e. atoms, lists, records and frames, and

- allow for explanation and trace facilities.

A typical image processing software we might develop is both backward and forward chaining. For example, consider an algorithm to scrutinize circular biscuits. The algorithm we develop has a *super-goal*, that is to scrutinize biscuits. But we have a number of sub-goals relating to this super-goal; we need to locate circular objects of the correct size, then measure properties and check for extent of damage if any. However, there may be differing numbers of biscuits in an image frame. Therefore having detected all the circular objects, we need to carry out the scrutiny task for each of the objects. Furthermore, if we find a biscuit is broken, then for that object the remaining scrutiny tasks need not be carried out. Therefore, within our *super-goal* we have data driven processing, i.e. the code executed depending on the input image.

However, there are a number of problems associated with traditional programming methodology. A major problem for very large tasks, the programming (the control structure as well as the task itself) becomes very involved; this makes it difficult to change or maintain the system. A possible solution considered is to build a *rule-based system* or a *knowledge based system* (KBS) or an *expert system*. The main idea behind an expert system is that human knowledge (which is often in a procedural form) should be made explicit in the form of *rules* which are then used by *inference engines*, a collection of which then forms an *expert system shell*.

Object recognition tasks consist fundamentally of searching and matching features; this task requires *a priori* knowledge about such entities as geometry, size, contrast, colour and texture. This knowledge can be represented either in procedural or declarative form. The procedural knowledge is most effective in representing how certain tasks may be carried out, such as preprocessing or segmentation.

A rule-based system (RBS) has the following properties:

1. incorporates practical knowledge in a production rule system,

2. knowledge is incremental, skill is proportional to the knowledge base,

3. solves a wide range of complex problems by choosing and combining results in an appropriate fashion,

4. scheduler – adaptively determines the execution sequence,

5. trace facility and natural language interface.

An RBS is therefore a *modularised know-how* system; with the knowledge being in one of the following forms:

1. specific inference from observation,

2. abstraction, generalisation and categorisation of given data,

3. necessary and sufficient condition for achieving a given goal,

4. likeliest position to look for relevant information,

5. arbitration.

A rule has the form:

$$\text{if} <\texttt{condition}> \text{then} <\texttt{action}>$$

where the *condition* is referred to the antecedent and the *action* the consequent. The interpretation of the rule is, if the antecedent can be satisfied then the consequent too. In such as case, if the consequent defines an action then the system executes the specified action; if the consequent defines a conclusion, then the system infers a conclusion. A simple system would then be a procedure which is an *interpreter* which iteratively applies rules to the initial database until the given goal condition is satisfied.

```
procedure interpretor:
      database := initial database; !database = blackboard
      WHILE not goal(database) do
      begin
            rule := choose(rules-applicable-to(database));
                                    !rule = <condition><action>
            database := perform(<action> , database);
      end;
```

Some example of rules are given below. These rules have been applied by Nazif and Levine [Nazif 84] in low-level image processing (image segmentation by split-and-merge methods). Typical rules in this system are:

RULE (801):
```
      IF:   (1) There is a LOW DIFFERENCE in REGION FEATURE 1
            (2) There is a LOW DIFFERENCE in REGION FEATURE 2
            (3) There is a LOW DIFFERENCE in REGION FEATURE 3
      THEN: (1) MERGE the two REGIONS
```

RULE (901):
```
      IF:   (1) The REGION HISTOGRAM is BIMODAL
      THEN: (1) SPLIT the REGION according to the HISTOGRAM
```

It is important to note that a RBS can only solve problems if it incorporates rules that use symbolic descriptions to characterise relevant situations and corresponding actions.

To summarise: $RBS = Knowledge\ base + inference\ engine$. A RBS [Hayes-Roth 85b] has a static memory which consists of facts and rules and a working memory (often referred to as *problem-solving state information*) where temporary assertions are kept. It should also be noted, rules are not universally valid, assumptions are valid in only certain reference frames. Therefore, one needs to ensure that rules are applied appropriately and in a meaningful context. PROLOG was the first general purpose logic-based programming language; it is essentially a RBS. A structure of a simple RBS is shown in Figure 7.12.

A traditional image processing program differs from a RBS in that it lacks a rule and selection sub-system. However, it often has a knowledge database containing model information (not rules). Instead of a rule interpreter, there is a user defined program consisting of one or more algorithms with control and dataflow designed into the system. However, we often have a common workspace which is used and updated by the algorithms. A program lacks the flexibility of a RBS. For a modification in its task execution, a new program may need to be written because the control and data flow may be changed by the addition of the new code. In a RBS, only the new rules need to be added to the knowledge base. However, for image processing, code execution efficiency is very important because of the need to work in real-time; this is a real advantage of a formal program over a RBS. A typical structure (to a first approximation) of a traditional image processing program is shown in Figure 7.13.

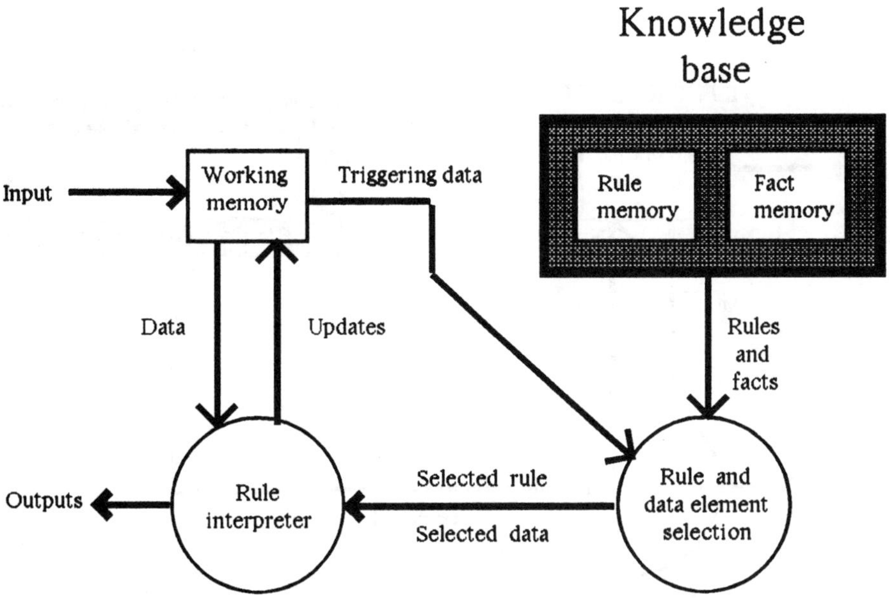

Figure 7.12: Architecture of a simple rule based system.

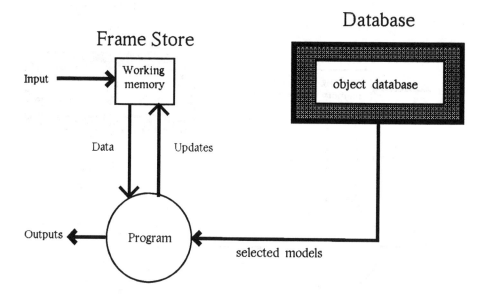

Figure 7.13: Architecture of a simple traditional image processing program.

To sum up RBS, we will describe its advantages and disadvantages and then indicate current research goals. The advantages of a RBS is:

1. it presents problem-solving method in a way which is suitable for computers,

2. it is modular,

3. it is incremental,

4. it is explainable,

5. it provides a framework for conceptualising computation,

6. it provides parallel method for problem solving, and

7. it makes distinction between analytic and imperative know-how.

Its main disadvantages are:

1. there is no analytical foundation for deciding which problems are solvable,

2. there is no methodology or technique to test consistency and completeness of rule set,

3. there is no theory of knowledge organisation,

4. there are no good rule compilers or specialised hardware, and

5. there are no easy way to integrate RBS into data processing.

So what are the current goals? Much of the work is directed by US Department of Defence (DoD) under the 'Strategic Computing Initiative'. They expect to build system with:

1. rule base with greater than 10 thousand rules,

2. increase speed by a factor greater then 2,

3. build more types of inference engines,

4. improve reasoning with uncertainty,

5. simplify the creation and expansion of knowledge bases, and

6. develop parallel methods.

Now, let us consider how we might represent knowledge about objects. To begin with, what is knowledge? By *knowledge* we mean "justified true belief" and *representation* allows us to encode this into a data structure. We will consider several methods by which we can encode knowledge.

Another similar method for knowledge representation is the *production system*. This is essentially the *rule based system* which we have already discussed.

These two methods have rules which are essentially independent of each other and therefore allow incremental development. A third method for knowledge representation is a *semantic net*, this is very different from the other two. A semantic net is an associational network model for representing the "objective" meaning of words; he also introduced the concept of inheriting attributes. In a semantic net (SN) the knowledge is represented in a highly interconnected structure which consists of nodes interconnected by a series of labelled arcs. The nodes symbolises an object and the arcs the relationships between objects. An example of a SN is given below, the interpretation of which is "monitor is on the table".

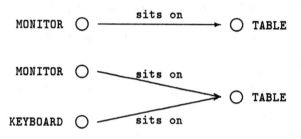

However, consider a slightly more complex situation. We have two propositions: (a) "monitor sits on the table" and (b) "key board sits on the table". Does this mean – the monitor and the keyboard sits on the same table, or two different tabless? To resolve this type of problem, at least two types of nodes are provided: *instance* and *class* nodes. We will use a ◯ and a △ respectively to denote these two types of nodes.

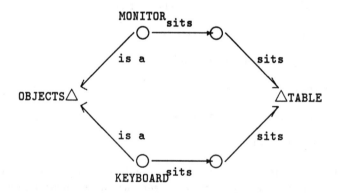

In a SN, an instance is allowed to *inherit* the properties of its originating class. A "is a" link which is usually associated with this property. An example of this is given below.

```
         sits on          is a           shape
   O  ──────────→  O  ──────────→  △  ──────────→  O
MONITOR                          TABLE      parallelepiped
```

The most important feature of a SN is its ability to associate concepts into a graph structure, this avoids the use of special inference rules. But this representation has several major drawbacks: it lacks expressive power (therefore the net has to be enriched making it difficult to handle), and the representation is often ambiguous.

The ideas of SN has been extended by Minsky, the terminology now used is a *frame*. The nodes are called frame and the arcs are relations. A frame is a named collection of information. This is similar to RECORD type structures in an Algol like language. The record fields are called *slots*; these have a name and a value, the values can be symbols, atoms, lists, names of functions, sets of production rules or links to other frames. Inheritance property of a frame is what distinguishes it from a RECORD. The inheritance property supplies a frame with implicit slots and values that are obtained (inherited) from other frames. Frames have very strong modelling expression. Frames by themselves can represent abstract data types as well as instances of these types.

7.5 Industrial Inspection

An industrial vision system must be able to carry out a few "simple" tasks: (i) It must process simple images. A simple image is one which has been captured under controlled illumination from a fixed view point such that 3D objects in some instances can be regared as being 2D. (ii) The system must operate at real-time (this implies use of specialised hardware). It is observed that pipelined processors provide a cost effective solution for speeding up the computation, although in many circumstances multiple pipe-lines may be required. Although use of pipelined systems leads to high data throughput, there is also a long delay between input and output data. This could be a problem for real-time control of processes. However, because processing behaviour in industrial inspection is so well known that problem of real-time control using pipelined processors do not usually arise. (iii) The inspection system must produce a low false alarm rate, i.e. the algorithm must be robust. The few false positives the system may produce could, if required, be checked manually. (iv) The system must be flexible so that it can accomodate easily changes in the inspection task or changes in the manufacturing process.

One of the reasons for the slow penetration of use of machine vision in industrial manufacturing has been the very instringent requirement on both the hardware and the software for the task. What is most often *wanted* is a

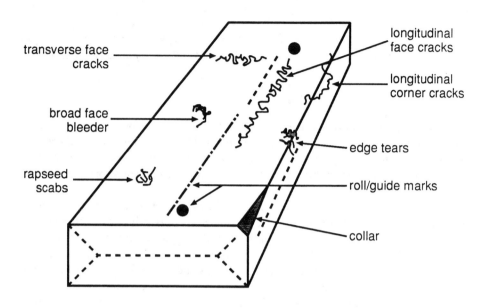

Figure 7.14: Typical fault found on a steel slab.

low-cost system, operating at a high-speed (working at the production rate of a product), high accuracy (in recognition, location and orientation of product), and flexibility in the machine vision system (it must accomodate changes in size of objects, must operate in a cluttered environment, and must be able to cope with changes in position and orientation of objects.

Most of the systems being developed for industrial inspection involve a model-based approach; however the models are often not geometrical ones. The model-based approach needs to address: (1) What features need to be extracted to desribe an object? (2) How do we represent and group features to describe classes of objects? and (3) How do we match the located features with the models?

7.5.1   Metal Surface Inspection

Several systems have been described in the literature [Sardis 79, Mundy 80, Suresh 83]. In steel-making a continous casting method is employed. This involves the molten steel being continually being poured into a water-cooled mould. As the steel solidifies, it is drawn out in a ribbon along the roll table. The ribbons are cut into sections of predetermined length to form slabs. However, these slabs have imperfections which need to be analysed before further processing of the slabs can be carried out. Typical types of faults found in a slab are shown in Figure 7.14.

A considerable amount of energy is used in the heating and cooling of the steel. Currently in manual inspection, the hot slabs have to be cooled to the ambient temperature before the manual inspection can be carried out. If the slabs are free from significant defects, then the slabs are reheated for further processing on a strip rolling mill. An automated system can remove this reheating process from the cycle. This would save energy hence reducing production cost and could lead to a fully automated steel mill which could involve in-line conditioning and hot rolling of the steel.

An automated system will have to locate the various defects and their size and width to determine the slab's disposition, whether is it satisfactory for further processing or requires further conditioning or has to be rejected. The slabs varies considerably in their dimensions: the length could be between 1 and 60 cm, the width could be between 55–190 cm, the length varies between 1.2 m and 12 m. The processing temperature is between 800 and 1100 C. The speed at which the slabs moves on the rollers is 10 cm/s. At this slab velocity, the system has to process some 550 kpixels/s. Even if the slabs have few imperfections, they tend to have many "features" which must be inspected.

Suresh describes a system built by Honeywell [Suresh 83] which works with images in the visible band; images captured using a camera with an infra-red blocking filter. Two cameras are used: The *data camera* views transverse to the slab motion; the picture is generated using a linear CCD array. The *position camera* works simultaneously with the data camera to determine the slab position. The images are segmented and the features extracted using an array processor. The processed data is passed to the host computer for the analysis of the defects. Because the faults are basically line-like features, the segmentation process consists of edge-enhancement and thresholding. These operations are carried out in a scan-line basis.

7.5.2 Lumber Processing

In a manual system, the log is sawn into different grades with the defects being randomly placed in the wood. The defective lumber is then resawn into smaller parts to remove these defects. This is both a labour-intensive process and involves waste of valuable lumber. In an automated process [Conners 83], the log is first scanned using an industrial photon tomograph to generate an image of a slice through the wood. Several slices are used to create a 3D description of the log which can then be used to determine the log geometry and locate the knots. Knowing the log geometry and the knot positions, an optimum sawing strategy can be computed to give maximum yield. Furthermore, using this information, a computer-driven saw can be used which will automatically position and turn the log and set the feed speed. However, this method will still leave some surface defects (such as knots, stains, and worm holes) in the lumber which are not detected by the tomography. The planks are left to dry and the surfaces are lightly worked. After this both surfaces are analysed using optical digital images [Conners 83]. The defect pattern found is used to determine optimal cutting of the planks.

The problem in processing lumber images is the great variability amongst them, even within the same species; no two knots are the same and each type of wood has a unique grain pattern. The defects in lumber arise from either biological or manufacturing defects. The biological defects include knots, steep or spril grain, worm or grub holes and the pith; the manufacturing defects arise from the sawing and the handling of the wood. There are a variety of techniques for detecting these defects. Laser scanners have been used to detect splits and knots of varying sizes [King 78, Mathews 76]. The tonal property of the lumber can be used to detect certain types of defects: for example, clear wood is lighter than knots, holes or cracks. These techniques need to be combined with pattern recognition techniques: ie. holes and knots are circular while checks and splits are narrow and long.

Parallel processing of lumber has been proposed by Conners by subdividing the image into a number of rectangles [Conners 83]. For each of the image patches, the tonal property of the lumber is measured by computing (i) the mean, (ii) the variance, (iii) the skewness, and (iv) the kurtosis. Also the coocurrance matrix is used in the texture analysis. The resolution of the images required is determined by the size of checks and cracks (which run parallel to grain) which have to be detected.

7.5.3 Integrated Circuit Inspection

We are continuing to observe the growth in density of both integrated circuits (ICs) and gate arrays. This has lead to problems in using traditional techniques for testing. The testing procedures for ICs include: (i) "pre-cap", this is visual inspection using various techniques such as stereo microscope, X-rays and electron beams, (ii) electrical testings which are at various stages in the manufacturing process, (iii) environmental testing, this include *burn-in*, temperature, humidity and shock tests, these are long and costly processes, and (iv) statistical testing of samples.

The ICs fail because of [Pau 83]: (i) corrosion or microcracks, these lead to open or short circuits, (ii) oxide breakdown by such processes as static discharge, (iii) surface defects such as dust, (iv) dirty photomasks, (v) die cracks, (vi) packing defects, and (vii) thermal mismatch.

7.5.4 Printed Circuit Boards

It is very difficult to manually check for defects in complex multilayer printed circuit boards (PCB) which are commonly in use today, for example IBM produce the TCM boards which are 600×700mm in dimension and have 20 layers and the circuits are about 0.081mm wide [Seraphim 82a]. It has been reported that humans can detect about 90% of faults in a single layer PCB. This detection rate goes down to about 50% for six layer boards [Yu88]. Even when fault free power and ground layers have been established using electrical testing, the sucess rate is no better than about 70%. It is said that the integrated ciruit is a $9 billion buisness but have a scrap $3 billion worth of product because of manufacturing faults [Bunze 85]. Therefore automatic optical inspection is required to determine that the line widths, the line spacing, voids and pin-holes are within the specification. These tasks cannot be carried out by electrical testing. Further advantages of optical testing are it can check the PCB against its CAD model and the method is non-intrusive, hence avoids mechanical damage.

Many of the defects on PCB, IC and photomasks can be detected by locating local features. Assumptions made are that good features, such as line tracks on PCBs and ICs are regular whilst defective features are irregular. These defective regions can be located by use of morphological operations (Figure 7.15). These operations will locate such defects as lines too thin or lines too close together, or protrusions, thin sprouts and micro-cracks (Figure 7.16).

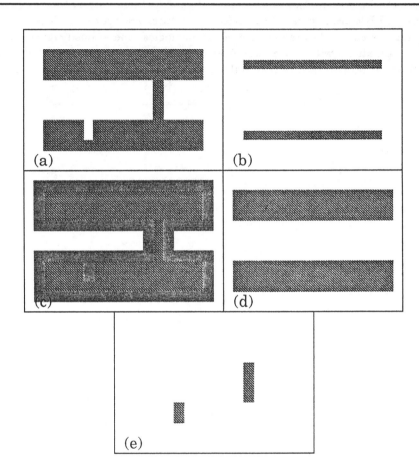

Figure 7.15: Using opening and closing morphology operations shorts and intrusions can be detected.

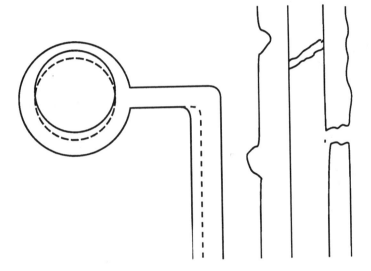

Figure 7.16: Some typical faults with pcbs and ics are shown. These faults include mis-placement, shorts, cracks and protrusions.

7.6   Biomedical Image Processing

Many researchers have used image-processing techniques to solve a particular class of problem, that of automatic processing of biomedical radiographic images. This has been achieved by developing systems which may be broadly classified into three categories: (1) the computer-aided diagnosis (CAD) systems, (2) the fully automatic computer diagnosis (FACD) systems, and (3) the interactive computer diagnosis (ICD) system. The word *diagnosis* is not being used in its clinical sense, we are only referring to the process of reaching a decision.

CAD – A system in which the computer makes the diagnosis based on information supplied by a physician.

FACD – A system in which the computer processes the image and arrives at its own diagnosis. Some advantages and disadvantages of a FACD system are listed below.

   (i) Computers are good at making quantitative calculations but still not capable of responding to the range of the human perception.

  (ii) Image processing is computationally complex and requires a large data throughput and therefore often requires fast computation and expensive hardware. A future solution may be VLSI-based SIMD array processors.

 (iii) Use of image processing and computers seems suitable where a significant amount of physician time can be saved, for example, a third of radiographic usage is in chest examination [Conners 82]. Consequently, use of image processing in chest radiographs is more widespread than its use in cephalometric radiography.

ICD – A system in which the physician aids the computer in the analysis of the image. This system is intermediate between CAD and FACD.

Chien and Fu have made studies of lungs in chest radiographs for known and unknown diseases [Chien 74]. Their method is to extract the lung boundary and make texture measurements. The image field consists of 128×256 pixels. The boundary is extracted by locating five 'corner' points using template matching in localised regions where the feature are expected to minimise computation. The templates are generated by averaging several images together. Joining up these points to form a polygon gives a good approximation to the true lung boundary. Diagnosis is made on basis of the lung texture measurements. Five texture measurements are made; these are (i) average grey level, (ii) number of edges – a measurement of coarseness, (iii) the inverse measure of edges, (iv) autocorrelation measurement and (v) conditional entropy which measures homogeneity. These measures are used to make one of three diagnostic decisions about the lungs: (i) normal, (ii) abnormal or (iii) no decision.

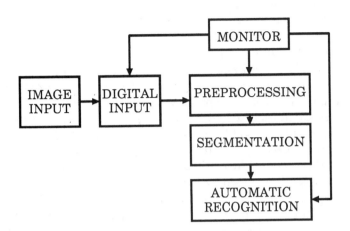

Figure 7.17: The image-processing system due to Harlow.

In developing a system for automatic analysis of chest radiographs for the heart, Harlow reports some of the problems encountered in cephalometric image analysis: (1) extreme variety and complexity within and between radiographic images and (2) small details, which are present in the image and which often lead to the detection of false positives. The system developed by Harlow et al., [Harlow 76] is shown in Figure 7.17. The monitor program controls and evaluates the results at each of the stages.

The preprocessing consists of projection of the sum of the grey-values in both the horizontal and vertical directions. Using the resulting histograms, the position of the lungs in the image can be located. Texture measurement of the lungs is carried out to determine the mean and the entropy of the grey-levels. This indicates an increase or decrease in vascular change, relating to heart disease. The segmentation process consists of a model-driven tree search. It begins by looking for large objects and, using these as a reference, it looks for smaller objects; therefore the search is context-sensitive. Using such a method, the heart outline is extracted and described using Fourier descriptors; ratios of area of the concavity between the lungs (i.e. the region where the heart resides) to its perimeter are also important. The distance measurements of features between the lung cavity are also performed.

Wechsler and Sklansky describes [Wechsler 77] a system for finding the rib cage in chest radiographs. To begin with, a 17 × 14 in. radiograph is reduced photographically to 5 × 4 in. and finally digitised using a drum scanner to a 256 × 224 pixel image. The analysis of image consists of four phases: (i) preprocessing, (ii) local edge detection, (iii) global boundary detection and (iv) linking ribs together. Image preprocessing consists of edge enhancement using the fast Fourier transform. Next edges are extracted by applying the Sobel and the Laplacian detectors in combination with a thresholding operation. The rib contours are found by matching the edge pixels for straight, parabolic and elliptical curve segments using the Hough transform. Finally, the dorsal and ventral ribs are linked together using a fourth-order polynomial curve to smooth the contours and fill in the gaps.

Sklansky *et al.* [Sklansky 80] describes a system for analysing nodules found in chest radiographs. The method consists of analysing a 256×256 pixel array of 8-bit precision. The images are preprocessed using the following steps: (1) the image is normalised using a zonal notch filter, (2) the edges are extracted and thresholded, and (3) straight lines are found and removed. The resulting image is fed to a circle detector to locate circular objects or arcs of prescribed radius. The output then enters a heuristic boundary tracker. The circle detector is a Hough transform as described by Kimme *et al.* [Kimme 75].

The Hough detected circle centers are used as a guide by the boundary tracking algorithm. The boundary following algorithm consists of a sequence of heuristic searches confined within a quadrant of a circle. This cuts down on the search required. The boundary information, such as its length, and number and length of straight segments, is used to classify the detected nodules into *tumour* or *nontumour* categories.

In summarising the techniques described above, we find, in many of the cases, the radiographic images have been preprocessed to remove noise using median filters or edges have been strengthened by applying extremum filters. Subsequently, in many of the systems, edges have been extracted. A variety of methods have been employed including Mero-Vassy, Sobel and the Laplacian. For feature recognition, a variety of boundary descriptors such as Fourier descriptors, central $(r - \theta)$ and edge-direction profile $(s - \psi)$ graphs and Hough transforms for lines, arcs and circles have been employed.

7.7   Cephalometric Analysis

Cephalometric analysis provides a means of measuring craniofacial form and structure in living subjects by use of standardised radiographic techniques. In 1931 H. B. Broadbent [Brodbent 31] introduced techniques for examining the skull by serial radiographs taken with standardised head fixation.

The clinician uses cephalometric radiographs to observe gross anatomic relations in order (i) to assess the physical structure of the cranium and the facial bones, (ii) to search for structural changes because of disease and (iii) to interpret physiological conditions. The second use of cephalometry is for measurement and description of features. By establishing a known population variation, comparison of one individual with another can be made. The third application of cephalometrics is to record and measure change in serial radiographs. Finally, radiographs are used for treatment planning.

The essence of cephalometric radiography is the standardisation of head fixation introduced by Broadbent. The patient's head position and the orientation of the X-ray beam are established in such a way that repeated exposures may be

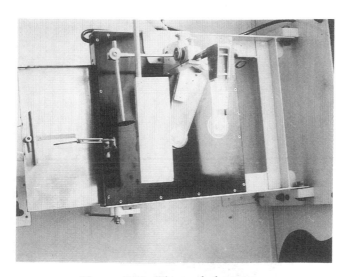

Figure 7.18: The cephalostat.

made on different occasions under essentially the same conditions. The instrument used to position the patient is called a "cephalometer" or a "cephalostat", shown in Figure 7.18. The ear posts are tightened into the external auditory meatuses (outer ears) sufficiently to steady the patient's head. The ear posts should not be so tightened as to become painful and in fact over-tightening can be counter-productive as the patient becomes increasingly uncomfortable. The patient is still able to rotate his head in the vertical plane and some procedure, such as use of a mirror to keep the eyes at a certain level, is required to ensure that all radiographs are taken in approximately the same position. By carefully positioning an aluminium wedge, it is possible to obtain a clear image of the soft tissue outline of the face.

Patient position for the lateral view is measured to the sagittal plane as determined by the midpoint between the ear posts positioning the patient. A lateral skull radiograph is shown in Figure 7.19.

Serial X-ray tracings have been used to assess the growth of the facial skeleton. Superposition upon various parts of the skull has been used to show growth at any particular point relative to another, and has encouraged its employment as an aid to diagnosis. This method of superimposition is made meaningful by locating each tracing by means of a fixed point, or a registration point. This point should be easy to identify with great precision and should be free as much as possible from the influence of growth. Although a number of registration points have been used in the past, it is now common practice to use the centre of the sella turcica and a line joining it to the nasion (see below for a definition).

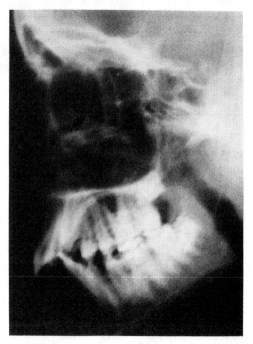

Figure 7.19: A lateral skull radiograph.

Broadbent and later workers have devised definitions for some of the principal terms used in cephalometrics, and these are given below [Krogman 57, Chaconas 80] (see Figure 7.20).

Nasion The anterior (foremost) point of the suture (seam-like articulation of two bones at their edges) at the junction of the frontal and nasal bones in the mid-sagittal plane (plane extending longitudinally to divide the skull in two similar halves).

Sella The mid-point of the sella turcica, a bony crypt, determined by inspection.

Menton The lowest point on the symphyseal shadow (shadow caused by the halves of the lower jaw meeting at the chin). (In some instances the term gnathion, the point on the chin determined by bisecting the angle formed by the facial and the mandibular plane, and menton have been used synonymously but there may be obvious differences between the two, as shown in Figure 7.20).

Pogonion This is the anterior point in the profile of the bony chin.

Gonion A point midway between the anterior and posterior (hind-most) points on the angle of the jaw, obtained by bisecting the angle formed by tangents to the mandibular lower border and the posterior margin of the ascending ramus (jaw bone).

Anterior Nasal Spine The tip of the anterior nasal spine seen on the lateral X-ray film (see Figure 7.20).

'A' Point (Sub-spinale) The deepest middle point between the anterior nasal spine and the prosthion (see Figure 7.20).

'B' Point (Supra mentale) The deepest point in the concavity of the bone between the infra dentale (bottom of the front teeth) and the pogonion.

Articulare The point of intersection of the dorsal contours of the condylar process (ellipsoid knob of bone, which articulates with a corresponding socket of temporal bone) and the external cranial base.

I The point where the long axis of the lower central incisor intersects the incisal edge.

Mandibular Plane Several mandibular planes are used, depending on the analysis. The most common ones are: a tangent to the lower border to the mandible (jaw bone); a line between gonion and gnathion; or a line between gonion and menton. It is not critical which one is used, as long as the same one is used consistently.

Palatal or Maxillary Plane Represented by a line drawn through the anterior and posterior nasal spines.

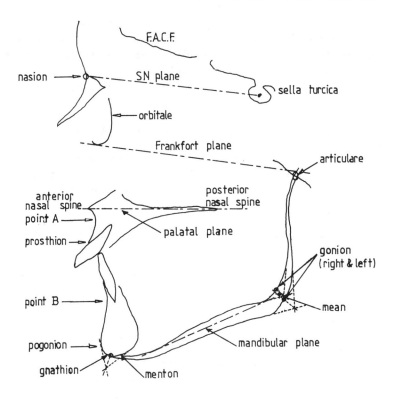

Figure 7.20: Tracing of a lateral radiograph indicating the standard landmarks.

An important point to be made with respect to reliability in measurement involves an individual's ability to locate a specified landmark on the radiograph. Some landmarks are more reliably located than others [Baumrind 71], furthermore, some landmarks are more reliably located at one age than another; the Bolton point (Figure 7.20) is relatively clear at a young age but becomes increasingly obscure with the growth of the mastoid process (excrescence of the human skull just behind the ears, containing air-spaces which communicate with the middle ear) until there are large discrepancies between observers as to its location on radiographs of older patients.

Furthermore, besides inter-radiograph differences, intra-radiograph differences are important in the longitudinal evaluation of growth or the effects of treatment. Although the cephalometer is well aligned with the central X-ray beam and the film, the external auditory meatuses are not so rigidly attached to the skull and they have considerable mobility. Therefore, films taken of the same patient on successive days are not identical; there are small shifts in position relative to the ear posts.

It is not easy to measure the quality of a radiograph. Assessment of quality is highly subjective and clarity often varies between anatomical sites on the same radiograph. However, one can assume that in a sample of radiographs there is a range of quality between the 'best' and the 'worst' films, and it is reasonable to assume that the level of tracing error is affected by this range.

There are three factors which determine what impact the error in location of a specific landmark will have on the linear and angular values involving that landmark: (1) the magnitude of the error involved in locating a specified landmark, (2) the linear distances between features which are connected in a measurement – the greater the distance, the smaller the error, and (3) the direction from which the line joining two landmarks intersects the envelope of error of each landmark.

Measurement of cephalometric radiographs has played an important part in orthodontic research and treatment planning. Measurements are used for two major purposes: description and prediction. The description involves (1) categorizing cases according to type (i.e. "a case with an $A\hat{N}B$ angle of 6 degrees" – the angle between point A, the nasion and point B), (2) defining the amount by which an observed case departs from some accepted norm, and (3) indicating the extent of change during treatment.

7.7.1 The Need for Automation

Baumrind *et al.* [Baumrind 71] have inferred that (i) even when one is replicating assessments of the same radiograph, errors in landmark identification are too great to be ignored; (ii) the magnitude of errors varies greatly from landmark to landmark, thus certain features such as the gonion, the lower incisor apex or the mesio-buccal cup (lower first molar) are unreproducible, and therefore of less significance in cephalometric measurements; and (iii) the distribution of errors for most landmarks is not random, but rather systematic – i.e. each landmark has its own error envelope. From this, one concludes that there is a systematic error in determining the landmark locations and the distribution and the magnitude of the error is dependent on the feature. Also, the features change over a period of time and therefore the systematic error envelope also changes. From Broadway's observations [Broadway 62], it can be seen that the error between observers in determining a feature location is greater than for an observer locating the same feature on different occasions, and the level of error is affected by training and experience. It should be stressed that tracing error is an important source of error in cephalometry, in fact it may be the major source of systematic error.

Several workers have demonstrated [Cohen 84, Richardson 66] that the use of high resolution digitising tablets (typical resolution 0.017 mm) produces little or no gain in reproducibility in locating the landmarks, the subjectivity of the human operator being thought to be the cause; an automated system would eliminate human subjectivity and hence play an important part in producing repeatable results.

Therefore, in determining any cephalometric landmark, there are two sets of errors: (i) systematic – associated with the shape of the feature and (ii) subjective. The latter, because of human interaction, can be further subdivided into two classes – (a) random error and (b) systematic error. There is a contribution from the latter since the level of error is observed to be affected by training and experience.

The systematic error associated with the landmark cannot be affected since it is intrinsic to the landmark. However, from the research of Broadway and Richardson, it may be assumed that a significant source is human subjectivity. Therefore, by automating the landmark location process involving human interaction, the subjectivity error is reduced and the overall error in feature location should also be reduced.

7.7.2   An Image Processing Solution to Cephalometry

Given that a possible solution to improving reproducibility in measurement of cephalometric radiographs both for growth study and for treatment planning may be to automate the system, a consideration to an 'ideal' system needs to be given.

What should an ideal system be capable of? It should do the three tasks frequently carried out by orthodontists: (i) recognise cephalometric features, (ii) measure facial structural changes because of either growth or treatment and (iii) indicate abnormal growth. Furthermore, we would want a system which would have the following specifications:

(a) It should reproduce results significantly more accurate than humans using tracings.

(b) It should operate as fast as or faster than a human operator. To achieve this we will have to consider parallel processing (at the low-level, pixel level, processing it is possible to gain N^2 speed up for many image-processing operations involving N^2 pixels).

(c) It would ideally be a fully automated system without expert intervention to free Orthodontists from the tasks so that the system can be operated by a non-expert with little training.

(d) It would include an expert system to interpret the output data and advise on diagnosis.

We will only concern ourselves with image-processing aspects of a possible ideal system.

If the annual growth rate between the menton and the sella is approximately 1–2 mm [Solow 86], then to observe any growth or change between the two landmarks, each of the landmarks needs to be detected to considerably better than ±0.5 mm. The effective resolution that may be attained is determined by the size of the penumbra on the radiographs and the film grain. For some typical values for effective anode size of 1.5 mm, the anode-patient distance of 1.52 m and the patient-film distance of 0.1 m, the penumbra is about 0.1 mm. The size of the film grain is also typically 0.1 mm [van Aken 62]. The area of the radiograph containing cephalometric features of interest is approximately 22×22 cm. Then to digitise the whole of this area with 0.1mm per pixel resolution an array greater than $2K \times 2K$ would be required. This is clearly impracticable.

However, we need only segment the regions containing a feature of interest, needing to digitise only a sub-image of the (22×22) cm of the radiograph. From the location of the features, we can calculate their spatial separation and the angle between the features. All this we could do using a motorised travelling stage. Using a model to drive the stage, only the areas containing features of interest need be digitised for subsequent processing. Furthermore, the stage will provide us with our measurement reference frame.

However, in the first instance, with current technology and cost of a large motorised travelling stage with 0.1 mm accuracy in movement, we do not expect to build an automated stage with the specifications as set above. In this section, we will attempt to describe a system which may be more reasonably built. Using the constraint placed upon us, we will describe our goals.

We wish to develop a system with the following characteristics:

1. there must be no operator interaction with the algorithms with the exception of the placement of the cephalometric feature in view of the camera,

2. the system must match human performance, more so in respect to reproducibility of results than in speed,

3. closely related to (2), the system must be robust, and

4. as a first step towards matching and bettering human operational speed we will develop parallel algorithms for CLIP4 machines [Duff 78]. Such a system is shown in Figure 7.21.

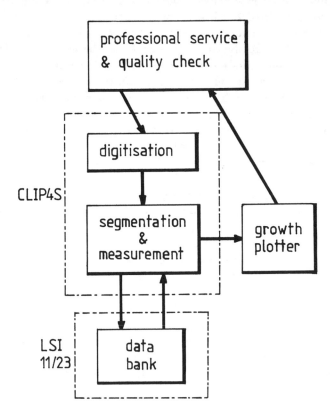

Figure 7.21: Our intermediate target image-processing system.

Figure 7.22: The stage for inputting cephalometric images.

The constraint on reproducibility along with the flexibility and the size of the processor array available imposes some limits on our choice of the segmentation algorithm.

The initial work done on CLIP4 with a resolution of 0.17 mm/pixel showed promise that an automated system to measure cephalometric techniques could be developed [Hussain 84, Hussain 85]. However, at such resolution, on CLIP4 with its 96 × 96 processor array, a single feature, such as the sella turcica, covered the whole field. The higher resolution CLIP4S computer with its virtual 512 × 512 processor array, provides the 0.1 mm per pixel resolution sought and a large enough field of view that the problem of CLIP4's resolution does not arise; the sella turcica at this resolution covers an area only about 1/9th of the pixel array. Therefore, to provide for an external reference frame is a much simpler task. The stage consists of a table with three sets of grooved plates, the grooves being 15 mm apart, see Figure 7.22. The film-holder can then be moved in steps of 15 mm in the x- or y-plane. Then calculating the number of steps required to move from one feature to another, and their position in the CLIP4S array, the spatial separation of the two cephalometric landmarks can be computed.

Normally, in a cephalometric analysis, the landmarks defined at the beginning of this section are used. However, because of the computational times, it was decided that only a subset of the features should be segmented to demonstrate that an automatic system could be developed for analysing cephalometric radiographs.

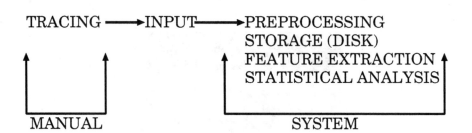

Figure 7.23: Method of Sugawara *et al.*

7.7.3   What type of system should we develop?

The use of image-processing techniques to extract information from cephalometric radiographs has been very limited. With the exception of three works
(one of them the subject of this section), the others have been exclusively
limited to manual tracings of radiographs and the use of computers has been
limited to merely calculating distance and angle between landmarks and for
database storage and retrieval. These systems are classified as being CAD.
Such manual methods for landmarking cephalograms have now become quite
common and have been reported by several authors [Baumrind 71, Baumrind
80, Solow 72, Walker 72, Chebib 76, Faber 78].

   A method with limited image-processing involvement has been reported by
[Sugawara 80].   While the techniques involved have been developed for the
study of the mandibular form, the method can easily be extended to study
other cephalometric features. The radiographs are manually traced and the
tracings are digitised, followed by some limited preprocessing before feature
extraction. The preprocessing of the images consists of:

(1) Thinning

(2) Node and line extraction

(3) Elimination of noise, spurious branches and minute loops

(4) Interpolation between two disconnected lines.

The overall method is summarised in Figure 7.23.

   The first extensive use of image-processing techniques to segment cephalometric lateral skull radiographs was applied by [Cohen 84, Ip 84]. Algorithms
to segment two features, the menton and the sella, were developed. In both
cases, the image was preprocessed by use of extremum and median filters to enhance the images, also an expectation window was manually generated within
which the feature was located. In the case of the menton segmentation, a modified version of the Ridler and Calvard threshold algorithm [Ridler 78] using

the values only within the expectation window was used to separate the object from the background. The menton was determined by picking out the middle pixel of the bottom row of the binary image within the expectation window; it is confidently expected that the bottom object contains the menton. For the sella, the image is again binarised using the Ridler and Calvard algorithm and subsequently thinned using the Arcelli masks [Arcelli 75]. The sella is determined by finding the concavities of the skeletonised object and taking the centroid.

The systems developed by Sugawara *et al.*, and Ip respectively are both ICD. In both cases, a human expert helps in the image segmentation process. For example, a physician has to make a tracing of the mandible outline in Sugawara's system and in Ip's system, an operator has to clean up the segmented image and generate expectation windows for subsequent processing. The systems we now review involve much less human interaction and can be broadly classified as being FACD.

A model-driven method to segment cephalometric images is given in [Levy-Mandel 86]. A radiograph is digitised to 256×256 pixels, with 8-bit precision, with the top-left octant of the image containing the forehead and the bottom-left quadrant the spine. Very briefly, the method consists of impulse noise removal by use of a median filter and edge extraction by use of a Mero-Vassy operator [Mero 75], a simplified version of the Hueckel operator. All the edges are extracted by use of a global line detection algorithm. Briefly, the method consists of setting up an upper threshold (S_1) limit, and scanning the image line by line until a point with grey-value greater than S_1 is found. The algorithm then performs a hill climb to find a local maximum (L_m) and tracks the object by following the neighbouring maxima while the maxima are greater than a lower threshold (S_2). This is performed per scanline for the complete image. Having found the lines, a model-driven line tracking algorithm is used to extract the relevant landmarks. A reference map is used; using a search area of 10×10 pixels in the reference map, lines are tracked given their start and end conditions and the general direction and length of the line. Having found the line, checks are made to confirm that the correct line has been found; if not, the feature is tracked once again. If, for any reason, one feature cannot be located, then the algorithm will fail to locate subsequent features since the previous landmark position dictates where the next landmark is to be searched for.

Any algorithm we devise for segmenting the cephalometric images must be (1) robust, (2) accurate, and (3) parallel. By the latter we mean that it should be SIMD parallel. Furthermore, any algorithm we use to extract a feature must conform as closely as possible to the clinical definition for the feature. Any algorithm which fails to meet these criteria will be rejected as being unsuitable.

It is noted that neighbourhood non-linear (rank) filters are generally better than linear filters because they blur the picture less than linear filters do. While the extremum filter does enhance the edge, it has been found that its combination with a median filter gives a better, less noisy, result and the convergence to a root-image is faster.

Using the criterion of parallelism, we can reject any algorithms which are essentially serial or not data parallel. Amongst these are the set of algorithms which describe and match objects using boundary information with variants of chain coding. These are the $s - \psi$, the $r - \theta$ graph and the Fourier descriptors. However, there are parallel template matching techniques such as the use of normalised central moments or use of two-dimensional templates. However, this only allows us to recognise objects and not to segment objects. Therefore, a suitable segmentation algorithm will need to be found.

We should consider the usefulness of parallel template matching for segmenting cephalometric images. This refers to both two-dimensional grey-level sub-image template matching and Hough transforms. One may consider that template matching may be a suitable method for segmenting the sella and the articulare features. To segment the palate or the mandibular line using 2D templates would require templates greater than (200×30). As indicated in [Wong 76b], in the presence of noise, correlation peaks are difficult to pick out. With cephalometric images, we have potentially even greater problems; those of scale and rotation. Using the sella turcica as an example, the feature cannot easily be described analytically. There is a large variation in size and shape of the feature within patients of the same age group. Furthermore, the feature changes in size and often in its shape during the periods of growth. This makes the task of choosing or generating a sub-image template very difficult.

The formulation of Hough transform for a circle [Kimme 75] and a straight line [Davies 86a] show that these algorithms can be implemented in parallel on an SIMD array computer.

Now consider the usefulness of optical flow in segmenting the cephalometric features. Optical flow gives us information regarding motion of objects in the image. The first problem is that the image sequences need to be registered. Otherwise, any change in the head position in the radiograph sequence will be interpreted as being motion; this may drown the motion because of the growth. In the present collection of radiographs, no registration information is present. Using cephalometric features as the registration points is not suitable since these points themselves may suffer from growth.

What is actually required is some simple method for measuring growth by measuring the spatial and angular separation between specified features. This requires us to locate a number of specified features in the radiograph and use a reference frame.

The types of operation which are efficient on CLIP4 are point-wise, local-neighbourhood, and global propagation operations. These operations fit directly into the architecture of the machine. Our segmentation techniques will make extensive use of these classes of operation. This still gives us flexibility in the way we interpret the segmented data. For example, it may be possible to segment the image such that the segments may be incomplete. Then a *linguistic* or a *syntactic* approach can be taken. Grammar can be used to label the segments and use of semantics can be made to give meaning to the labelled segments allowing us to understand the scene. However, we can make life much easier for ourselves. We do not need to assign meaning to our image scene. We know that a given scene contains a specified feature. Therefore, we can approach the problem *descriptively* or *algorithmically*. Provided we can extract complete segments, then we can, in many cases, distinguish between objects by computing some of the basic shape parameters proposed in [Kitchen 83b], such as area and perimeter length.

In our system each segmentation window, chosen by an operator, contains a distinguishable feature of interest, so a specific algorithm can be implemented to extract the feature. This increases our resolution and reduces the number of noise objects within each segment window. While this increases the computational cost, this is compensated by use of an SIMD array processor. Both Harlow [Harlow 76] and Levy-Mandel have used a model-driven segmentation technique. Initially, they segment large features and use their location as a guide to subsequent fine-level processing; this technique is not necessary in our case.

In this section, segmentation algorithms for three point-like structures and three line-like features are presented. The point-like features are the sella, the menton and the articulare, and the line-like structures are the palate line, the mandibular plane and the floor of the anterior cranial fossa (FACF).

It should be emphasised, while the image quality of the radiographs is generally poor, the segmentation processes have been simplified by arranging that the images have fixed orientation and resolution. The fixed orientation has meant that the line-like features which are either ridges or edges on the radiograph have a predetermined gradient. The resolution imposes a limit on the size of the feature which is being processed. Whilst the features grow with the size (and age) of the patient, these changes are small compared with the variability in dimensions observed between different patients. Assumptions of a minimum size for the palate line have been made. For the mandible line, no size assumptions need be made. Once the orientation is fixed, the mandible line is the bottom edge of the mandibular shadow inclining downward from west to east.

The segmentation processes described below have been considerably simplified because a large body of *a priori* knowledge is available for the segmentation. Some of the *a priori* knowledge which is available will be discussed here. To begin with, (as stated above) the orientation of the radiographs is fixed. The radiographs are viewed in their normal orientation, i.e. the chin is at the bottom and the sella turcica towards the top. Where *a priori* knowledge of the shape, size, orientation and location of the objects is available, this has been used in the segmentation. At other times, some values, such as edge gradients or threshold values, have either been determined experimentally or fashioned in an ad hoc manner. Of the *a priori* knowledge that is available, the shapes and orientations of the structures are known. The sella turcica is an upright ∪-shaped ridge, the articulare is an inverted ∨-shaped cavity and the menton is the lowest point on the chin (and the chin is the bottom-most object in the image). The width of the sella turcica ridge is known not to exceed 25 pixels and the edge parameter of the articulare cavity is known to exceed 200 pixels.

We begin by noting the preprocessing algorithm common to the segmen-

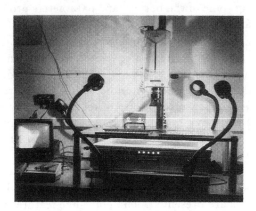

Figure 7.24: The apparatus for inputting an image into CLIP4S.

tation of all the six features under discussion. Then an outline of the six
segmentation algorithms is given.

7.7.4 Common Segmentation Preprocessing Procedures

Three of the features we are segmenting appear topologically as grey-level ridge-
like structures. These features are the sella turcica, the palate and the floor
of the anterior cranial fossa (FACF). The sella turcica appears as a ∪-shaped
object while the palate and the FACF are near-horizontal lines. The ridge-like
features are segmented in all three cases using high-pass filters. The high-pass
filtered image is obtained by subtracting the original image from its low-pass
filtered image. The low-pass operation is carried out using repeated grey-scale
expand and shrink operations [Goetcherian 80]. However, for the palate and the
FACF, a directionally sensitive high-pass filter is used. The filters are made
direction-sensitive by limiting the neighbourhood in the expand and shrink
operations. The number of iterations required of the neighbourhood function
is determined by the width of the ridges. See individual algorithms for details.

For the sella and the palate, having found the correct ridge lines, the central
line of the ridge, the arête, is determined using the MAXINOBJ routine. These
arête points are used in the case of the sella to compute the concavity; and for
the palate, the palate line is determined by fitting a least-squares line to its
arête data points. A least-squares technique is also used to fit a straight line
to the segmented FACF data.

7.7.5 Segmenting the Sella Turcica

The Sella Point ('S' point, Figure 7.20) is stated to be the midpoint of the sella turcica by inspection [Krogman 57]. This is a very subjective definition of a cephalometric landmark and as such requires redefining so that an algorithm may be devised to automatically extract the coordinates of the sella point. The definition of the sella to be used in this book is that the sella is the centroid of the concavity of the sella turcica. The sella turcica is a ∪-shaped bony crypt.

Our task in segmenting the sella turcica is to find all the ∪-shaped ridge objects by an area measurement technique and having found the correct ridge, locate the sella point by computing the centroid of the cavity of the ∪-shaped ridge. A method to accomplish this task is presented below in Algorithm 7.1.

Algorithm 7.1 Segmenting the Sella Turcica

1. *Input an average of the same image field containing the sella turcica feature and store result in* I_0, *Figure 7.25.*

2. *Make a copy of image* I_0 *to* I.

3. *Filter* I *for salt-and-pepper (spike) noise with a low-pass filter and store result in* I_{lp}.

   (a) *A spatial filter that is known to remove noise smaller than the filter window but which preserves edges is the median filter. Since edge preservation is not an absolute necessity at this early stage, a faster method for image filtering, such as a weighted averaging or a low-pass filter may be chosen. For our purposes, both methods were found to give satisfactory results. However, the low-pass filter is faster.*

4. *Next obtain the ridges in the image* I_{lp} *using a high-pass filter, and store result in* I_{hp}.

5. *Threshold image* I_{hp} *using edge strength information, §5.2.1 Algorithm 5.1, to choose only the strongest ridges, which are also the most probable sella ridge. Store result in* I_{ridge}. *The result of thresholding the high-pass filtered image using the THR_EDGE algorithm is shown in Figure 7.26.*

   (a) *One of the most prominent features of the sella turcica image is the strong gradients because of the ridges in the image. Experimentally we have determined that at least 30% of the steepest gradient pixels form the ridges. The threshold value is chosen by calculating the average grey value of the* $(30 \pm 5)\%$ *of the pixels with the strongest gradients in* I_{hp} *using a binary tree search. In 95% of the cases attempted, a threshold value chosen from the grey-value of the 30% (call it p_tile) of the strongest edge pixels has given satisfactory segmentation. If the segmentation is not satisfactory, then the operator can change the p_tile value interactively. Increasing the p_tile value decreases the threshold value and vice versa.*

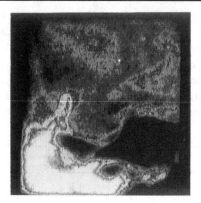

Figure 7.25: The input image of a sella turcica on CLIP4S.

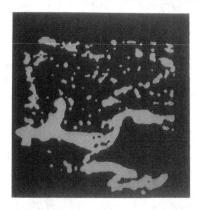

Figure 7.26: Result of thresholding the high-pass filtered sella turcica image using the THR_EDGE algorithm.

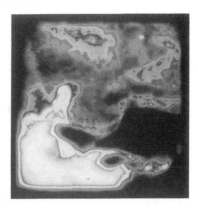

Figure 7.27: A sella image filtered using Beta-spline.

6. *Filter image* **I** *again, this time using Beta-spline technique with the the tension parameter* (β_2) *set to 50 within the ridges found in step (5). Store the result of Beta-spline filtering in image* **I**$_{spline}$.

(a) *In step (3) a low-pass filter was used to remove spike noise from the image. It was not important, since only the basic outline of the ridges was later required, whether the resultant image was smooth or edges were preserved accurately. Use of a median filter was rejected because while it preserves edges better, it is computationally expensive. A fast method for filtering images has been found to be fitting Beta-spline to the data. Furthermore, results have shown that a Beta-spline type filter gives very similar results to both median and weighted average filters.*

(b) *The Beta-spline has two very useful parameters, bias and tension. The tension parameter in particular is very useful in the image filtering process; the amount of "smoothing" is varied by the tension parameter. Furthermore, different tension parameters could be set within the image. For example, at the edges or ridges, the tension parameter could be set high to preserve the original edge and ridge shape while smoothing strongly all others. Where there are ridges, the tension parameter* (β_2) *has been set at 50, elsewhere* $\beta_2 = 0$*; also,* $\beta_1 = 1$ *everywhere (i.e. no bias). The result of the sella image filtered by a Beta-spline is given in Figure 7.27*

7. *Filter image* **I**$_{ridge}$ *to remove ridges which are only horizontal or only vertical (i.e. not* ⊥- *or* ∪-*shaped) using a combination of directionally sensitive high-pass filters. Ridge segments, which are only vertical or only horizontal, of width less than 21 pixels are removed as shown in Figure 7.28.*

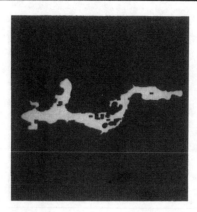

Figure 7.28: Ridge segments of width less than 21 pixels are removed using shrink and labelling operations.

8. *Determine in parallel the horizontal concavities in the I_{ridge} image. This is to distinguish between the ⊥-shaped and ∪-shaped ridges.*

9. *Choose the sella turcica concavity by a size and shape discrimination algorithm using eccentricity measurements.*

   (a) *At this stage, computationally more complex, and more importantly serial-over-objects, i.e. computing for each object at a time, shape discrimination calculations need to be carried out to find the correct sella turcica concavity. More often than not, only one object is present after the size discrimination calculations.*

   (b) *There are a number of simple shape measures. Amongst these are measures of aspect ratio. A method for measuring the aspect ratio has been proposed by [Tang 83] using moments. More simpler than computing aspect ratios is calculating the ratio of the circumference to the area of the object.*

10. *Now mark the maxima (Figure 7.29) with the new sella turcica ridge using I_{spline} as the label, see Algorithm 7.3.*

11. *Finally, determine the new concavity (Figure 7.30(a)). The sella (Figure 7.30(b)) is determined by calculating the centroid of the concavity, Algorithm 7.6.*

Concavities of the sella turcica are simple shapes ranging between an ellipse and a circle. We observe, most of the sella turcica concavity to be near circular; however, some are more elliptical. If one assumes a = 2b for one extreme in

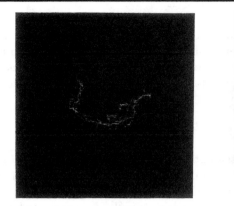 

Figure 7.29: (a) The maxima within the sella turcica ridge. (b) The thresholded image of (a).

 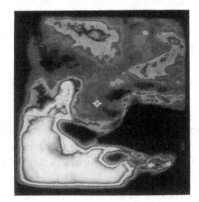

Figure 7.30: (a) The concavity of the sella turcica and (b) the sella (centroid of the concavity marked by the circle).

the sella turcica shape (typically this ratio is observed to be smaller), then the range for $(C/A)^2$ measure can be calculated

$$\frac{4}{a^2} \leq \left(\frac{C}{A}\right)^2 \leq \frac{10}{a^2}$$

The semi-major axis radius is measured by constructing the bounding rectangle for the shape. Applying shrink operations and keeping a note of the number of iterations required for the rectangle to disappear. The width of the body is less than $2n + 1$ if the number of iterations required are n. The object is accepted as being the concavity of the sella turcica if

$$\frac{n}{2} \geq \left(\frac{A}{C}\right) \geq \frac{n}{\sqrt{10}}$$

This measure has required many assumptions about both the diameter of the major-axis and the eccentricity of the body. A confirmation is required if the above measure is near the limits. This is achieved by calculating the eccentricity of the body [Ballard 82]. Eccentricity of a large number of segmented sella turcica cavities was computed to find the range of this value. If the eccentricity of the body ($\mathbf{R}$) is $\epsilon < 0.6$, the object is accepted as being an eligible candidate for the sella turcica concavity.

However, there is a problem because a part of the floor of the anterior cranial fossa is also being segmented. Thus when the horizontal concavity is calculated, a "lip" is also formed which is not part of the sella turcica concavity. This lip needs to be eliminated. This is achieved by shrinking the concavity eight times. The resultant concavity is shown in Figure 7.31. The number of shrink operations is sufficient because the lip is thin but could be long. Then a "bounding rectangle" is formed so that it covers the ∪-shaped ridge by expanding the shrunk concavity. The number of iterations of the expand operation chosen is 28: eight iterations is required to counteract the shrink operations so that the old concavity minus the lip is recovered and an additional twenty expand operations are applied to cover the ridge. A new ridge image, containing only the sella turcica ridge is obtained by logically ANDing this image with $\mathbf{I}_{ridge}$ image. There are two major sources of error in computing the centroid from the concavity of the sella turcica. One is because of the "lip". Although we have attempted to correct for this, there always some lip apparent. This biases our centroid to the right. The other error is due to computing the concavity using an eight-connected array. Because the left side of the sella turcica is usually higher than its right side, the tops of the ∪-shape of the sella turcica is not normally joined with a straight line but rather by a *concave* one. This biases the centroid position towards bottom-left. This latter problem can be overcome by using the parallel convex hull algorithm presented in Chapter 5.

Choose the ∪-Shaped Ridge of the Sella Turcica

We are at a stage in our segmentation of the sella turcica where the ridges of certain width have been extracted using a high-pass filter and subsequently

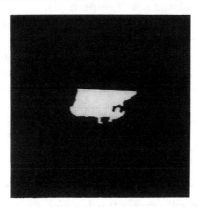

Figure 7.31: The horizontal concavity of the sella turcica.

thresholded (Algorithm 5.1). However, because all directions were enabled in the high-pass filtering of the sella turcica, the ridges extracted contain not only the ∪-shaped sella turcica but also many horizontal and vertical ridges. To choose the correct ∪-shaped ridge and reduce the serial search that might be required later to find the sella object amongst others, straight line segments will be removed.

Our task is to remove all ridge segments which are only horizontal or only vertical and do not have a ∪-shaped component. By a horizontal ridge we mean a ridge with a $|gradient| \leq 1$; a vertical ridge has a $|gradient| > 1$. The proposed method for removing these unwanted ridges is by applying a binary spatial low-pass filter, using shrink followed by expand operations; Serra refers to this as "opening" [Serra 82]. The change we propose is that the sequence of expand operations be replaced by a single global propagation operation, which we will refer to as *LABEL*.

Before describing the *LABEL* operation, we show how to obtain a speed advantage over the use of the expand operation. If a global propagation operation takes t_{gp} and an expand operation t_{nh}, then global propagation is faster when

$$t_{gp} < n \cdot t_{nh}$$

where n is the number of iterations. Now $t_{gp} = t_{pw} + N \cdot t_{pg}$, where $t_{pw} = t_{nh}$ (a point-wise operation takes same execution time as a neighbourhood operation), N is the array size and t_{pg} is the time for the signal to propagate through a processor. Then,

$$t_{nh} + N \cdot t_{pg} < n \cdot t_{nh}$$

50

$$n > N \cdot \frac{t_{pg}}{t_{nh}} + 1$$

Then substituting the appropriate values for CLIP4 [Otto 84] and CLIP4S (see Table 2.3.1), gives $n > 5$ and $n > 3$ respectively.

The *LABEL* operation uses a *seed* (the shrunk object in this case) to mark the pixels within an *object* (the ridge image before being shrunk) which are topologically connected to the seed. The object is defined by the connectivity we assign to it. In essence this is what the expand operation does in a low-pass filter. The *LABEL* function is algebraically defied as

---

$$\mathbf{object}(propagation\_in \cup \mathbf{label\_obj}) \ \Rightarrow \ propagation\_out$$

$$\mathbf{object} \cap propagation\_in \ \Rightarrow \ \mathbf{Labeled\_Object};$$

As a shorthand we will express this function as

$$\mathbf{Labeled\_Object} \ \longleftarrow \ LABEL(\mathbf{object}, \mathbf{seed});$$

We begin by low-pass filtering the ridge image in the vertical direction, i.e., we enable directions 2 and 6. This removes any ridge with vertical width less than $2n$ where n is the number of iterations. Similarly, we carry out a low-pass filter in the horizontal direction on the previous result. This removes ridges with horizontal width less than $2n$. The result is that the remaining ridges must be ∪-shaped or at least ⊥-shaped. From our observation of the cross-sectional width of the sella turcica ridges, we have chosen $n = 12$; this is a minimum value which has been found to remove most non-∪-shaped ridges. Any non-∪-shaped ridge not removed will be removed in the next phase of the processing.

Our next task is to look for upright ∪-shaped ridges. A simple method for looking at such objects is to observe if they are concave to propagation signal from the edge of the array in a horizontal sense. We can compute the horizontal cavity for each object serially, or attempt to compute all the cavities in parallel using a single global propagation operation. We choose to do the latter. However, to prevent concavities being found between non-connected objects, global propagations need to be restricted within the bounding rectangles, each containing a single object. It is not important how we compute the bounding rectangles, however, a method for it is presented in Chapter 5 where we discuss computation of convex hulls. The function for computing the horizontal concavities is given in step (5) of Algorithm 7.2. The propagation signals are gated to only the 4 and 8 directions. The propagation occurs through the background within the bounding rectangles. If any region within the **B_Rect** does not receive a propagation then it is a concavity in the gated 48 direction. However, all the regions outside the **B_Rect** also have not received any propagations and consequently have been set to one. The true concavities are restricted to within the **B_Rect** and therefore a masking operation is required to choose those **Concavities** which lie within the **B_Rect** (step 6).

Algorithm 7.2: Detecting an ∪-shaped ridge.

(1) tmp ⟵ S_{026}^n(**All_Ridge**);
(2) **All_Ridge** ⟵ *LABEL*(**All_Ridge**, *tmp*);
(3) tmp ⟵ S_{048}^n(**All_Ridge**);
(4) **All_Ridge** ⟵ *LABEL*(**All_Ridge**, tmp);
(5) $\begin{cases} \overline{\textbf{All\_Ridge}} \cap (\textbf{B\_Rect} \cap prop\_in) & \Rightarrow prop\_out \\ \overline{prop\_out} & \Rightarrow \textbf{Concavities} \end{cases}$
(6) **Concavities** ⟵ **Concavities** ∩ **B_Rect**;

Finding the Maximum and Minimum values within Objects

One of the method for finding the biggest label within objects is to conduct bit-plane searches and discard points which are not maxima. The algorithm given below is due to G. P. Otto and details of which can be found in [Otto 82b, Otto 84].

To illustrate this algorithm consider the following example. To find the maximum of the following five 4-bit numbers

$$1100 \quad 1010 \quad 0101 \quad 1101 \quad 1000$$

Compare, first, their most significant bit (m.s.b)

$$1 \quad 1 \quad 0 \quad 1 \quad 1$$

Since some of the numbers have a one for their m.s.b, the number with the m.s.b of 0 cannot be a maxima and therefore can be discarded. Then the next m.s.b of the eligible maxima candidates are

$$1 \quad 0 \quad 1 \quad 0$$

Again the numbers with the next m.s.b of zero can be discarded for the same reason. This leaves two numbers

$$1100 \quad 1101$$

The next m.s.b are

$$0 \quad 0$$

Therefore none of them can be discarded. Then the final bits are

$$0 \quad 1$$

Discard the zero, leaves with the maxima which is

$$1101$$

On CLIP4 and CLIP4S, these operations can be done in parallel to find the maxima with in each object.

Algorithm 7.3: Determine the maximum in objects.

for $i = p$ **down** **to** 1 **do**
      marker $\Rightarrow$ **carry**;
      $\neg$**indexp**$_i$ $\cap$ **marker** $\Rightarrow$ **temp**;
      objects $\cap$ (**temp** $\cup$ *propagation_in*) $\Rightarrow$ *propagation_out*
      objects $\cap$ $\neg$*propagation_in* $\Rightarrow$ **n_indexp**$_i$;
      marker $\cap$ $\neg$(**n_indexp**$_i$ $\cup$ *carry*) $\Rightarrow$ **marker**;
      carry $\leftarrow$ 0;
od

Algorithm 7.4: Determine the minimum in objects.

for $i = p$ **down** **to** 1 **do**
      marker $\Rightarrow$ **carry**
      $\neg$**indexp**$_i$ $\cap$ **marker** $\Rightarrow$ **temp** ;
      objects $\cap$ (**temp** $\cup$ *propagation_in*) $\Rightarrow$ *propagation_out*
      objects $\cap$ $\neg$(**object** $\cap$ *propagation_in*) $\Rightarrow$ **n_indexp**$_i$;
      marker $\cap$ $\neg$(**n_indexp**$_i$ $\cup$ *carry*) $\Rightarrow$ **marker**
      carry $\leftarrow$ 0 ;
od

Given an image **indexp** and an image **object** within which the maximum values are to be found, initially an image **marker** need to be set equal to objects since all points in **object** are possible candidates. However, **marker** will be modified as false candidates are discarded. The algorithm (MAXINOBJ) requiring 2 point-wise and 1 global propagation operations per bit-plane is given below.

A complement to this algorithm is the one which finds the minimum label (MINIMUM) within objects, the algorithm for which is given in Algorithm 7.3.

Choose the Correct Cavity

In Algorithm 7.2, we found a horizontal cavity which we expect to be the sella turcica cavity, but this needs confirmation. However, there is a problem because of a part of the floor of the anterior cranial fossa often being segmented. Thus when the horizontal concavity is calculated, a "lip" is also formed which is not part of the sella turcica concavity. This lip needs to be eliminated. Furthermore, by restricting the propagation signal to the 48 direction, only a part of the total concavity is determined. Therefore, not only will the lip have to be removed but the concavity recalculated.

First, we will discuss the removal of the lip. This is achieved by shrinking the concavity eight times. This has been found to be sufficient since the lip is thin but could be long. Then a "bounding rectangle" is formed by expanding the shrunk concavity so that it covers the ∪-shaped ridge. The number of iterations of the expand operation chosen is 32; eight iterations are required to counteract the shrink operations so that the old concavity minus the lip is recovered and an additional 24 expand operations are applied to cover the ridge. Then we use this bounding rectangle to "clip" the **All_Ridge** image (refer back to Algorithm 7.2) to lie within the bounding rectangle.

Second, the concavity is recalculated. The peaks of the ∪-shaped ridge need to be found and the concavity formed by these peaks will then be determined. The peaks are determined using the MAXINOBJ algorithm. Basically, the subroutine requires three inputs, the image **Filtered_Image**, the object **All_Ridge** within which the maxima are to be calculated and the connectivity, dir, of **All_Ridge**. For example, given only vertical connectivity, a maximum per column within **All_Ridge** of the image **Filtered_Image** is found. The image **Filtered_Image** is the beta-spline filtered version of the original sella image (**Original_Image**).

A point in the ∪-shaped ridge is accepted as being on the arête (peak of the ridge) if, and only if, it is a peak under at least two connectivity conditions; this is to prevent choosing some noise points. Unfortunately, the image of the arête points is not a single continuously connected body since not all peak points are maxima under at least two connectivity conditions. Therefore a connected body is formed by expanding the image **line** until all points join to form a body and thinning [Arcelli 75]. Now the concavity [Reynolds 83] can be calculated. The complete algorithm to remove the lip and recalculate the concavity is detailed in Algorithm 7.5.

Algorithm 7.5: Detecting the maxima within the sella turcica ridge.

All_Ridge $\longleftarrow$ **All_Ridge** $\cap$ **bounding_rectangle**;
line $\longleftarrow$ 0;
for $dir = \{E - W, N - S, NW - SE, NE - SW\}$ **do**
    peak $\longleftarrow$ $MAXINOBJ$(**Filtered_Image, All_Ridge,** dir);
    line $\longleftarrow$ **line** + **peak**;
od
line $\longleftarrow$ $THRESHOLD$(**line**, 2);
concavity $\longleftarrow$ 0;
for $dir = \{E - W, N - S, NW - SE, NE - SW\}$ **do**
$$\left\{ \begin{array}{ll} \textbf{line} \cap \overline{propagation\_in_{dir}} & \Rightarrow propagation\_out \\ \overline{propagation\_in_{dir}} & \Rightarrow \textbf{tmp} \end{array} \right.$$
    concavity $\longleftarrow$ **concavity** $\cup$ **tmp**;
od

Note that the MAXINOBJ routine is also used in calculating the arête points within the palate ridge. In this case, only the north and south directions are enabled.

Calculating the Centroid of a body

Once a 'solid' body, such as the concavity of the sella turcica has been extracted using Algorithm 7.5, its centre of mass, the centroid, can be found using the following equation:

$$\mathbf{R} = \frac{\sum m_i \cdot r_i}{\sum m_i}$$

where m_i is the mass of a unit element at a distance r_i from the origin and $\mathbf{R}$ is the vector position of the centre of mass of the object.

For a digitised image, the mass of an element m_i can be taken to be its grey-value and the element is at a distance r_i from the origin of the array (bottom left corner). For a binary image, m_i are labeled 1 within the object, and 0 in the background. Thus summing over all n pixels in the object gives the area of the object.

This is implemented in parallel in CLIP using Algorithm 7.6.

Algorithm 7.6: Calculating the centroid.

$\sum m_i \equiv AREA \longleftarrow VOLUME(\text{image});$

$\sum m_i \cdot x_i / \sum m_i \equiv \underline{x} \longleftarrow VOLUME(\mathbf{R_x} \cdot \text{image}) \, / \, AREA;$

$\sum m_i \cdot y_i / \sum m_i \equiv \underline{y} \longleftarrow VOLUME(\mathbf{R_y} \cdot \text{image}) \, / \, AREA;$

$\mathbf{R_x}$ and $\mathbf{R_y}$ are ramp images in x and y directions. The ramps are generated using global propagation operations [Otto 84]. Since we wish to determine the centroid of a binary object, the multiplication can be replaced by a point-wise boolean AND operation, ANDing the object with each bit-plane of the ramp, this simplifies the arithmetic in CLIP. The VOLUME function sums the grey-value within an image.

7.7.6 Segmenting the Menton

The menton is defined to be the lowest point of the symphyseal shadow, and the gnathion is defined to be the point on the chin determined by bisecting the angle formed by the facial and the mandibular plane [Krogman 57]. From these two definitions and their locations, in some instances the terms gnathion and menton have been used synonymously but there are obvious differences between the two (Figure 7.20). The **mandibular plane** is represented by a line connecting the two most dependent points of the mandible or a line from menton to gonion and the **facial plane** represented by a line drawn from the orbitale (eye socket) at right angles to the Frankfort plane. However, determinations of these planes are very subjective and thus have both large systematic errors and random errors associated with their measurement. At each step of their determination there would be a propagation of these errors leading to a large final error on determination of any of these landmarks. For this reason, it was decided not to choose to determine either the menton or the gnathion but instead, the bottom point of the mandible (lying between the menton and the gnathion).

From our knowledge of the cephalometric radiograph and the 'windowing' operation, we know that the bottom-most object in the image is the chin and from our definition, the bottom-most point on the chin is the menton. Then to locate the menton point we need to extract the edges in the image, choose the bottom-most contour in the image field and determine the bottom-most point on this boundary.

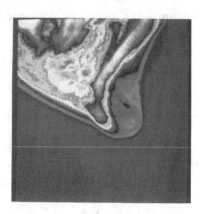

Figure 7.32: (a) An 8-bit image of the menton.

Algorithm 7.7 Segmenting the Menton

1. *Input menton image, $\mathbf{I}$, Figure 7.32.*

2. *Filter the image $\mathbf{I}$ for salt-and-pepper noise and enhance the edges in the images using sequences of rank filters and store result in $\mathbf{I}_{fil}$, Figure 7.33.*

3. *Extract edges in the image $\mathbf{I}_{fil}$ and store result in $\mathbf{I}_{edge}$.*

4. *Threshold edge image.*

    (a) Some penumbra edges are also extracted in the process. They generally have a lower contrast than the proper chin edge. This is a compromise between removing the weak penumbra edges and possibly fragmenting the chin edge by choosing too high a threshold value. This implies that some penumbra edges may be left in the image. The edges, furthermore, may not be single pixel wide. Therefore, the edges are skeletonised using Arcelli mask operations [Arcelli 75], Figure 7.34. However, since the penumbra are generally considerably weaker than the chin edges, the threshold leaves the penumbra edges as small strands or "hairs". These small strands or hairs can be removed by "chewing" [Reynolds 83]. The chewing process consists of applying near-neighbourhood operations to the binary skeletonised edge image to remove the "free-ends". This leaves objects such as in Figure 7.35, the lower border of the mandible.

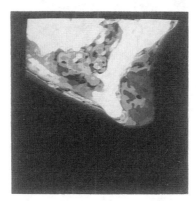

Figure 7.33: The menton image edges enhanced using a sequence of extremum and median filters.

Figure 7.34: The thresholded and skeletonised edges in the mandibular image.

Figure 7.35: The lower edge of the mandible, the chin.

5. *Extract the bottom point of the bottom-most object in image $\mathbf{I}_{edge}$, the lower border of the mandible, and store result in $\mathbf{I}_{chin}$.*

6. *Note that there may not be a single lowest point on the mandibular border extracted. Futhermore, the finite width, at the given resolution, of the menton point may not be a simple straight line but may have concavities. In such an image we will find more than a single bottom point or if we insist on a single point we may bias our points. To prevent such problems, we need to fill in the concavities only in the horizontal sense, Figure 7.36, and then determine the menton by computing the central pixel in the bottom-most line, §7.7.6. Figure 7.37 shows that the menton point has been found.*

Extracting the Bottom Point

For the segmentation of the menton, we wish to extract the bottom point of the chin; once the image has been correctly registered by rotating with respect to the palate line (this is a task for future research). Only two global propagation operations are required to extract the bottom point [Otto 84].

$$\begin{cases} \textbf{Chin\_Line} \cup propagation\_in_{456} & \Rightarrow propagation\_out \\ \textbf{Chin\_Line} \cap \overline{propagation\_in_{456}} & \Rightarrow \textbf{bottom\_point;} \end{cases} \quad (7.6)$$

$$\begin{cases} \textbf{bottom point} \cap propagation\_in_{678} & \Rightarrow propagation\_out \\ \textbf{bottom\_point} \cap \overline{propagation\_in_{678}} & \Rightarrow \textbf{bottom\_point;} \end{cases} \quad (7.7)$$

The image containing the thresholded edge image of the chin is **Chin_Line**, and the result, the bottom-right point is stored in **bottom_point**.

Figure 7.36: The horizontal concavity in the **Chin_Line** image.

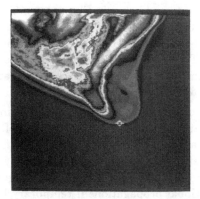

Figure 7.37: The image shows the menton point found by the algorithm.

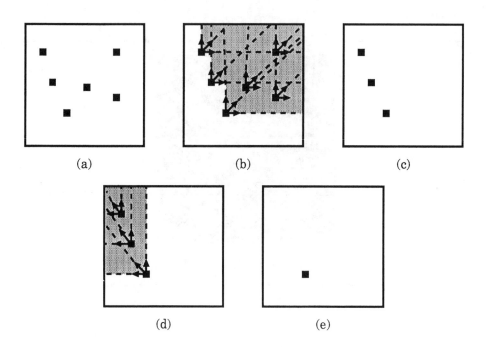

Figure 7.38: The operations required to determine the bottom-most right-most point in the array.

This operation is illustrated below in Figure 7.38. This algorithm will fail if the initial image is all 0. If the image is non-zero, then the algorithm will find the bottom-most right-most point (bottom-most left-most if the instructions are swapped round). The first operation will pick out at least one point from the bottom left (Figure 7.38(b)) and the second operation will pick out the point which is the bottom-most right-most point (Figure 7.38(d)).

By choosing another pair of direction lists, other important features can be extracted. For example, if the direction lists enabled are [234] and [128] respectively, then the above algorithm determines the first point in the array. This is the method that has been proposed in Chapter 5 to determine the four extreme points of the convex hull using a total of eight global propagations (one only needs four global propagations if temporary images are generated). For the other two extreme points, the directions of the propagations required are [128],[678] and [234],[456] respectively.

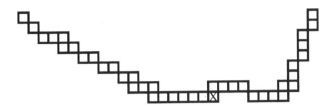

Figure 7.39: A schematic diagram of a possible bottom row of the chin in a segmented menton image.

However, extracting the bottom-right point of the chin does not determine the menton. The bottom of the chin may not consist of a single point but rather a row of points. The middle point of the row is defined to be the menton. However, a further problem may arise. The bottom row may not be continuous as shown in a schematic diagram (Figure 7.39).

There are several solutions to this problem. A solution is given below. The method in essence is the concavity determination problem already encountered in segmenting the sella turcica. However, here it is required not only to find the concavity of the chin, the regions which do not receive propagations from the horizontal directions, but to add the concavity region to the chin line that already exists. This can be achieved by a simple modification to the concavity algorithm. First, using the **bottom_point**, a horizontal line is generated (step 1 of **Algorithm 6.6**), this is used later to mask the bottom row of the chin. Then using the **bottom_point** as a seed, we label the chin. Thus we now have only the chin outline and no other structure in the image. Next, the horizontal concavities are computed. The bottom of the image, the row containing the menton, can now be extracted by ANDing the images **Filled_Chin_Line** and **bottom_line**.

The middle point of the **menton_line** which is the menton is obtained by chewing the object by fitting the following masks.

0	0	0
1	1	0
0	0	0

mask 1

0	0	0
0	1	1
0	0	0

mask 2

Where the masks fit the objects, the central pixel is deleted. This process of applying the two masks iteratively is continued until the image stabilises, i.e. until there is no change on application of the masks.

If **menton_line** consists of an odd number of pixels, then the central pixel is chosen. However, if even, then the left of the central two pixels is chosen.

Algorithm 7.8: Extracting the bottom point of the menton.

(1) $\begin{cases} \textbf{bottom\_point} \cup propagation\_in_{48} & \Rightarrow propagation\_out \\ \textbf{bottom\_point} \cap propagation\_in_{48} & \Rightarrow \textbf{bottom\_line}; \end{cases}$

(2) $\begin{cases} \textbf{Chin\_Line} \cap (propagation\_in \cup \textbf{bottom\_point}) & \Rightarrow propagation\_out \\ \textbf{Chin\_line} \cap propagation\_in & \Rightarrow \textbf{Chin\_Line}; \end{cases}$

(3) $\begin{cases} \overline{\textbf{Chin\_Line}} \cap propagation\_in_{48} & \Rightarrow propagation\_out \\ \textbf{Chin\_Line} \cup \overline{propagation\_in_{48}} & \Rightarrow \textbf{Filled\_Chin\_Line}; \end{cases}$

(4) $\textbf{menton\_line} \longleftarrow \textbf{Filled\_Chin\_Line} \cap \textbf{bottom\_line} \ ;$

7.7.7 Segmenting the Articulare

The articulare is the point of intersection of two bones (Figures 7.19 and 7.20), the jaw bone and the external cranial base [Krogman 57]. This shape can be approximated by an inverted V-shape. Our segmentation task is to search for concavities formed by the inverted V-shape and then extract the apex of the concavity. The edges are very sharp and therefore they are easy to extract in the image.

Briefly, the segmentation method employed has been to preprocess the image of the articulare field to extract sharp edges. Having extracted the edges, a model is used to find the articulare concavity. The model used is approximating the articulare concavity to an equilateral triangle of a certain size. If the model fits a concavity, then the feature is recognised as being the articulare concavity and the articulare is chosen by calculating the apex of the concavity.

The major problem encountered in extracting the articulare feature has been the poor quality of the radiographs. One of the problems is because of the plastic bars with ear-posts, which are needed to keep the head steady and at a certain orientation during the exposure of the radiograph, casting a shadow on the radiograph. Since the articulare feature lies behind the ear, the ear-post shadow covers the articulare feature and reduces its contrast. However, the main problem arise when the ear-post shadow only covers half the articulare feature. In such circumstances, the algorithm determines two concavities since the edge of the post shadow is segmented and cuts through the articulare feature. The other problem occurs with some radiographs where the patient is wearing earrings. The earrings, being often of metal, cast shadows which are much stronger than those cast by bones and again obscure part of the articulare.

Algorithm 7.9 Segmenting the Articulare Image

1. *Input an articulare image,* **I,** *which is an average of eight images of the same field. A typical articulare input image is shown in Figure 7.40.*

2. *Filter the image using a median filter and edges enhanced using a sequence of extremum and median filters, and store result in* **I**$_{fil}$. *The result of filtering and edge enhancement of the articulare image is shown in Figure 7.41.*

3. *Extract the edges in image* **I**$_{fil}$ *and threshold, Algorithm 5.1; store result in* **I**$_{edge}$. *The result of the edge operation is shown in Figure 7.42(a).*

   (a) *The consequence of sharpening the edges in the articulare image before extracting them is that the articulare cavity edges constitute discontinuities. Furthermore, from the profile image of the sharpened edges, Figure 7.41(b), the gradient of the articulare edges is known to be over 35 grey-values per pixel. Therefore, any pixel in the* **I**$_{edge}$ *image with values greater or equal to 35 is a candidate for*

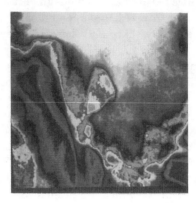

Figure 7.40: A typical articulare input image.

Figure 7.41: The result of filtering the articulare image using extremum and median filters. (a) The result of the filter and (b) profile across the articulare concavity showing that the edges have been enhanced.

 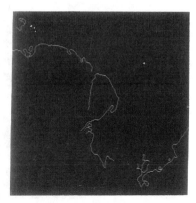

Figure 7.42: The edge image of the articulare. (a) The edge gradients greater than 35 and (b) strands longer than 200 pixels.

an articulare concavity edge point. Also a consequence of sharpening the edges before extracting them is that the thresholded edge image has edge contours which are single pixel wide. A priori knowledge which is available is that the contour of the articulare concavity is at least 200 pixels long. Thus any strands less than 200 pixels long are removed from the thresholded edge image, leaving only those strands which are still possible candidates for the articulare concavity, Figure 7.42.

4. *Compute the concavity for each remaining object in image I_{edge}. The concavity chosen by the algorithm, Algorithm 7.10, fitting the model is shown in Figure 7.43.*

   (a) *Of course, the choosing can be done at an earlier stage. If the test for the concavity size proves successful, then a record of the (A/C^2) value can be kept, where A is the area of the cavity and C is its perimeter length. If at a later stage, another concavity passes the area test and its (A/C^2) value is greater, then this new concavity is now the articulare candidate. This process is continued for all possible candidates.*

5. *Having found the articulare concavity, the apex – the point of intersection of the jaw bone with the external cranial base – is calculated in the same fashion as for determining the menton point after the chin edge had been determined, Algorithm 7.7.6. The articulare found is shown in Figure 7.44.*

Figure 7.43: The articulare concavity.

Computing and Fitting a Model to the Articulare Concavity

We have chosen to segment the articulare by searching for inverted ∨-shaped concavities and choosing the apex of the concavity to be the articulare. A model matching scheme is used to confirm the segmentation of the articulare concavity. The model we have chosen is to approximate the concavity to a triangle. This is justified on the basis of our observation of the concavity shape in a large number of radiographs. A shape measure often stated [Kitchen 83a] is the area (A) over square of the parameter length. For an equilateral triangle, this is

$$\frac{A}{C^2} = \frac{\sqrt{3}}{12}$$

An algorithm based on the model described above is detailed in Algorithm 7.10. It is to serially search the **edge_image** and generate concavities only in the horizontal direction and then choose the concavity best conforming to the articulare concavity model devised.

Having found the articulare concavity, the apex, the point of intersection of the jaw bone with the external cranial base, is calculated in the same fashion as for the determining the menton point after the chin edge had been determined.

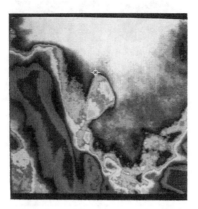

Figure 7.44: The articulare.

Algorithm 7.10: Computing and fitting a model to the articulare concavity.

cavity ⟵ 0;
<u>while</u> (edge_image NOT EMPTY) <u>do</u>
      tmp ⟵ $FIRST\_OBJECT$(edge_image);
      tmp_cavity ⟵ $HORIZONTAL\_CAVITY$(tmp);
      C ⟵ $\sum EDGE$(tmp_cavity);
      A ⟵ $\sum$ tmp_cavity;
      $test$ ⟵ $(A > 4000) \wedge \left(\left(\frac{A}{C^2} \right) \approx \frac{\sqrt{3}}{12} \right)$;
      <u>if</u> ($test$) <u>then</u>
                cavity ⟵ cavity $\cup$ tmp_cavity;
      <u>fi</u>
      edge_image ⟵ edge_image $\cap$ $\overline{\text{tmp}}$;
<u>od</u>
$test$ ⟵ NO_OF_OBJECTS(cavity) > 1;
<u>if</u> ($test$) <u>then</u>
          choose object with largest (A/C^2) value;
<u>fi</u>

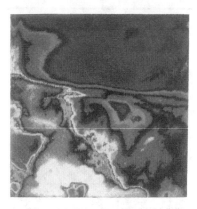

Figure 7.45: An input palate image.

7.7.8   Segmenting the Palate Line

The palate line (Maxillary plane) is defined as a line drawn through the anterior and the posterior nasal spines (Figure 7.20) [Krogman 57]. However, this definition has not been used here to segment the feature. The feature segmented is the hard palate and the palatal line is generated by fitting a straight line to the segmented palate data using a least-squares technique. The reason for this change in definition is because of the difficulty of segmenting the posterior nasal spine. The posterior nasal spine is often obscured by the molars and consequently is impossible to segment in many radiographs. However, most of the hard palate is visible and therefore more reliable in segmentation.

In the skull, the palate line is a plane of bone. When looking at a lateral projection of a skull on a radiograph, this plane of bone appears as a white ridge because of the radio-opaqueness of the bone. Above the palate is the nasal cavity and below is the mouth cavity. These are filled either with air or muscles; both air and muscles are radio-translucent and therefore appear as black on the radiograph which is a negative print. The palate then topographically appears as a ridge. Therefore, to extract the palate, we need to extract all (approximately) horizontal ridges. Then we need to check the length of the segmented ridges to choose the correct ridge. Once we have the palate ridge, we need to compute a best-fit line, by a least-squares mean, to the arête points.

Algorithm 7.11 Segmenting the Palate Line

1. *Input palate image, Figure 7.45(a).*

2. *Spatial filter image for salt-and-pepper noise with a 5 × 5 median filter.*

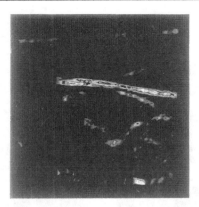

Figure 7.46: High-pass filtered palate image.

3. *Extract the horizontal ridges by applying a directionally high-pass filter.*

   (a) Palate ridges run basically from west to east in a slight incline across the array. Thus a directionally sensitive high-pass filter can be used to extract those ridges that are horizontal or at small inclines to the horizontal and reject those which are vertical. The result of the high-pass filter is shown in Figure 7.46.

4. *Threshold the ridge image.*

   (a) From the profile image of the palate ridge, it can be seen that once the relevant ridges have been extracted, only the steep ridges are likely to be the palate, or at least the candidates for the palate are reduced. Thus at an early stage many of the ridges found can be eliminated from the search by choosing a suitable threshold. Furthermore, the threshold may be chosen to be quite high. However, the profile was chosen at a particularly advantageous position. At other positions, parts of the palate may be obscured by weak shadows cast by other features. To remove other segments while leaving the fragmented palate behind would require considerable intelligence and rules to be embedded into the segmentation process. Such an algorithm is presented § 7.7.8 collinear.

   (b) A considerably simpler method is to choose a much lower threshold value, for the directional high-pass filtered image. This then segments the palate such that it forms a continuous line. But this method also retains many noise features which requires removal.

5. *During the ridge-extraction process, many horizontal ridges beside the palate may have been found. The palate lines are typically about 350 pixels long therefore we wish to remove ridge segments much smaller than this. We remove ridge segments smaller than 70 pixels long, this operation*

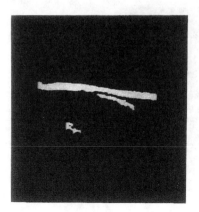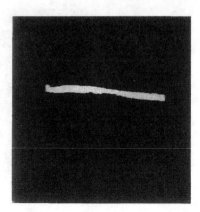

Figure 7.47: The palate ridge regions: (a) all the ridges and (b) the longest ridge is the palate ridge.

retains the palate line but removes structures which can be regarded as being noise. This operation is carried out by skeletonising the segments and chewing away the end points until strands smaller than 70 pixels are removed. The objects are relabeled using the "chewed" strands. Thus objects of length smaller than 70 pixels have been removed, but bigger objects are unaltered.

6. *The palate consists of the hard palate, the feature which is being segmented, and the soft palate; this is a structure which slopes downward from the hard palate (Figure 7.47). If the threshold value is chosen low, then in many cases the hard and the soft palates have been found to be joined and need to be separated. However, it is rather simple to separate these two objects since they have very different gradients.*

   (a) *The consequence of having different gradients is that their width (in the horizontal sense) is very different. The hard palate due to its gentle gradient has a very large horizontal width while the soft palate with its steep gradient has a width which is no greater than 20 pixels and often considerably less. The algorithm being applied is a binary version of a high-pass filter. By gating the local neighbourhood propagation signals to only the horizontal directions (i.e. directions 4 and 8), shrink (erosion) operations can be iterated 10 times to remove objects whose horizontal width is less than 21 pixels. However, this not only removes the soft palate but also modifies the hard palate. If one is not concerned about the loss of information at the extremities, then much of the ridge information can be retained by relabelling. The loss of information at the extremities is of little*

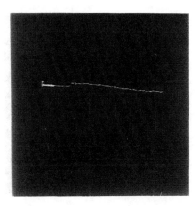

Figure 7.48: The maxima in the palate ridge.

consequence since the ridge objects have been segmented to allow the calculations of the ridge peaks which run centrally along the ridge.

7. *Now get the ridge peak, the arête, which runs (approximately) central to the ridge segments.*

   (a) *For the palate line gradient to be calculated, the arête of the ridge needs to be extracted. This is done using the MAXINOBJ routine [Otto 82b, Otto 84]. By restricting the connectivity of the ridge to only the north and south pixels (directions 2 and 6 respectively), the MAXINOBJ routine can be used to determine the maxima per column within the ridge image and thus the peaks in the ridge. The result of using the MAXINOBJ routine to determine the peaks within the ridge is shown in Figure 7.48.*

   (b) *This is not the only method to extract the arête points. A method employing local-neighbourhood operations is discussed in §7.7.8.*

8. *Finally fit a straight line to the maxima points in the ridge.*

   (a) *A least-squares technique by minimising moments is used to determine the gradient; the result of the least-squares fit is shown in Figure 7.49.*

Extract Collinear Objects from a Segmented Image

One of the problems which on occasion arises with the palate segmentation is that the palate gets fragmented if an incorrect threshold value is chosen. An algorithm is proposed which attempts to label the complete palate segment

Figure 7.49: The best-fit line for the palate ridge.

by looking for collinear objects in the segmented image, and if such segments are found, they are merged into a "single body". The rules require that the gaps between the fragmented segments must be small, say up to a maximum of 50 pixels; the fragments must of course be collinear, and the palate must be the biggest ridge object in the image, or have some other readily recognisable feature.

To prevent difficulties in determining the gradient, to extract the collinear objects from small segments, **object1** is chosen to be a subset of **object2**, the image containing all ridge segments. In **object1**, segments smaller than 50 pixels in length have been removed. Despite this, one may have to search through typically 6-8 candidates. Thus, the method can be computationally slow.

Having found the first object in the search image **object1**, its gradient is determined, and the gradient line (**line**) is used to label objects in the image **object2**. The objects which are labeled are co-linear. Next the length of the segments found (length1) and the gaps in between the segments (length2) are calculated. The total length of the collinear object is length1 + length2. If the total length is greater than 250 pixels, then the object found is a possible candidate for the palate. However a further check needs to be made; the number of segments making up the collinear object are counted. Even at the high threshold value chosen, the number of segments for the palate should not exceed 4. If the number of segments does exceed 4 then the object found can be rejected.

Get the arête points within a ridge

In this book, a method for extracting the maxima using global propagation operations to mark them within objects, i.e. the ridges, have been used extensively for the sella turcica and the palate features, Algorithm 7.5.

Algorithm 7.12: Extracting collinear objects.

bar ⟵ Set top row to 1;

<u>while</u> **object1**¬EMPTY <u>do</u>

 im ⟵ $FIRST\_OBJECT$(**object1**);

 line ⟵ $GRADIENT$(**im**);

 line ⟵ $GRADIENT$(**im**);

 im ⟵ $LABEL$(**object2**, **line**);

 tmp1 ⟵ $SPREAD$(**im**, *vertically*);

 $length1$ ⟵ $\sum$(**tmp1** ∩ **bar**);

 tmp2 ⟵ $CAVITY$(**tmp1**, *horizontal*);

 tmp2 ⟵ AND(**tmp2**, **bar**);

 $length2$ ⟵ $\sum$ **tmp2**;

 $no\_objects$ ⟵ $NO\_OBJECTS$(**tmp2**);

 object1 ⟵ **object1** ∩ $\overline{\text{im}}$;

<u>od</u>

There are other methods which do not use global propagation to detect maxima within objects but rather use serial search or local neighbourhood operations. Amongst the serial search methods are those proposed by [Levy-Mandel 86] and [Ashkar 78]. The Levy-Mandel's method has some similarities with a parallel method described below and therefore is considered in some detail. First, the edges in the image are found using the Mero-Vassy edge operator [Mero 75]; the line follower algorithm described below is used to extract the ridge line.

---

1. Define two threshold values
   - S_1 – upper threshold
   - S_2 – lower threshold

2. Scan the image in a raster fashion until a pixel (point A) is found with a value higher than S_1.

3. Check all the neighbours of A for the steepest direction and climb to the local maxima (point C).

4. Follow along the maximum contour until the pixel value falls below S_2.

5. Return to C and track in the opposite direction again until the pixel value falls below S_2.

6. Goto (2) and continue until the end of image.

The Levy-Mandel's method, however, does not find the ridge maxima but rather since it searches for maxima within an edge image, it will detect two lines per ridge.

The parallel method to extract maxima ridge points described below uses fuzzy logic.

$$\textbf{Edge} \leftarrow |\textbf{im} - S_{026}(\textbf{im})|; \tag{7.8}$$

$$\textbf{maximum} \leftarrow \textbf{Edge} - S^n_{026}(E^n_{026}(\textbf{Edge})); \tag{7.9}$$

$$\textbf{line} \leftarrow THRESHOLD(\textbf{maximum}, t); \tag{7.10}$$

$$\textbf{line} \leftarrow SKELETON(\textbf{line}); \tag{7.11}$$

In the example given, only the 026 directions are enabled. This is because, in the case of the palate ridge, only the horizontal edges are of importance. However any direction list could be enabled as the need arises. For example, if the direction list 048 was chosen, then only the vertical ridges will be found. Furthermore, all the ridges may be found by enabling propagations from all directions. The method is illustrated in Figure 7.50. Figure 7.50(a) shows an ideal cross-section of a ridge and its gradient. In Figure 7.50(b), the absolute value of the gradient is shown. The maxima in the ridge have zero gradient and therefore appear between the gradient maxima. The "zero-region" between the edge maxima can be found using a high-pass filter, Figure 7.51(a). The feature which would be detected by the high-pass filter is the shaded region in Figure 7.50(b).

By choosing a suitable threshold value only the strong ridges can be picked out, Figure 7.51(b). This is also achieved in Levy-Mandel's algorithm by suitably choosing S_1 and S_2.

The Levy-Mandel's method is most suitable for detecting step edges; for ridges two lines (one for each gradient) will be detected. The parallel method described above is suitable for detecting only the maxima within ridges. For detecting step edges, the second step needs modification as given below.

$$\textbf{maximum} \leftarrow \textbf{Edge} - E^n_{dir}(S^n_{dir}(\textbf{Edge})); \tag{7.12}$$

7.7.9   Segmenting the Mandibular Line

Several mandibular plane definitions have been used depending on the analysis. In the skull, the mandibular plane is an imaginary plane constructed between two halves of the lower border of the jaw bone, the mandible. On the radiograph, a strong shadow is cast by the mandible and the mandibular plane is a line along the bottom of the mandible shadow. The most common definitions of the mandibular plane used by clinicians are: a tangent to the

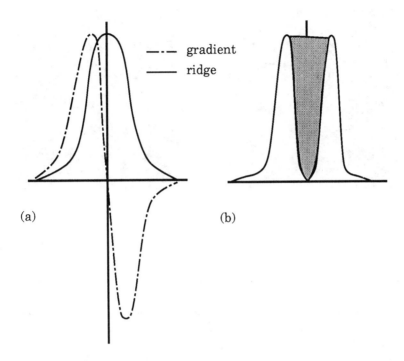

Figure 7.50: Extracting the maxima within the ridge: (a) the cross-section of the ridge and its gradient, (b) the shaded region represents the maxima in the ridge.

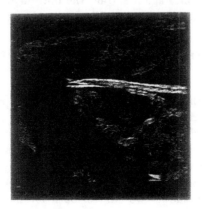

Figure 7.51: The edge image.

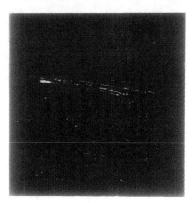

Figure 7.52: (a) The maxima within the ridge, (b) the maxima after thresholding.

lower border of the mandible; a line between the gonion and the gnathion (Figure 7.20); or a line between the gonion and menton. The definition which will be used here is the lower border of the mandible, between the gonion and menton. Having extracted the lower border of the mandible, the mandibular plane will be constructed by generating a line by a technique of least-squares fit to the segmented data. Since the lower border of the mandible is required to be segmented, there are at least two readily available options to segment the feature. The two options which will be considered are (i) threshold the image to separate the mandible from the background and then choose the bottom edge of the mandible, and (ii) sharpen the edges as for segmenting the menton and the articulare and once again extract the bottom edge. However, a better method found for segmenting the mandibular line is the Hough transform.

7.7.10   Segmentation by Thresholding the Mandibular Plane

Three thresholding techniques, (i) maximising entropy [Kapur 85a], (ii) relaxation [Ridler 78] and (iii) a local adaptive thresholding [Reeves 82] have been implemented and tested on a number of images containing the mandible to determine their suitability for segmenting the given images. A suitable threshold value given by the grey-level (s) for which the entropy $\psi(s)$ is a maximum. However, we notice from Figure 7.53, that the graph does not have an obvious maximum. The graph shows two curves, both for the same radiograph but the images have different amounts of background. As a consequence, it is observed that there is a difference of 20 levels in the threshold values found for

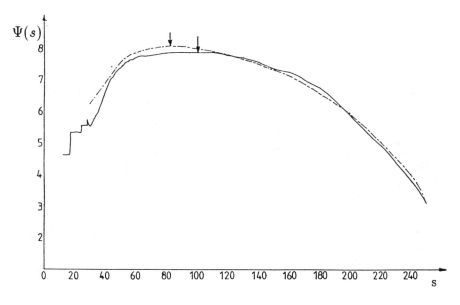

Figure 7.53: Graph of $\Psi(s)$ against s.

the two image. In Figure 7.54, a comparison of results using the entropy and the Ridler and Calvard relaxation method is given below. It is observed that the histograms do not have a binomial distribution and therefore there is no obvious position for grey-level partition. For the three examples observed here, the entropy method, despite the absence of a sharp peak, is more robust than the relaxation method and consistently chooses a lower threshold value than the relaxation method. The result of applying the local adaptive threshold operation on a mandible image is shown in Figure 7.55. A window size found to give good results is 17×17 pixels. Since the method is sensitive to noise, the noise in the image needs to be reduced by (i) averaging several images and (ii) applying a large window median filter (a 9×9 window has been found useful).

A conclusion to be drawn is that the relaxation method is not suitable for segmenting the mandible images. While the local thresholding technique does extract the border of the mandible, because of the local operations, structures within the background are also extracted. These "noise structures" will need to be removed. We will need to explore possible methods, such as increasing threshold level or texture analysis, to remove the noise regions. For these given images, the entropy method has proven suitable for extracting the lower border of the mandible.

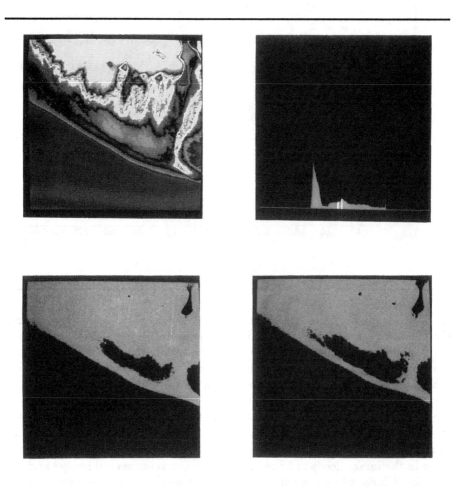

Figure 7.54: (i) Results of thresholding using the entropy and the iterative method: (a) the 8-bit mandible image, (b) the histogram of the image, (c) thresholding using the entropy method and (d) thresholding using Ridler's iterative method.

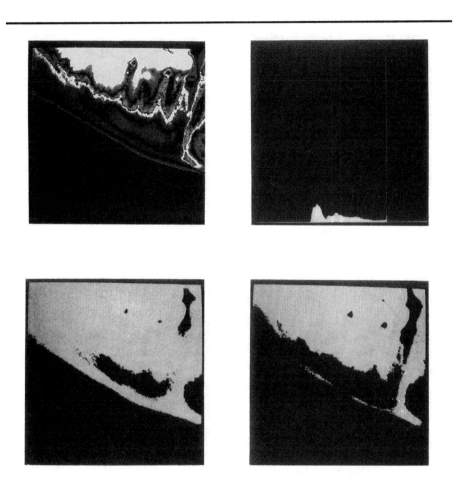

Figure 7.54: (ii) The images are as in Figure 7.54(i).

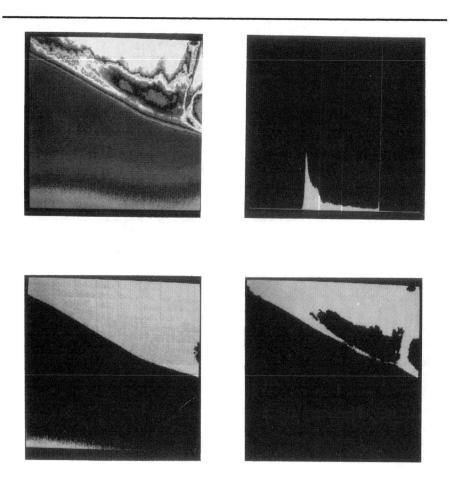

Figure 7.54: (iii) The images are as in Figure 7.54(i).

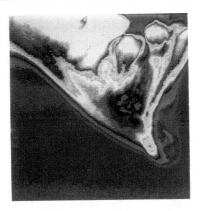 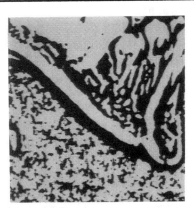

Figure 7.55: The result of thresholding a mandible image using the max-min scheme.

7.7.11 Segmenting the Floor of the Anterior Cranial Fossa

The floor of the anterior cranial fossa is the bone structure running upward from the sella turcica (Figures 7.19 and 7.20). A relief map of the FACF show that the structure is not symmetrical, therefore the arête points cannot simply be extracted by skeletonising the ridge. Any thresholding conducted to extract the ridge in the image (Figure 7.56) will result in an asymmetrical structure being extracted. Extraction of such a feature in different images will consequently be unreliable. Furthermore, choosing a threshold value to segment the image and separate the background from the foreground (i.e. the ridges) is very difficult. The histogram for the image is not bimodal; there is more than one class of objects in the image. This problem will be even more acute when a bigger field of a cephalometric radiograph is viewed on CLIP4S. Segmenting the FACF is the process of extracting one line-like feature amongst many other line-like structures from a 2D shadow cast by 3D objects in the skull in forming the radiograph.

There are at least two options: (i) to segment the maxima constituting the ridge by calculating the zero-crossings after differentiating the FACF ridge function, which is computationally difficult for real images because of integer arithmetic on CLIP4, or (ii) to extract strong edges, which is computationally simple since simple edge operations (such as Sobel or Prewitt edge operators [Rosenfeld 82]) and thresholding are sufficient; then choose the desired lines.

The cross-section of the FACF ridge has a triangular shape. By use of some simple non-linear rank filters, the given ridge function may be further approximated to a truncated triangle. It is then easier to segment the sharper edge on the top side of the ridge-like structure than search for the line constituting the FACF. Having managed to extract this edge, thresholding on this structure will be more stable, and therefore more reproducible. The method employed to segment this edge is described below using a directionally sensitive high-pass filter, followed by extremum and median filters (Figure 7.57). The ridge is now more prominent, and an edge can easily be extracted.

Algorithm 7.13 Segmenting the FACF

1. *Input an image,* im, *containing the FACF.*

2. *Apply a directional high-pass filter to the raw input image* im, *and store result in* $\mathbf{I}_{hp}$*. The result of the high-pass filter can be seen in Figure 7.58(b).*

3. *Enhance the edges in image* $\mathbf{I}_{hp}$ *by applying a sequence of median and extremum filter.*

4. *Extract edges in image.*

   (a) *The edge image is thresholded. The image is now filled and skeletonised to remove any closed loops in the image that may result from the extraction of the edge image.*

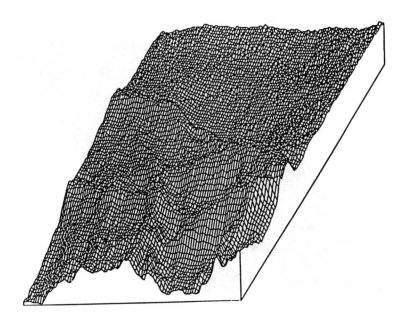

Figure 7.56: A topographical image of a FACF.

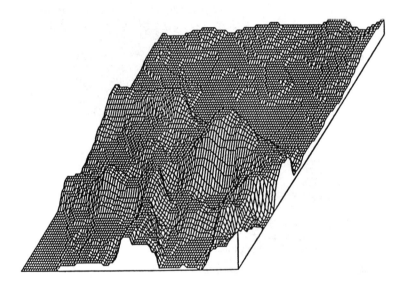

Figure 7.57: A topographical representation of a high-pass filtered FACF image.

 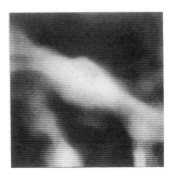

Figure 7.58: Typical images of the FACF: (a) a raw input image and (b) after high-pass filtering.

> *(b) At the chosen resolution, the FACF ridge is confidently expected from our knowledge of the image field to be long enough to occupy much of the array horizontally. Small structures, smaller than 20 pixels long, are removed by chewing and labeling [Otto 82b, Otto 84] with the thresholded image. The first object in the image is now confidently expected to be the floor of the anterior cranial fossa (Figure 7.59).*
>
> *5. Fit a straight line to the segmented FACF object using a linear least-squares fit technique (Figure 7.59(b)).*

A question we wish to ask: can the FACF be approximated to a straight line? A quadratic least-squares fit was applied to a number of FACF images to establish whether a linear least-squares fit was adequate to define the best-fit line.

If the line is part of an arc of a circle, the radius of curvature can be calculated. If a point on the arc lies at the point $O(x', y')$ at a radius r from the centre, then another point $P(x, y)$ on the arc is then given by

$$x = r \sin \theta \qquad y = r(1 - \cos \theta)$$

For small angles this can be approximated to

$$x = r\theta \qquad y = \frac{r\theta^2}{2} \approx \frac{x^2}{2r}$$

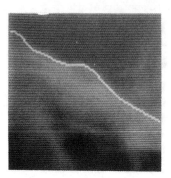 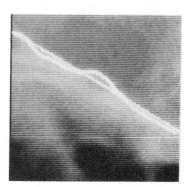

Figure 7.59: Segmented FACF images: (a) the line indicates the segmented edge and (b) the least-squares line is shown.

An indication of the radius of curvature is $y \approx ax^2$ therefore

$$r \approx \frac{1}{2a}$$

If the radius of the arc is greater than the size of the array, this is an indication of the line being linear. But before concluding that the linear least-squares fit is adequate to describe the line, it would be desirable to fit a higher order least-squares fit. Higher order terms leading towards zero would indicate a suitability for a first order fit.

7.7.12 Results of Cephalometric Analysis

In this section, a set of algorithms have been presented to segment three point-like features and three like-like features on a lateral skull radiograph.

For two of the features, the methods discussed are for locating concavities. For the sella, the search is for a U-shaped concavity, while for the articulare, the search is for an inverted V-shaped concavity. Having found the respective concavities, the sella point is determined by computing the centroid of the concavity and the articulare is the middle point of the first-row of its concavity.

The menton segmentation is different to the other two in that it involves modelling the edges. The edge shadows forming the radiograph have been cast by rounded surfaces. However, this is approximated to shadows cast by a sharp edged structure whose edges have been consequently blurred by low-pass filtering during the image formation stages. The consequence of this assumption is that an extremum filter can be applied to transform gently sloping edges to step-edges. The sharpened edges are then easier to extract. The bottom edge object in the image is the chin. The mid-point of the bottom-row of the chin object is the menton.

In segmenting the cephalometric landmarks significant use of *a priori* information has been made. For example, the orientation of the features is known along with their shapes and sizes.

A single gradient measurement is, of course, of no use since no knowledge of the orientation of the head at the time of the exposure of the radiograph is known and, furthermore, a new orientation is imposed when the radiograph is digitised. Therefore, angle measurements which are likely to be of consequence in growth measurements are the relative angles between three point-like features or two line-like features. Therefore, by developing the segmentation algorithms for the mandibular plane and the FACF, two relative angles can be measured; that between the palate and the mandible, and between the palate and the FACF.

The FACF segmentation is based on modelling the ridges, as it appears on the radiographs, by a combination of a polynomial and an exponential function. This function is further approximated to a "truncated"-triangle structure, with one steep slope and a gentle slope. The steep slopes are easily extracted from the image by applying an edge operation and choosing the strongest gradients. A straight line is then fitted to the segmented data by a linear least-squares line fitting technique.

For the mandibular plane, two possible methods of segmentation have been proposed: (i) using thresholds to separate the mandible from the background and (ii) using the edge-enhancement method of iterative application of extremum and median filters. The latter method was used for the segmentation for both the menton and the articulare features. A possible advantage of using this method is that the mandible and the menton segmentation may be integrated and thus save computational time. A problem which may arise for some radiographs is that the total feature may be too large to fit within the image field completely.

Using entropy to choose a suitable threshold value has shown promise, and the method is reasonably fast. There are improvements to this method which can be made: one such method is a search program to find the most suitable value for s by searching $\Psi(s)$ and exploring if the method is improved by incorporating the edge enhancement method.

Furthermore, results have been found to be good for segmenting the mandible line using the Hough transform (see Chapter 5). In any future implementation other than on CLIP4 or CLIP4S, this method should be considered before the other methods discussed in this section.

Reproducibility of Segmentation Results

Having developed algorithms for segmenting several cephalometric features, the accuracy of the algorithms needs to be tested. In this section, reproducibility of three of the features, the sella, the menton and the palate line will be discussed. Once the radiograph has been "locked" into the film holder, the film-holder can

Table 7.1: Comparing the reproducibility in sella location using CLIP4S and manual tracings.

	CLIP4S	Manual
Δ_x(mm)	0.27 ± 0.09	0.44 ± 0.14
Δ_y(mm)	0.31 ± 0.12	0.46 ± 0.14

be repositioned on the stage and therefore the experiment can be repeated. Any change in the position of the segmented feature is because of (i) the effect of noise on the algorithm and (ii) small movement in repositioning the film-holder on the stage in repeating the experiment.

The reproducibility test has been carried out using nine separate radiographs. For each radiograph, the algorithm was run five times and the results recorded. The mean and the standard deviation value of the x- and y-coordinates were calculated for the point-line feature's segmentation or the gradient and the intercept with the left-hand edge of the array for the line-like feature. To measure the reproducibility of the segmentation, one needs to note the standard deviation on the measurements.

One of the objectives of automating the segmentation process of cephalometric landmarking is to reduce the subjectivity in measurements. By eliminating subjective error it should be expected that the method proposed for segmenting the sella should be better or at least comparable in reproducing results with the manual methods of tracing currently used. The reliability of measurement of three cephalometric features with CLIP4S and a manual tracing method [Baumrind 71] are compared in Tables 7.1 and 7.2 for the sella and the menton respectively. The figures given are the mean of the standard deviation for repeated measurements. The error on the mean standard deviation is the standard error.

Sella Turcica

From the data in Table 7.1, it is observed that the segmentation algorithm proposed for the sella and the use of the stage provide means for increasing the reliability in measurement of the sella beyond that possible by manual tracing methods. Errors of ± 3.0 and ± 3.4 pixels are observed in the x- and y-axis respectively. There is approximately 36% improvement in reliability in locating the sella using CLIP4S over that using manual tracings (comparing with results obtained by Baumrind [Baumrind 71]. Thus our objective of improving the reproducibility in measurements of the sella feature by minimising human subjectivity has been met. Is this about the best which might be expected using CLIP4S and current resolution of the images? Much of the error in segmentation is because of the concavity forming portion of the algorithm.

Table 7.2: Comparing the reproducibility of the menton using CLIP4S and manual tracings.

	CLIP4S	Manual
Δ_x(mm)	0.30 ± 0.09	1.26 ± 0.36
Δ_y(mm)	0.05 ± 0.02	0.50 ± 0.36

Depending on the amount of the "lip" segmented and the variations in orientation, varying by a small amount, when the radiographs are repositioned, the centroid of the concavity varies. The cut-off at the top determines the error on the y-coordinates and the lip determines the error on the x-coordinate.

Menton

A comparison of the reproducibility in segmentation of the menton using CLIP4S against using manual tracings [Baumrind 71] is given in Table 7.2. The figures quoted are the mean of the standard deviation obtained in repeated measurements. Errors of ± 3.3 and ± 0.5 pixels are observed for the x- and y-coordinates respectively. This is an 80% improvement in locating the menton using CLIP4S compared with using tracings. The result of reproducibility on CLIP4S is observed to be significantly better than manual methods of making tracings and inputting the coordinates into a computer using a digitising table [Baumrind 71]. By reducing the subjectivity in the landmark identification, reproducibility has been increased in CLIP4S.

Palate Line

Now consider the reproducibility of the palate line segmentation. The mean value of the gradient of the palate line has little significance, with the exception that the gradient of the palate is small (less than $7°$) and is negative. The gradient, to a large extent, is determined by the operator in placing the radiograph on the film-holder and the orientation of the head when the radiograph was taken. The reproducibility of the algorithm is indicated by the size of the standard deviation (Δ) of the gradient and the intercept.

$$\Delta_{gradient} = 0.0078 \pm 0.0061$$

$$\Delta_{intercept} = 2.1 \pm 1.6$$

Thus the mean standard deviation of the gradient is less than $0.5°$, and the standard deviation on the intercept is better than 2.1 pixels (1 pixel $\equiv 0.09$ mm).

Unfortunately, there are no other independent data for single gradient measurements. Much of cephalometric measurement involves angle measurement between two line features such as, for example, the Frankfort and the Mandibular plane. The errors in angle measured between these two lines are typically between $\pm(3 - 4)°$ [Gravley 74]. If the FACF or the mandibular plane can be segmented with the same accuracy as the palate line, then for example, the mandible-palate angle may have an error of $\pm 1°$. This is better than the manual measurements using tracings.

Furthermore, a simplistic argument can be proposed to indicate the extremes in error that may arise in the gradient of the palate measurement using manual tracings. The typical gradients measured are approximately -0.12. Then if the length of the palate line is 50 mm [Solow 86], the vertical displacement of the anterior and the posterior nasal spine (Figure 7.20) is 6 mm. If errors of ± 0.5 mm are assumed for determining either of these two points (the assumptions are reasonable from observation of the manual measurements of the sella and the menton points [Baumrind 71]), the vertical displacement can then be 7 mm. This gives rise to a gradient of -0.14. The angular difference between these two gradients is $1.1°$. No account of the errors in horizontal displacement has been taken.

Therefore, it is reasonable to assume that angular measurements between features using CLIP4S are at least as reliable, if not more so, than manual measurements using tracings, as was observed for the sella and the menton.

Using the stage as a global reference frame, the method is not only allows one to measure the distance $\vec{SM}$ but also to measure the distances $\vec{SA}$ (d_{sa}), and $\vec{AM}$ (d_{am}) between the features. Furthermore, the angles between these features can also be calculated to give indication of angular growths. Algorithms have also been proposed to segment three line-like features, the FACF, the palate and the mandible. Once their gradient and the point of interception with an axis is known, the global coordinate system can be used to measure angles between the features as indicated in Figure 7.60.

To test that the stage is capable of providing a system with sufficient accuracy in measuring distances such that growth trends may be observed, the d_{sm} distance has been measured for five radiographs of a single patient taken over a period of about seven years. Each distance measurement d_{sm} was measured four times and the mean and the standard deviation of the measurements calculated. Results of which is given in Table 7.3 and is also shown in Figure 7.61.

Comparison with Manual Measurements

The sella–menton distance has also been measured manually using tracings of the features and measuring the distance using a ruler with 1 mm spacing. The result is given in Table 7.4.

The measurements were made by two human operators (A and B). Operator A made the measurement twice, separated by one month. Four sets of measurements for each radiograph were made. Since a ruler with 1 mm spacing was used, measurements were attempted to the nearest half millimetre.

Looking at the values for observers A1 and A2, it is observed that the values are consistent with one exception. However, the values for two different

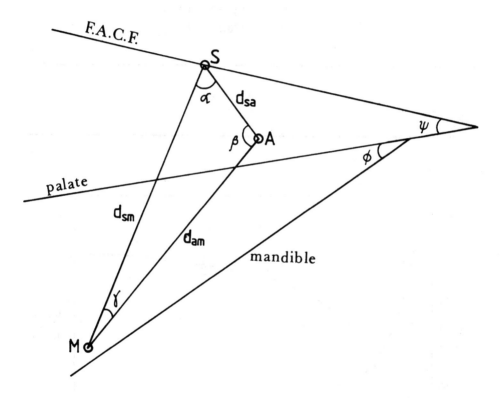

Figure 7.60: The spatial and angular measurements possible by segmenting the six features.

Table 7.3: The result of the measurement of the sella–menton distance using CLIP4S.

Age (yr:mnt)	d_{sm}(mm)
13:00	125.57 ± 0.13
15:04	130.69 ± 1.15
16:01	131.89 ± 0.37
17.05	132.40 ± 0.42
19:08	133.12 ± 0.17

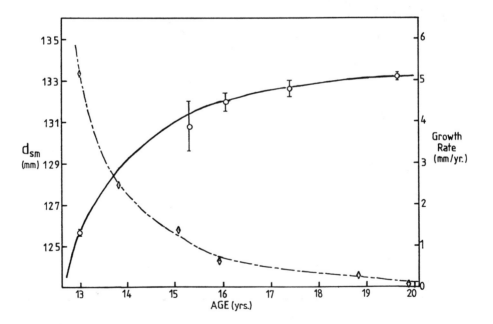

Figure 7.61: The growth of the distance between the menton and the sella for a single patient.

Table 7.4: Results of manual measurement of the sella–menton distance.

Age (yr:mnt)	A1 (mm)	A2 (mm)	B (mm)
13:00	125.0 ± 0.5	125.0 ± 0.0	124.5 ± 0.5
15:04	130.5 ± 0.5	130.5 ± 0.5	130.0 ± 0.5
16:01	131.5 ± 0.5	131.5 ± 0.5	130.5 ± 0.5
17:05	132.5 ± 0.5	132.0 ± 0.5	131.0 ± 0.5
19:08	132.5 ± 0.5	132.5 ± 0.5	131.5 ± 0.5

operators are quite distinct. This indicates that the inter-tracer errors are smaller than intra-tracer errors as a result of subjectivity applied by humans in measuring the distances. Although the same definition for the features was available to both the operators, different judgements have been made for the same data. This type of error is not present using CLIP4S since definitions are fixed once the algorithm has been given. Only changes that occur are due to inputting the images afresh, lighting conditions and camera response change with time. Since the reason for the one large error using CLIP4S is known, results indicate that reliability of the d_{sm} measurement is as good as manual measurements and generally better.

One further point, the d_{sm} measurement has been found to be greater for CLIP4S than manual methods. There are two possible explanation for them: (i) the human eye possibly responds better to the brighter edge whilst CLIP4S is responding to the penumbra edge and (ii) use of the extremum filter propagates the edge outward and thus a slightly lower edge, possibly 2 or 3 pixels lower, is extracted.

From the outset, the aim has been to develop algorithms which would allow automatic segmentation of certain cephalometric landmarks on lateral skull radiographs. These segmented features can then be used to measure distance and angles between the features and thus allow the study of the growth of skull with time. Furthermore, the system may be used in treatment planning since changes in size because of the treatment may be observed.

What has been achieved? First, the segmentation process has not been fully automated. The process of choosing the required grey-range has not been automated and requires some interaction. However, the interactions are simple and an operator with one hour's training will be competent to operate the system.

On the positive side, algorithms have been developed which are capable of segmenting the desired features. Automating a process is useful if (i) it is cost-effective, (ii) it is more reliable and accurate than human operators or (iii) it speeds up the process. The cost-effectiveness of using CLIP4S for cephalometric measurement is difficult to judge since it is an experimental machine and the only one available. Furthermore, CLIP4S is not fast — segmenting a single feature takes approximately 3–5 minutes. A human operator would be expected to measure all three distance measures (d_{sm}, d_{sa} and d_{am}) and the two angles (ϕ and ψ) within a period of about 10 minutes. However, the exercise has been to show that image-processing techniques are suitable in cephalometric measurement and an intermediate system could be developed using CLIP4S which is capable of matching human performance. The algorithms developed on CLIP4S should allow the measurement of the features with reliability comparable to or better than that of a human operator.

Reliability tests for segmentation of three of the features (the sella, the menton and the palate line) have shown that CLIP4S is more reliable in extracting these features than is possible manually — approximately 36% improvement in locating the sella and over 80% for the menton relative to the situation for manual tracings.

Once the features can be extracted accurately, then the distance between the features can be measured. Using CLIP4S, the distance between the sella and the menton, d_{sm} has been measured for a series of five radiographs of a single patient taken between the age of 13 years and 19 years and 8 months. The results have been given in Table 7.3. The same feature was also measured by two human operators manually using tracings. Operator A measured the values on two separate occasions, separated by a period of 1 month. The results are given in Table 7.4. The measurement on CLIP4S, after looking at the standard deviation on d_{sm} which is a guide to measurement reliability, is certainly as good as for those obtained by manual methods.

Since the annual growth between many of the cephalometric landmarks of interest is between approximately 1 and 2 mm (but greater during puberty), to observe growth trends, the reliability in distance measurement must be better than ± 0.5 mm. Thus measurement of d_{sm} indicates that CLIP4S can be used for growth study using a sequence of radiographs.

7.8 Automated Cytogenetics

Analysing chromosomes consists of three stages: (1) taking tissue specimen and producing slides, (2) detecting cells which are in the metaphase of cell division, and (3) carrying out measurements on the chromosomes. The resolution and size of images one requires to carryout the later two tasks is governed by the size and seperation of the chromosomes. Chromosomes usually vary in length from about 1-15 μm. They have a width of about 1 μm. This means that typical pixel size used is about 0.1 μm. The slides also have considerable amount of debris from broken blood cells and metaphases.

Many of the systems developed for chromosome analysis involve the following common steps [Piper 80]:

(i) location and digitisation of each metaphase cell;

(ii) segmentation by simple thresholding using histogram density [Granum 80]; and

(iii) classification by feature measurement.

The chromosomes features used are its area, length, banding patterns and the centrometric index. For these features to be measured, the chromosome's medial axis (the axis of symmetry) has to be obtained and then the location of the centromere (i.e. polarity) has to be determined. Once the feature measuremets have been carried out, *karyotyping* (classification) of chromosomes into its 24 classes can be carried out. karyotyping allows the detection of abnormalities in the structure of chromosomes or additional or missing chromosomes. Manually, karyotyping is carried out by taking photographs and then using pairs of scissors to cut the chromosomes and then paste them together into groups as if "hanging them on a clothes line"; this form of display is called a *karyogram*. This is a long and laborious process which is suitable for automation. Lloyd *et al.* comment that manual searching of metaphases on slides takes upwards of 5 minutes. Therefore automation leading to 2 minutes for machine searching is important because this can enhance the productivity of the operator. The operator can then process and correct segmentation and do classification of a previous slide as the machine operates on the current one.

One of the earliest machines which could threshold and measure area, perimeter length, mean-density and aspect ratio at video rate was built by Cambridge Instruments [Fisher 71]. Since then a number of machines for automatic cytogenetics have been commercially available from Joyce Lobel (Magiscan2 [Graham 83]) and Image Recognition systems Ltd (Cytoscan 110). The Cytoscan 110 was developed from FIP (Fast Interval Processors) an MIMD computer [Piper 86, Lloyd 87a]. FIP consists of a master processor and upto ten slave processors (each processor is a Motorola MC68000) connected to each other using a VME bus. Each of the processors have 0.5 Mbytes of local memory and 2 Mbyte pool of global memory. In addition, there is 0.25 Mbytes of display RAM. All memory is dual-ported so that it can be accessed by other processors over the VME bus. The operating system used on the master processor is OS9. The typical amount of data involved with a chromosomes is about 1 kbyte. The VME bus has a bandwidth of about 40 Mbytes/s, therefore I/O overhead is negligible. The performance of the system is determined by the need for comparative intensive operations in determining the orientation of the chromosomes, then rotating it to vertical and then feature measurement.

The global memory acts as a *blackboard* on which a list of data and the operations which need to be carried out on them is written. A free worker processor inspects this *blackboard* for any task it can carry out. It copies the data to its local memory, carries out operations and writes the result back to the blackboard.

However, the segmentation by thresholding has a number of limitations. Too high a threshold value could be chosen; this leads to fragmentation of the chromosomes into smaller pieces. Another problem, some chromosomes either touch or overlap each other; leading to the formation of clusters. Clusters are easily detectable if a convex hull is generated and displayed for each connected data set. An operator can then interactively draw a line to separate touching chromosomes and overlapping chromosomes can be treated in a similar way.

Also, this operator interaction can be overlapped with the machine classifying the other chromosomes. This ensures a better utilisation of both the operator and the hardware.

Many of the chromosomes are bent, so invariant measures are made with respect to its medial axis. Also, as mentioned previously, the orientation needs to be determined so that the chromosomes are hung "vertical" in the karyogram. The orientation is determined from the minimum bounding rectangle for each of the chromosomes. If assumption is made that the chromosomes are almost straight, then an approximation to medial axis can be found by taking the midpoint of each horizontal line through the chromosomes once its been rotated to vertical. If the chromosomes are bent, then a more conventional thinning algorithm is used [Hilditch 69, Piper 85]. The axis obtained is smoothed using a low-pass filter and extend to the ends of the chromosome shapes.

Once the medial axis has been determined, profile measurements transverse to this axis is carried out. The profile measurement consists of carrying out grey-level sum or histogram along the profile lines.

Next the centromere is computed from the analysis of the profile density. However, this method only works for unbanded chromosomes. For banded chromosomes, the profile value computed is $\sum md^2 / \sum m$ where d is the distance along the profile line from the medial axis and m is the pixel density. A centromere is the minimum value in the profile [Groen 85, Graham 87].

The features measured are grouped into four levels [Piper 80]. Level one features include area, grey-level density and convex hull of the chromosomes. For level two and higher features, the medial axis has to be computed. The level two features include the length and profile density of the chromosomes. These features are used to discriminate between normal chromosomes classes [Lundsteen 86]. It is important to note that while global features can be computed more reliably, they are usually not suitable for discriminating variation or abnormalities within classes; for this local feature computation are required (such as the profile measurements).

Once the features have been measured, each chromosome is classified (in parallel) using a maximum likelihood classifier which has previously been trained on a large data set [Piper 87]. Piper has shown that the classifier can be improved further if one uses the knowledge that there are usually two chromosomes in each class [Piper 86].

7.9   Concluding Remarks

Apart form the Cephalometric analysis system which was developed for an SIMD array processor and the Cytogenetic systems for which specialised parallel hardwares have being developed, most of the systems are still being developed to run on von Neumann architectures. Although increasingly small transputer based MIMD systems are being used to increase the processing speed. In this chapter we have shown how a MIMD system could be used for object recognition through the use of a *blackboard* processing model. This type of system could be effectively implemented using a Transputer array or Transputer farms. Although it should be noted that implementation of a *blackboard architecture* is probably more efficient on a shared memory machine rather than a message passing machine. For specialised applications it is probably more cost-effective to build a pipelined architecture using DSPs. However, such systems tend to be quite inflexible. With a pipelined architecture there is the problem of its use in a real-time system because of the pipeline delay which could lead to problems of real-time control. However, it is suitable for many industrial applications because of its high data throughput, and the pipeline delay is not a problem because the industrial process it is monitoring is "fixed" and the system behaviour known in advance.

Chapter 8

Future Trends

8.1 Introduction

The progress in image processing and image understanding depends on three factors: (1) semiconductor technology; (2) computer architecture; and (3) AI techniques and algorithms. These are the enabling technology for the success in the field. We expect that with VLSI and WSI, the gate speed will continue to increase with use of materials such as gallium-arsenide. Furthermore, use of wafer-scale integration will allow for fault tolerance. The von Neumann architecture is based on the control flow of instructions. This has been enhanced by the Harvard architecture where instruction and data can be fetched in parallel. This book has covered many varied architectures which are in popular use for image processing.

In this book, we have explored the special hardware (SIMD, MIMD and MISD) developed for computer vision. Many of then motivated by our limited understanding of the biological vision system. We have also presented parallel vision algorithms and the various means by which parallelism can be exploited. Based on these hardware and algorithms, we have presented a number of successful applications of machine vision. In many instances, accuracies in measurement obtained can only be achieved through machine vision.

Despite all these, use of computer vision is not as widespread as one would wish. We wish finally to examine, after 30 years of research and development in image processing and image understanding, the question: what are the major problems still to be solved to produce more varied and robust applications? Towards this end, we will examine the question what have been the fundamental problems in IP/IU? What are the limitations of currently available computer architectures which prevent development of real-time applications? Finally we will explore the direction in which IP and IU are moving to make further progress in the field. This would include a look at artificial neural networks as exploiting massive parallelism to solve many of the ill-defined vision problems. We also explore optical computers as a means of attaining greater speed because the circuit can operate at the speed of light and its inherent parallelism.

8.2   What Has Been the Problem with IP?

Before the emergence of specialised and parallel hardware, the major hindrance to progress in image processing/computer vision was image acquisition and processing. Often small, low-resolution (often 64×64 or smaller) images were used. Since then computing power has grown considerably. We now commonly use 512×512 or larger images. Raw processing power is now often not the problem except for real-time applications.

Despite this, progress in IP/IU has remained slow compared to the other branches of computer science, such as computer architecture, compiler technology, computer graphics and theorem proving. The large advances in the last 20 years we have seen in IP/IU has been in the support technology: the processors have become more powerful and cheaper, memory has grown bigger, faster and cheaper, storage device capacity has increased (now long image sequences can be digitised and stored on magnetic discs or optical discs. These advances in technology will be discussed further in the next section.

Attempts are afoot to make computer vision into a *real science*. There is an urgent need for coherent theories describing the visual processes and what needs to be computed for successful application of machine vision; and these theories will have to be extensively tested against real-world data. Many have attempted to utilise theories directly into applications rather take a more scientific approach with the consequence of failure and limited success. A scientific approach should be to observe the visual processes and from the observation to derive theories, which then can then be used to predict certain behaviours. Experiments could then be devised to test the predictions. Currently there is a chasm between theory and application.

It is arguable whether we currently require a general-purpose computer vision system; and certainly we are currently incapable of building one. Our goal must be to create vision systems which are good at carrying out certain tasks. To make the construction of such systems possible we need to constrain the problem we need to solve by using general knowledge and expectation behaviours of objects; however, we should avoid building systems which are too domain-specific. One of the biggest hindrance to testing algorithms and how they perform is the lack of *ground-truth*. For much of the data we need to process, it is often difficult to establish the ground-truth. However, without this, we cannot verify the performance of our algorithms.

Although we do not know or understand about most of the processes involved in a biological vision system, we do know that visual information such as motion, shape and the intrinsic image information are extracted [Hubel 62, Julesz 75, Gibson 79, Marr 82]. Although the algorithm used in computer vision need not be the same as in biological vision, understanding of which information are important and how they get used would be helpful in formulating algorithms for their computation. There needs to be interaction between computer vision, biological vision and computer architecture. This interaction has been evident in the field from the beginning and more of the finding ought to be used by computer vision practitioners.

In Chapter 5 we presented various algorithms for image segmentation. It is without doubt that segmentation is important in scene interpretation. However, segmentation based on intensity is not very successful; for different applications different threshold values have to be set and these have to be determined either empirically or in an ad hoc fashion. This should not be surprising. Objects are characterised by other information besides intensity. Many of the techniques discussed in Chapter 6 (such as stereo, texture, shape from X) have to be combined to obtain successful segmentation. This then raises the issue of control. How do we combine the information from these different intrinsic images?

Currently, we are moving way from the bottleneck in computing power to a bottle neck in data acquisition. Special-purpose hardware can now be built which will carry out certain operations at video rate. However, factory process speed has also increased. Now it is not uncommon for object on conveyor belt to move at between 500 and 1500 objects per minute (and for continuous processes such as rolling of steel in mills, the speed can be 300 m per minute). One method to reduce the data rate would be capture and process regions interest at high resolution and the remainder of the image in low resolution. Also, currently we are too often having to customise the application; this is proving very expensive. We require algorithms which are adaptable and robust so that we do not require to rewrite the algorithms if they were developed to recognise apples but now we need to recognise oranges or potatoes. This requires that techniques have to be robust and not dependent on illumination and noise.

8.3 Limits of Current Computer Architecture

Although which parallel algorithms are required has not been firmed up (especially because most of the algorithms have been developed on serial computers), there are four major aspects to image processing and image understanding: (i) preprocessing, (ii) feature detection, (iii) iconic-symbolic transformation, and (iv) scene interpretation. For all of these processes, there are different computational requirements and data representation. Image processing involves such operations as spatial filtering, image enhancement and image restoration. These consist mainly of local operations. Feature detection once again involves largely local operations. Although some global operations (such as histogramming, moments and image statistics) may be necessary. Because low- and intermediate-level image processing largely involve local operations, they are massively data-parallel and are efficiently mapped onto SIMD array processors.

The data throughput required for image processing is given by the following relationship $In^{2r}fN^2$, where I is the number of instructions per pixel, n is the window size of the operation (typically 3×3 or 5×5), r is the order of the computation (typically 1 for convolutions and ≥ 2 for median filtering, moments etc.), f is the frame rate (for video rate, this is 25) and N^2 is the number of pixels in the image (typically 512×512 images are processed). If we assume that at least 100 pixel instructions are required to carry out some simple processing on images, then we typically require 8×10^9 image operations per second. This gives us some indication of the speed of the computer required for a rudimentary real-time image processing system. If we use a pipelined computer with p processors in the pipe, the computation can be speeded by a factor p, and using a N^2 array processor, the speedup could be by a factor N^2. This reduces the complexity and the speed at which each of the processors needs to operate in an SIMD array computer.

We are likely to see improvement in price/performance for all aspects of the semiconductor technology such as for processors, memory, dsp and communication. Many have attempted to empirically determine the trends in the semiconductor industry [Noyce77, Myers 91]. According to Moore's Law: (i) the number of transistors on a chip doubles every 2 years (this was first stated in 1975 and is still valid). Currently 4 Mbyte RAMs are available, we expect 16 Mbyte RAMs in 1992 and 256 Mbyte RAMs by 2000. Lundstrom states that the transistor density increase by a factor 10 every 5 years, and this trend is likely to continue to the end of this century [Lundstrom 85]. On this basis by year 2000 we should expect to see 50 million transistor on a chip. Intel expects that it will release i686 in 1996 with 20 million transistors and there will be 100 million transistors in the i786 by 2000. Since the size of processors are also getting smaller we should expect DSP and supporting processors on a single chip; this would of course improve performance.

8.4   Direction of IP and Computer Vision

As we have discussed in the previous section, soon we should expect 200 MIPS RISC processors and 1 gigabit RAMs. Performance of SIMD array machines such as the Connection machine and MPP are already impressive. As we will discuss later, greater computational speed is likely to become available with the use of optical computers. We are beginning to see increased use of parallel hardware for image processing (such as Transputers, IPSC, CM); however, with the exception of CM, they are not the massive arrays we discussed as being most efficient for low-level image processing (see Chapters 4, 5 and 6).

The direction computer vision is generally taking is in the greater use of geometric models for 3D scene analysis. Although much work still needs to be done to map 2D segmented data to the 3D geometric models efficiently (we have to infer what the model is from the 2D information available). Also, the use of active vision is becoming more prominent within society. We have found that general scene recovery is very difficult; however, active vision, by making the ill-posed problems in computer vision (such as optical flow and shape from X techniques) well-posed [Aloimonos 88b], could allow for more successful general recovery although the task would remain computationally expensive. This also requires that more than a few image frames are processed. This has led to the *purposive vision paradigm* as opposed to the *reconstructionist paradigm* arguing that recovery of full structures is not necessary; we need to use certain vision modules in certain parts of the image to solve particular tasks.

Application of IP/IU is now mature for biomedical analysis and certain industrial automation tasks such as surface inspection. We are likely to see new applications in:

(1) traffic automation / law enforcement;

(2) safety critical monitoring of vehicles;

(3) Security systems;

(4) document processing;

(5) teleperception, progress in robot vision alone has been a disappointment; and

(6) food products and pharmaceuticals.

As well as computer vision forming a basis for science as discussed in the previous section, greater experimentation on calibrated data will lead to it also becoming an engineering subject. The good points that we now observe in IP/IU are a greater use of a combination of methods; and more people willing to combine ANN, traditional IP/CV and AI to produce successful applications.

8.5 Artificial Neural Networks

Many problems, especially those involving vision are to-date better solved by humans than digital computers. The primary reasons for these are: (1) the vision problem is fundamentally ill-defined, the recovery of depth (shape) of three-dimensional surfaces from two dimensional images is leads to ambiguity, and (2) there is a very large volume of data to be processed.

Progress in computation vision has undeniably been slow. Many argue that this problem has arisen through the use of the von Neumann computer architecture [Widrow 90]. The primary relationship between biological neural network and artificial neural network (ANN) is that both uses large number of simple processing units (the neurons) and the network is capable of solving large complex problems by learning.

Neurons are analogue devices; they take input from synapses and output continuous valued results to other neurons. In mammalian nervous system there are over 10^{10} neurons. This is a very large number; however, they are quite slow devices in comparison to current digital computers; to integrate the values from the connecting neurons takes over a few milliseconds. However, we can reason about quite complex visual problems in tens of milliseconds therefore they must work in a very highly parallel method. Many neurons respond to a single stimulus.

In the visual area of the cerebrum, local features (such as lines at different orientation) are detected [Hubel 02]. Higher up in the visual pathway, these features are grouped together. This hierarchy of operations we have already discussed in Chapter 1; however, we wish to re-emphasise that these processes uses both bottom-up (afferent, feed-forward), which is used in pattern recognition, and top-down (efferent, feed-back) signals, which is used selective attention, pattern segmentation and associative recall.

ANN consists of neuron-like analogue cells. Stimulus is applied in a two-dimensional input layer. The data is feed-forward and processed up to the recognition layer, which is the last stage. The recognition layer consist of gnostic cells, one cell is activated for a given input pattern. A connection network uses the ANN computation model, not so much because it emulates faithfully the biological processes but because of the computational power and convenience of the system. The short-term memory of a network is the state of the units, and the long-term memory of the system are the weights, these being modified by learning. There are three types of learning algorithms: (i) *supervised* learning, the input and output pattern are specified (*clamped*); (ii) *unsupervised* learning, it learns by looking at the regularities in the input vectors; and (iii) *reinforced* learning. At the heart of these three learning paradigms is *matrix-by-vector* multiplication, which we know to be a highly parallel operation.

In a connectionist model the neurons are like processing elements (PEs) called units which have states determined by the input. The "total input" of an unit j is usually a linear function of the output of the other units.

A specific method used for supervised learning is the *backpropagation* method [Rumelhart 86]. The method which uses an iterative gradient-descent technique is a generalisation of the delta-learning rule [Widrow 60] used in a feed-forward network with hidden layers. The deficiencies of this method is that there is no biological evidence that a synapse can send messages bidirectionally. Furthermore, gradient-descent techniques tend to have slow-convergence and often get stuck in local-minima.

Backpropagation is a generalisation of the least squares method. With hidden layers (Figure 8.1), it is a more powerful network but is more slower in the training phase. The major problem with this network is that the learning time scales poorly for a large network [Hinton 89]. Also there is no biological evidence that synapses could be used in both directions. Furthermore, it suffers from the general problems faced by gradient-descent techniques: the energy minimisation can lead to the network being stuck in local minima and the slow convergence rate.

A measure of how well a network is performing with currently set weights is given by

$$E = \frac{1}{2} \sum_{j,c} (y_{j,c} - d_{j,c})^2$$

where $y_{j,c}$ is the output and $d_{j,c}$ is the desired state of unit j for input-output case c. The error can be minimised by starting with random weights and

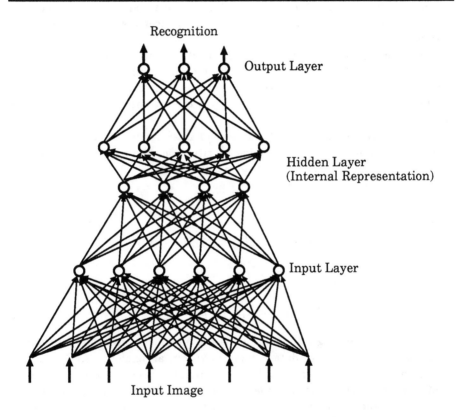

Figure 8.1: A multi hidden-layer artificial neural network.

changing them by an amount

$$\Delta w_{ji} = -\epsilon \frac{\delta E}{\delta w_{ji}}$$

This is called the least-squares learning procedure.

A network consists of N nodes (neurons) each with K connections (synapses), i.e. each node receives input from the output of K other nodes. In many networks, for each neuron, the input from K other nodes may be used; the connectivity is arrived at randomly using equal probability for connection to any node. The network which consists of neurons and synapses are the hardware of the system and the weights connecting the neurons is the software of the system.

The cerebellum in the brain acts as associative content addressable memory (ACAM) and it is trainable. The edge information in images is extracted in the retina by lateral inhibition between retinal neurons. In the cortex, the lateral excitation process computes the brightness of the regions enclosed by the edges it previously found.

An artificial neuron network has two phases: retrieving information and learning new information. In a retrieval situation, the network is in a feedforward mode and the network can be expressed by

$$u_i(l+1) = \sum_{j=1}^{N_l} w_{ij}(l+1)a_j(l)$$

$$a_i(l+1) = f_i(u_i(l+1), \theta_i(l+1))$$

where $1 \leq i \leq N_{l+1}$ and $0 \leq l \leq L-1$.

In the learning phase

$$w_{ij}(l) = w_{ij}(l) + \eta \Delta w_{ij}(l)$$

where η is the updating rate and Δw_{ij} is the increment of the weight change.

Because the volume of data is large in vision, an analogue ANN can play important role in preprocessing. For later stages of processing, one requires a digital network. This is because the lower accuracy of the computation supplied by the analogue circuit is adequate for the preprocessing stages which often require more parallelism and more speed. Analogue circuits are usually smaller and therefore can be more densely packed; however, they also tend to be fixed in their configuration and therefore cannot be used to solve problems where the circuit has fewer neurons than are required to solve the problem. This limitation does not usually exist with digital neural networks where there is usually the flexibility to partition the problem and the ability to implement different networks on the same hardware. The problem can be partitioned by allowing the results of multiple pass through the system with different data to be combined. Also, digital circuits are more accurate than analogue ones and digital networks have learning capabilities. They also scale well to handle larger problems because they are not affected by noise as analogue ANNs are, in which

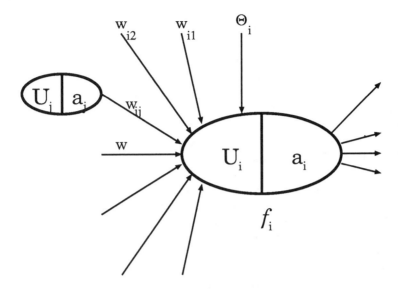

Figure 8.2: An artificial neuron.

extra noise is added by the size of the circuits. Furthermore, digital ANNs can be designed as hierarchies of networks which act cooperatively. However, digital ANNs, even when implemented in VLSI, take up large silicon area; they tend to be much slower than analogue ANNs and there is the greater cost for the network interconnection.

There are primarily two different neural network models: (1) *Feed-forward* [Rumelhart 86] – there are layers of artificial neurons set up as a directed graph (i.e. no loops), an input is applied to the first layer and the output result is at the last layer. These networks can be described as a special case of combinational circuits. (2) *Feed-back* [Hopfield 82] – the network can be described by an undirected graph; the nodes are the neurons and the connections are the synapses. Typically the neurons are connected by "wires" which are bidirectional (i.e. have symmetric weights). The feed-back network operates like an asynchronous computer. Given some initial stare (the input), the network is capable of carrying out some computation and eventually reaches a state state (the result).

Both these types of networks performs the same type of functions, the neurons acts as threshold logic units; the output of a neuron is a functions of the sum of the inputs which are multiplied by weights. Such neurons as these were introduced by McCulloch and Pitts [McCulloch 43], they can used to implement linearly separable functions. A single layer "MP neuron" is created from an array of n neurons

$$y_j = \mathcal{F}\left(\sum_i w_{ij} x_i - \theta_i\right)$$

Such a network is capable of computing the three boolean functions AND, OR, and NOT by using threshold (θ) of 2, 1 and 0 respectively. In the last case w_{ij} has to be -1. Such a network works synchronously in discrete instances of time. Rosenblatt showed that the MP network with adjustable synapses could be "trained" to classify certain pattern types; such a network is called the "Perceptron" [Rosenblatt 58]. Minsky and Papert showed that such a network would not be able to distinguish between many simple patterns because a such single-layer network could not compute the XOR function [Minsky 69]; therefore the perceptron is not an universal computing machine (i.e. Turing machine). Such limitation does not exist in symbolic algebra hence this led to the general demise of the perceptron. Despite these limitations, many Perceptron-like machines, such as WISARD I [Aleksander 84] have been shown to be good at pattern recognition. Furthermore, certain networks such as the Boltzmann machine [Hinton 84] do not suffer the same limitations as the Perceptrons and is actually better at finding certain solutions to computational problems than using algorithms which involves exhaustive search methods.

Learning in Perceptrons, Hopfield Net and Boltzmann machines is fixed. In humans, we are continually learning new facts and new actions and these new learning can revise old facts and beliefs and filter out irrelevant information. In symbolic processing we would refer to this as the *truth maintenance system*. Therefore, in human system there is both a stable and plastic (adaptable) mode for learning so that we do not over write our old knowledge. Carpenter and Grossberg have addressed this question of how can a system continually learn and yet be stable and prevent old memory from destruction in the process of learning; a solution they propose is the *adaptive resonance theory*, ART [Carpenter 88]. The ART model is an adaptation of the *competitive learning model* [Hinton 89].

In competitive learning, the set of input vectors are divided into disjointed clusters such that for each cluster the input signal is similar. It is referred to as competitive learning because because the hidden layers compete with each other to become active. A unit which receives most input becomes active and the others are turned off, this is the "winner-take-all" rule. The winner then increments is weight vector, thereby receives even more input in the future phase.

The most fundamental difference between ART and other ANNs (such as Perceptrons, Boltzmann machines, and back-propagation) is that ART is designed for real-time learning of transitory patterns using an unlimited number of inputs until its memory capacity is full. In the other systems, new learning washes away old memory if there are signals from too many inputs. This problem is often prevented by either limiting the number of inputs or shutting down the learning process if memory capacity catastrophe occurs. The other main differences between ART and other ANNs are: (i) learning is unsupervised in

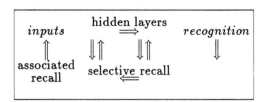

Figure 8.3: Fukushima's neocognitron model. In the hidden layers it hierarchically extracts and groups features.

ART, (ii) it maintains its own plasticity, (iii) it does not use passive learning, instead the learning is regulated by active attentional focus, (iv) while Boltzmann machines and back-propagation use noise to perturb the system out of local energy minima by simulated annealing, ART globally reorganizes the energy function by self-regulatory hypothesis testing using an approximate match mode and buffering against external noise [Carpenter 88].

Grossberg *et al.* have described methods for visual pattern recognition using ART. The input images to ART are preprocessed by first segmenting the data into foreground and background using a range detection technique. Then a neural network is used for enhancement and filtering and completing the boundaries. Finally, the objects are made invariant to translation, scale and rotation using the Fourier-Mellin filter [Casasent 76]. The later two preprocesses stages are said to be analogous to operations carried out in the visual cortex [Schwartz 80, Carpenter 88].

Fukushima refers to an ANN model called cognitron [Fukushima 75] for pattern recognition; the ANN uses unsupervised learning akin to the competitive learning paradigm. However, the recognition process is not invariant to translation or scale; Fukushima has extended the cognition model, *neocognition*, to overcome these deficiencies [Fukushima 88]. The functionality of neocognitron is further enhanced by having both feed-forward and feed-back connections which give it selective attention functionality as well as pattern recognition functionality. The selective attention methodology is claimed to work better than associative recall because it is invariant to translation and scale. The system learns to recognise and classify objects without a teacher using a similarity in shape test; this is reinforced learning using both bottom-up and top-down processing (Figure 8.3).

A review of hardware for artificial neural networks is provided in [Hecht-Nielson 88]. While we have generally observed some difficulty of programing and mapping algorithms to massively parallel machines, ANN provides massively parallel computation with the added benefit of learning and thus avoiding the difficulties of algorithm development.

If the voltage on the input wire is held at ground, then the signal from other

neurons is
$$I_{ij} = V_{outj}T_{ij}$$
where the output voltage at cell j is V_{outj} and T_{ij} is the connection strength. The output voltage is a function of the total current input at that cell. The function basically acts as an amplifier which has a non-linear transfer characteristics. The multiplications can be carried out based on Ohm's law using registers and the summation of the current in the circuit is given by Kirchhoff's law.
$$V_{outi} = f(\sum_{j=1}^{N} I_{ij}) = f(\sum_{j=1}^{N} V_{outj}T_{ij})$$

One method for achieving significant speed in computation is by using VLSI architectures. This method tends to be about a magnitude faster than using digital accelerator boards. Because each neuron cell is simple, a high packing density on the VLSI chip can be achieved. Several VLSI based neural networks have been built. Graf *et al.* describe a 54 PEs VLSI chip which is fully connected using a programmable connection matrix [Graf 88]. Because the weights need to be more than 8-bits, the back-propagation is not suitable for analogue implementation [Graf 89]. However, they state that 5-bit weights are of sufficient resolution for the feed-forward part of the network. If the function to be computed is known in advance and it is not likely to change, then the weights can be defined using fixed value resistors. This leads to the network being physically smaller. However, it is more flexible for the network to be trainable, this then requires the circuit to have storage cells for weights and connections. Also, if the input signal and the weights are binary, then the multiplication can be reduced to the use of AND and XOR functions rather than have full adders or even multipliers. Graf *et al.* describes a *content addressable memory* (CAM) chip which can recall 46 different templates each 96 bits long (Figure 8.4), thus each feature can be described using up to 96 bits. In a CAM, the content of a cell is recalled by word matching, where as in random access memory (RAM), words are recalled by supplying a particular address. Therefore, RAMs typically need to be addressed serially but all cells containing a particular value can be recalled in parallel using a CAM. The CAM cells are analogue but the input and output is digital. The chip has 4416 connections, a computation cycle takes 100 ns, and therefore it is capable of making more than 44×10^{10} connections per second. A CAM chip can compute inner products and is therefore efficient at template matching.

Another analogue circuit is described by Hutchinson *et al.*. They have developed a 48×48 silicon retina using a resistive network of hexagonal cells which work in real-time. The input to the network is the output of logarithmic response photoreceptors. In such a network, the potential drop across the registers give a good approximation to the Laplacian operator.

Artificial neural networks are being developed for automatic target recognition (ATR) because the functions need to have the following characteristics:

- The algorithms need to be robust against changing target and environ-

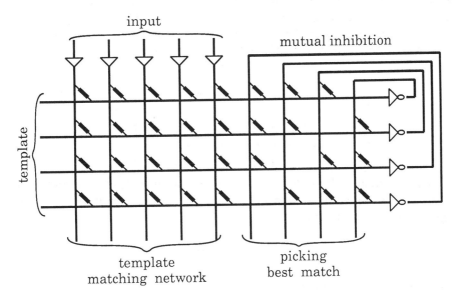

Figure 8.4: A typical diagram of a content addressable memory.

ment, on conventional computers, this leads to algorithms being computationally very complex and expensive.

- It is difficult to make the algorithms adaptable. ANN provides the means of training for new environment and target patterns.

- The algorithms need to find a limited number of good features.

- The algorithms will need integrating with *a priori* knowledge.

ANN provides automatic learning and continuous refinement to these using back-propagation learning rules by choosing weights in the hidden layers. This saves on deriving very complex rules of behaviour and recognition patterns, which then have to be integrated into algorithms.

Hopfield showed that ANN with threshold logic units can be used in solving complex optimisation problems and minimisation problems [Hopfield 82]. These computationally capable networks consist of non-linear recursive units which update themselves in a random, asynchronous and recursive manner. Hopfield further showed, that by using symmetric weight networks, the network can reach stable states which have energy minima. This ANN model is further improved for computation cost [Hopfield 84] by making the neurons act in a synchronous, continuous and deterministic manner using a sigmoid response function. He further showed that the network converges to a global energy minima rather than a local one. A Hopfield ANN is a single-layer, time iterative feed-back network which has associative recall properties. The network consists of N binary valued $\{-1, 1\}$ artificial neurons which are linked to other neurons by symmetric weights $\{w_{ij} = w_{ji}\}$. Thresholding units are added to linear associators for iterative feed-back autoassociation tasks.

For information retrieving, the network can be expressed as

$$u_i(l+1) = \sum_{j=1}^{N} w_{ij} a_j(l)$$

$$a_i(l+1) = \begin{cases} 1 & \text{if } u_i(l+1) > -\theta_i \\ 0 & \text{if } u_i(l+1) < -\theta_i \\ a_i(l) & \text{if } u_i(l+1) = -\theta_i \end{cases}$$

For the extended Hopfield ANN, the input and the output have continuous values and we integrate time delay as the capacitance of real neurons using sigmoid activation function.

$$f(x) = \frac{1}{1 + \exp(-x/u_0)}$$

then the retrieving phase can be expressed as

$$u_i(l+1) = \sum_{i=1}^{N} w_{ij} a_j(l)$$

$$a_i(l+1) = \frac{1}{1 + \exp(-u_i'(l+1)/u_0)}$$

where $u_i'(l+1) = K_i(u_i(l) + \theta_i) + K_2(u_i(l+1) + \theta_i)$ and K_1 and K_2 are proper constants. When $u_0 \to 0$, then $f(x)$ is a step function.

The behaviour of the Hopfield ANN is determined by the global energy function

$$E = -\sum_{i<j} s_i s_j w_{ij} + \sum_{J} s_j \theta_j$$

where s_i and s_j are the state of two units in the range $\{+1, -1\}$. The states are chosen such that the energy of the network is minimised.

$$\Delta E_j = E(s_j = -1) - E(s_j = +1) = -2(\theta_j + \sum_i s_i w_{ij})$$

where ΔE_j is the energy gap at unit j for a change from a $+1$ state to a -1 state.

$$E = \begin{cases} +1 & \text{if } \Delta E > 0 \\ -1 & \text{otherwise} \end{cases}$$

A fully connected Hopfield ANN with N units can store $0.15N$ random vectors [Hinton 89], therefore its memory capacity is approximately 0.15 bits per weight even though each of the weights are represented using an integer value.

Hopfield ANN has found its use in both pattern recognition [Farhat 85] and computational vision [Koch 86].

In a Boltzmann machine [Ackley 85, Hinton 84] which is a generalised Hopfield network, the sigmoid coefficient u_0 is substituted by a temperature control parameter T

$$a_i(l+1) \quad = f(u_i(l+1), \theta_i)$$
$$= 1/(1 + \exp(-(u_i(l+1) + \theta_i)/T))$$

In a Boltzmann network, the global state of the network is determined by the energy difference. The states in the range $\{0, 1\}$ are updated using a stochastic decision rule. The probability of an unit adopting state 1 is

$$p_j = \frac{1}{(1 + \exp(-\Delta E_j/T))}$$

If this rule is applied iteratively, the network reaches a "thermal equilibrium". At high "temperature", the network approaches the equilibrium point fast but then each unit has an equal probability of attaining high or low energy state. If low temperature is chosen, then the approach to the equilibrium point is slow but there is a greater chance that each of the units is in a low energy state. Therefore, the network is initially trained using high T value then as it approaches equilibrium, T is lowered. This form of training is called "simulated annealing" [Kirkpatrick 83]. A VLSI chip has been fabricated for simulated annealing which is given stochastic properties using noise [Alspector 87]. While the retrieving phase of the Boltzmann network is similar to the Hopfield network, the learning phase is different; it is closer to a two-step Hebbian learning.

ANN also provides massive parallelism for feature extraction computation; work has been reported for feature extraction in noise [Roth 89] and computation of the Hough transform [Oyster 87]. Uses of ANN for solving various ill-posed problems have been reported: shape from shading [Koch 86], optical flow [Hutchinson 88, Bulthoff 89] , image restoration [Zhou 88], image segmentation [Cortes 89, Dailey 89], Marr-Poggio stereo [Gmitro 87].

8.6 Optical Computing

The interest began in optical computing with the development of the matrix-vector multiplier (OMVM) at Stanford [Goodman 78] which is capable of carrying out a series of analogue multiplications. Subsequently others have developed or proposed architectures for inner-product, systolic and outer-product processors [Sawchuk 84, Sawchuk 86]. The advantages of an optical computer is that it has a very large fan-in and fan-out (i.e. large space and time bandwidth), it is inherently two-dimensional and parallel, and it is reconfigurable. Because of the large bandwidth, it can operate on many independent channels and these channels can propagate through each other with little or no interference or cross-talk. However, on the downside, it is an analogue device: it has little flexibility, these are the problems of noise accumulation in analogue systems and problem with deterministic noise. There are also the limitations in the input-and-output devices. Some of the problems are alleviated by development of digital optical or optoelectronic computers [Farhat 87, Psaltis 85].

Optical techniques can be used for image operations such as addition, subtraction, multiplication, two-dimensional Fourier transforms, correlation and convolution to name but a few [Goodman 68, Hausler 77, Casasent 78, Jahns 80, Lee 81, Stark 82].

In the pre-LSI and -VLSI era, the gates and gate interconnection on chips was expensive. This is one of the fundamental reasons for the development of the von Neumann architecture which have separate CPU and memory. However, with the development of the VLSI technology, gates have become cheap but now the communication is expensive and this has pushed the development of systolic architectures. With optical systems, the situation is different once again, In an optical system, the gates can be cheap or moderately expensive but the connection is cheap. Therefore, one can attempt to minimise the hardware redundancy and make more use of connectivity. Thus we can have several processors and memory units which are interconnected by either a fixed network or dynamic crossbar switch; it is very much like a shared-memory system. The memory can be accessed in parallel and therefore have parallel communication between the CPU, the memory and the I/O system. These three subsystems can become a unit in a larger optical gate array. This could give rise to cellular logical arrays and associative memory constructed of optical parallel processors.

An optical neural network arrangement used in implementing a 32-neuron network which behaves as associative memory is described in [Farhat 85]. In an optical implementation (see Figure 8.5, the signal can only take positive values. Therefore, two channels are used, one acting as the positive carrier and the other the negative carrier. However, this leads to the requirement for twice as many sources and detectors as from digital ANN, and hence an increase in the size of the matrix. An alternative to using two channels is to use a bias to a mean of unity.

An optical ANN can be programmed using holographic memory. The advantage of such optical computing is the very fast switching speed, the reduced interference and the greater fan-out of the circuit. Typically, an optical computer can have communication bandwidth of 1–100 Gbytes per second as opposed to 10–1000 Mbits/s for electronic circuits.

More generally, greater use of optical computing and optical transforms can be used along side the digital circuits for the pre-processing of the image data. These could involve the optical computation of the Hough transformation. The key point to optical implementation is that the Hough transform is equivalent to the Radon transform [Deans 81]. Therefore, it can be implemented by performing a series of spatial Fourier transforms [Steier 86]. A single Fourier transform is proportional to a single line of Hough transform for a given θ. Transform for different θ can be obtained by rotating the scene with a Dove prism and carrying out the Fourier transforms.

Optical systems are also capable of computation of geometric moments [Kumar 86]. The key point of the implementation is the relationship between the intensity of the Fourier transform and moments.

Multilayer optical learning network is described in [Wagner 87]. Description of optical cellular logic processors of image processing can be found in [Jenk-

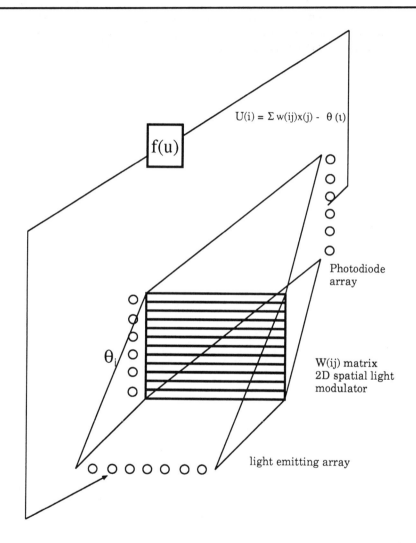

Figure 8.5: An example of an artificial neural network using an optical computer.

ins 85, Tanida 85, Yatagai 86, Caulfield 86]. They found that many image processing algorithms use cellular logic and automata or mathematical morphology. As with others who have developed special digital architecture for image processing, they conclude that these operations are best implemented on array processors. As optical processors are spatially parallel and capable of implementing image algebra (mathematical morphology), they are suitable vehicle for high-speed image processing. A 4/8-connected optical SIMD machine which can implement binary image algebra is the Digital Optical Cellular Image Processor (DOCIP) [Huang 87, Huang 89].

Bibliography

[Abdou 79]   I. E. Abdou and W. K. Pratt. Quantitative design and eval-
                 uation of enhancement/thresholding edge detectors. *Proc.
                 IEEE*, 67:753–763, 1979.

[Abramson 63]   N. Abramson. *Information Theory and Coding*. McGraw-
                 Hill, New York, 1963.

[Ackley 85]   D. H. Ackley, G. E. Hinton, and T. J. Sejnowski. A learning
                 algorithm for Boltzmann machines. *Cognitive Sci.*, 9:147–169,
                 1985.

[Adams 68]   D. A. Adams. *A Computational Model With Data Flow Se-
                 quencing*. Technical Report CS 117, Computer Science Dept.,
                 School of Humanities and Sciences, Stanford University, 1968.

[Ahuja 86]   S. Ahuja, N. Carriero, and D. Gelernter. Linda and friends.
                 IEEE Computer, 19(8):26–34, 1986.

[van Aken 62]   J. van Aken. Geometrical errors in lateral skull x-ray projec-
                 tion. *Trans. European Orthodontic Soc*, 74–86, 1962.

[Akl 78]   Selim G. Akl and Godfried T. Toussaint. Efficient Convex
                 Hull Algorithms for Pattern Recognition Applications. In
                 Proceedings of the 4th Int. Pattern Recognition Conference,
                 pages 483–487, Kyoto, Japan, 1978.

[Aleksander 84] I. Aleksander, W. V. Thomas, and P. A. Bowden. WISARD
                 – A radical step forward in image recognition. *Sensor Review*,
                 120–124, July 1984.

[Allan 80]   S. J. Allan and A. E. Oldehoeft. A Flow Analysis for the
                 Translation of High-Level Languages to a Data Flow Lan-
                 guage. *IEEE Trans. Computer*, C-29(9):826–831, 1980.

[Aloimonos 88a] J. Aloimonos. Visual Shape Computation. *Proc. IEEE*,
                 76(8):899–916, 1988.

[Aloimonos 88b] J. Aloimonos, I. Weiss, and A. bandopadhay. Active vision.
                 Int. J. Computer Vision, 2(1), 1988.

[Alspector 87]   J. Alspector and R. B. Allen. A neuromorphic VLSI learning
                  systems. In P. Losleben, editor, *Advanced Research on VLSI*,
                  pages 313–349, MIT Press, Cambridge, MA, 1987.

[Amdahl 67]   G. Amdahl. Validity of the single-processor approach to
                  achieving large-scale computer capabilities. In *AFIPS Conf.
                  Proc.*, pages 483–485, 1967.

[Andrews 74]   H. C. Andrews. Digital image restoration: a survey. *IEEE
                  Computer*, 7:36–45, 1974.

[Andrewds 77]   H. C. Andrews and B. R. Hunt. *Digital Image Restoration*.
                  Prentice-Hall, Englewood Cliffs, NJ, 1977.

[Annaratone 87]   M. Annaratone et al. The Warp Computer: Architecture,
                  Implementation, and Performance. *IEEE Trans. Computers*,
                  C-36:1523–1538, 1987.

[Arcelli 75]   C. Arcelli, L. Cordella, and S. Levialdi. Parallel thinning of
                  binary pictures. *Electronics Letters*, 11(7), 1975.

[Arnold 78]   D. Arnold. Local context in matching edges for stereo vision.
                  In *Proc. Image Understanding Workshop*, pages 65–72, May,
                  1978.

[Arvind 78]   Arvind and K. P. Gostelow. Some Relationships Between
                  Asynchronous Interpreters of a Dataflow Language. In E. J.
                  Neuhold, editor, *Formal Description of Programming Con-
                  cepts*, pages 95–119, North-Holland, Amsterdam, 1978.

[Arvind 83]   Arvind and R. A. Iannucci. A Critique of Multiprocessing von
                  Neumann Style. In *Proc. 10th Symp. Computer Architecture*,
                  pages 426–436, Stockholm, 1983.

[Ashkar 78]   G. P. Ashkar and J. W. Modestino. The Contour Extraction
                  Problem with Biomedical Applications. *Computer Graphics
                  and Image Processing*, 7:331-355, 1978.

[Ataman 80]   E. Ataman, V. K. Aatre, and K. M. Wong. A fast method
                  for real time median filtering. *IEEE Trans. Acoust., Speech,
                  Signal Processing*, ASSP-28:415–420, 1980.

[Attneave 54]   F. Attneave. Some informational aspects of visual perception.
                  Psychol. Rev., 183–193, 1954.

[Attneave 66]   F. Attneave and M. D. Arnoult. The quantitative study of
                  shape and pattern perception. In L. Uhr, editor, *Pattern
                  Recognition*, pages 123–141, Wiley, New York, 1966.

[Backus 78]   J. Backus. Can programming be liberated from the von Neumann style? A functional style and its algebra of programs. *Communications of the ACM*, 21(8):613–641, 1978.

[Bacon 82]   D. Bacon, H. Ibrahim, R. Newman, A. Piol, and S. Sharma. *The* NON-VON PASCAL. Technical Report, Computer Science Dept., Columbia Univ., May 1982.

[Bajcsy 88]   R. Bajcsy. Active Perception. *Proc. IEEE*, 76(8):996–1005, 1988.

[Baker 80]   H. Baker. Edge-based stereo correlation. In *Proc. Image Understanding Workshop*, pages 168–175, Apr., 1980.

[Bal 89]   H. E. Bal, J. G. Steiner, and A. S. Tanenbaum. Programming languages for Distributed Computing languages. *ACM Computing Survey*, 21(3):261–322, 1989.

[Baldock 88]   R. Baldock and S. Towers. First Steps Towards A Blackboard Controlled System for Matching Image and Model in the Presence of Noise and Distortion. In J. Kittler, editor, *Pattern Recognition, 4th International Conference (Lecture Notes in Computer Science - 301)*, pages 429–438, Springer Verlag, Cambridge, U.K., 1988.

[Ballard 75]   D. H. Ballard et al. Automatic analysis of human haemoglobin fingerprints. In *Proc. 3rd Meeting, International Society of Haemotology*, London, 1975.

[Ballard 81]   D. H. Ballard. Generalizing the Hough Transform to detect Arbitrary Shapes. *Pattern Recognition*, 13:111–122, 1981.

[Ballard 82]   D. H. Ballard and C. M. Brown. *Computer Vision*. Prentice-Hall, Englewood Cliffs, N.J, 1982.

[Ballard 87]   D. H. Ballard. *Eye movements and spatial cognition*. Technical Report 218, Comp. Sci. Dept., Univ. of Rochester, 1987.

[Barlow 82]   R. H. Barlow and D. J. Evans. Parallel algorithms for the iterative solution to linear systems. *The Computer Journal*, 25(1):56–60, 1982.

[Barnard 80]   S. T. Barnard and W. B. Thompson. Disparity analysis of images. *IEEE Trans. Pattern Analysis and Machine Intelligence*, PAMI-2:334–340, 1980.

[Barrow 71]   H. G. Barrow, and R. J. Popplestone. Relational descriptions in picture processing. In R. Meltzer and D. Michie, editors, *Machine Intelligence 6*, Elsevier, NY, 1971.

[Barrow 81]   H. G. Barrow and J. M. Tenenbaum. Interpreting Line Draw-
                   ings as Three-Dimensional Surfaces. *Artificial Intelligence*,
                   17:75–116, 1981.

[Barsky 81]   B. A. Barsky. *The Beta-spline : A Local Representation
                   Based on Shape Parameters and Fundamental Geometric
                   Measures*. PhD thesis, Univ. of Utah, Salt Lake City, 1981.

[Barsky 83]   B. A. Barsky and J. C. Beatty. Local control of bias and
                   tension in beta-splines. *ACM Trans. Graphics*, 2(2):109–134,
                   1983.

[Barsky 85]   Brian A. Barsky and Tony D. DeRose. The Beta2-spline: A
                   Special Case of the Beta-spline Curve and Surface Represen-
                   tation. *IEEE Computer Graphics and Applications*, 46–58,
                   1985.

[Batcher 68]   K. E. Batcher. Sorting networks and their applications. In
                   AFIPS Conf. Proc., pages 307–314, 1968.

[Batcher 80]   K. E. Batcher. Design of a massively parallel processor. *IEEE
                   Trans. on Computers*, C-29(9):836–840, Sept 1980.

[Baumrind 71]   S. Baumrind and R. C. Frantz. The reliability of head film
                   measurements - 1. landmark identification. *American Journal
                   of Orthodontics*, 60(2), 1971.

[Baumrind 80]   S. Baumrind and D. M. Miller. Computer-aided head film
                   analysis: the university of california san francisco method.
                   American Journal of Orthodontics, 78(1):41–65, 1980.

[Beaudet 78]   P. R. Beaudet. Rotationally invariant image operators. In
                   Proc. 4th IJCPR, pages 579–583, 1978.

[Bedner 84]   J. B. Bedner and T. L. Watt. Alpha trimmed means and their
                   relationship to median filters. *IEEE Trans. Acoust., Speech,
                   Signal Process.*, ASSP-32:145–153, 1984.

[Bergland 69]   G. D. Bergland. Fast Fourier transform hardware implemen-
                   tations — an overview. *IEEE Trans. Audio Electroacoust.*,
                   AU-17:104–108, 1969.

[Berman 85]   S. Berman, P. Parikh, and C. S. G. Lee. Computer Recogni-
                   tion of Two Overlapping Parts Using a Single Camera. *Com-
                   puter*, 18:70–80, March, 1985.

[Bezier 74]   P. E. Bezier. Mathematical and practical possibilities of
                   unisurf. In R. E. Barnhill and R. F. Riesenfeld, editors,
                   Computer Aided Geometric Design, pages 127–152, Academic
                   Press, New York, 1974.

[Bhattacharya 83] B. K. Bhattacharya and G. T. Toussaint. Time- and storage-efficient implementation of an optimal planar convex hull algorithm. *Image and Vision Computing*, 1(3):140–144, 1983.

[Billingley 70] F. Billingley. Applications of digital image processing. *Applied Optics*, 9:289, 1970.

[Binford 82] T. O. Binford. Survey of model-based image analysis systems. *Int. J. of Robotics Research*, 1:18–64, 1982.

[Blackman 58] R. B. Blackman and J. W. Tukey. *The Measurement of Power Spectra*. Dover Publications, New York, 1958.

[Blackmer 81] J. Blackmer, P. Kuekes, and G. Frank. A 200 MOPS systolic processor. In *Proc. SPIE, Vol. 298, Real-Time Signal Processing IV*, Society of Photo-Optics Instrumentation Engineers, 1981.

[Blake 82] A. Blake. Fixed Point Solutions of Recursive Operations on Boolean Arrays. *The Computer Journal*, 25:231–234, 1982.

[Bolles 83] R Bolles and R Cain. Recognising and Locating Partially Visible Objects: The Local-Feature-Focus Method. In A Pugh, editor, *Robot Vision*, chapter 2, pages 43–82, IFS (publications), UK, 1983.

[de Boor 78] C. deBoor. *A Practical Guide to Splines*. Volume 27 Applied Mathematical Sciences, Springer-Verlag, New York, 1978.

[Bracewell 84] R. N. Bracewell. The fast hartley Transform. *Proc. IEEE*, 72:1010–1018, 1984.

[Brice 70] C. Brice and C. Fennema. Scene analysis using regions. *Artificial Intelligence*, 1:205–226, 1970.

[Brigham 74] E. O. Brigham. *The Fast Fourier Transform*. Prentice-Hall, Englewood Cliffs, N.J, 1974.

[Brodbent 31] H. B. Broadbent. A new x-ray technique and its application to orthodontics. *Angle Orthodont.*, 1:45, 1931.

[Broadway 62] E. S. Broadway, M. J. R. Healy, and H. G. Poyton. The accuracy of tracings from cephalometric lateral skull radiographs. *Trans. of the B.S.S.O. The Dental Practitioner*, 455–459, August 1962.

[Bronson 90] E. C. Bronson, T. L. Casavant, and Leah H. Jamieson. Experimental Application-Driven Architecture Analysis of a SIMD/MIMD Parallel Processing System. *IEEE Trans. Parallel and Distributed Systems*, 1(2):195–205, 1990.

[Brooks 81]   R. A. Brooks. Symbolic Reasoning Among 3-D Models and 2-D Images. *Artificial Intelligence*, 17:285–384, 1981.

[Brooks 87]   R. A. Brooks. A hardware retargetable distributed layered architecture for mobile robot control. In *Proc. IEEE Int. Conf. on Robotics*, pages 106–110, 1987.

[Brown 89]   C. Brown and M. Rygol. Marvin: Multiprocessor Architecture for Vision. BMVC'90, Oxford, 1990.

[Bulthoff 89]   H. Bulthoff, J. Little, and T. Poggio. A parallel algorithm for real-time computation of optical flow. *Nature*, 337:549–553, Feb., 9, 1989.

[Bunze 85]   V. Bunze. Automated optical inspection: Quality and profit pick-up for PWBs. *Conductor*, 2–3, June 1985.

[Burt 81]   Peter J. Burt. Fast Filter Transforms for Image Processing. *Computer Graphics and Image Processing*, 16:20–51, 1981.

[Burt 83]   P. J. Burt and E. H. Adelson. A multiresolution spline with application to image mosaics. *ACM Transactions on Graphics*, 2(4):217–236, 1983.

[Buxton 83]   B. F. Buxton and H. Buxton. Monocular depth perception from optical flow by space time signal processing. *Proc. R. Soc. Lond.*, B 218:27–47, 1983.

[Cannon 89]   S. Cannon. Concurrent file system — Making highly parallel mass storage transparent. In *Proc. Supercomputer '89*, St. Petersberg, FL, 13-20, May 1989.

[Canny 86]   J. Canny. A Computational Approach to Edge Detection. *IEEE Trans. Pattern Analysis and Machine Intelligence*, PAMI-8:679–698, 1986.

[Cantoni 85a]   V. Cantoni, C. Guerra, and S. Levialdi. Towards an evaluation of an image processing system. In S. Levialdi, editor, *Integrated Technology for Parallel Image Processing*, pages 43–56, Academic Press, London, 1985.

[Cantoni 85b]   V. Cantoni, M. Ferretti, S. Levialdi, and R. Stefanelli. PAPIA: Pyramidal Architecture for Parallel Image Analysis. In *IEEE Proc. of 7th Symp. on Computer Arithmetic*, pages 237–242, 1985.

[Cantoni 87]   V. Cantoni and S. Levialdi. PAPIA: A Case History. In L. Uhr, editor, *Parallel Computer Vision*, pages 3–13, Academic Press, London, 1987.

[Carpenter 88]   G. A. Carpenter and S. Grossberg. The ART of Adaptive Pattern Recognition by a Self-Organizing Neural Network. *IEEE Computer*, 77–88, March, 1988.

[Casasent 76]   D. Casasent and D. Psaltis. Position, Rotation, and Scale Invariant Optical Correlations. *Applied Optics*, 15:1793–1799, 1976.

[Casasent 78]   D. Casasent, editor. *Optical Data Processings: Applications (Topics in Applied Physics, Vol. 23)*. Springer Verlag, Berlin, 1978.

[Case 81]   J. B. Case. Automation in Photogrametry. *Photogram. Eng. Remote Sensing*, 47(3):335–341, 1981.

[Catmull 78]   E. Catmull and J. Clark. Recursively generated b-spline surfaces on arbitrary topological meshes. *Computer-Aided Design*, 10(6):350–355, 1978.

[Caulfield 86]   H. J. Caulfield. Systolic Optical Cellular Array Processors. *Optical Engineering*, 25(7):825–827, 1986.

[Chaconas 80]   S. J. Chaconas. *Orthodontics*. PSG Publishing Company, Litteton, Massachusetts, 1980.

[Charniak 85]   E. Charniak and D. McDermott. *Introduction to Artificial Intelligence*. Addison-Wesley, Reading, MA, 1985.

[Chebib 76]   F. S. Chebib, J. F. Cleall, and K. J. Carpenter. On-line computer system for the analysis of cephalometric radiographs. *The Angle Orthodontist*, 46(3):305–311, 1976.

[Chen 75]   T. C. Chen. Overlap and pipeline processing. In H. S. Stone, editor, *Introduction to Computer Architecture*, Science Res. Assoc., Chicago, IL, 1975.

[Chen 82]   C. H. Chen. A study of texture classifications using spectral features. In *Proc. 6th Int. Conf. on Pattern Recognition*, pages 1064–1067, Munich, 1982.

[Chen 88]   Chin-Tu Chen, Jin-Shin Chou, Wei-Chung Lin, and C. A. Pelizzari. Edge and surface searching in medical images. In *Medical Imaging II*, pages 594–599, SPIE, 1988.

[Chen 90]   S. S. Chen, J. M. Keller, and R. M. Crownover. Shape from Fractal Geometry. *Artificial Intelligence*, 43:199–218, 1990.

[Cheng 90]   H. D. Cheng, C. Tong, and Y. J. Lu. VLSI Curve Detector. *Pattern Recognition*, 23(1/2):35–50, 1990.

[Chien 74]   Y. P. Chien and K-S. Fu. Recognition of x-ray picture pat-
                terns. *IEEE Trans. System, Man, and Cybernetics*, SMC-
                4:145–156, 1974.

[Chow 72]   C. K. Chow and T. Kaneko. Boundary detection of radio-
                graphic images by a thresholding method. In S. Watanabe,
                editor, *Frontiers in Pattern Recognition*, pages 61–82, Aca-
                demic Press, New York, 1972.

[Chow 77]   W. K. Chow and J. K. Aggarwal. Computer Analysis of Pla-
                nar Curvilinear Moving Images. *IEEE Trans. Computer*, C-
                26:179–185, 1977.

[Chung 85]   H. Y. H. Chung and C. C. Li. A systolic array processor
                for straight line detection by modified Hough transform. In
                *IEEE Workshop. Comput. Arch. Pattern Analysis Database
                Mgmnt.*, pages 300–303, 1985.

[Clark 86]   K. L. Clark and S. Gregory. PARLOG: Parallel programming
                in logic. *ACM Trans. Program. Lang. Syst.*, 8(1):1–49, 1986.

[Clarke 84a]   K. A. Clarke. *Reconstruction of Nuclear Medicine Images
                on the CLIP4 Computer*. PhD thesis, University of London,
                1984.

[Clarke 84b]   M. J. Clarke and C. R. Dyer. Curve Detection in VLSI. In
                King-Sun Fu, editor, *VLSI for Pattern Recognition and Image
                Processing*, pages 157–173, Springer Verlag, Berlin, 1984.

[Clocksin 80]   W. F. Clocksin. Perception of surface slant and edge labels
                from optical flow: a computational approach. *Perception*,
                9:253–271, 1980.

[Clowes 71]   M. B. Clowes. On seeing things. *Artificial Intelligence*,
                2(1):79–116, 1971.

[Clune 87]   E. Clune, D. J. Crisman, G. J. Klinker, and J. A. Webb.
                *Implementation and performance of a complex vision system
                on a systolic array machine*. Technical Report CMI-RI-TR-
                87-16, Robotics Institute, Carnegie-Mellon Univ., Pittsburgh,
                PA, 1987.

[Cohen 80]   E. Cohen, R. Riesenfeld, and T. Lyche. Discrete b-splines
                and subdivision techniques in computer-aided geometric de-
                sign and computer graphics. *Computer Graphics and Image
                Processing*, 14(2):87–111, 1980.

[Cohen 84]   A. M. Cohen, H. H-S. Ip, and A. D. Linney. A preliminary
                study of computer recognition and identification of skeletal
                landmarks as a new method of cephalometric analysis. *British
                Journal of Orthodontics*, 11(3):143–154, 1984.

[Conners 80]   R. W. Conners and C. A. Harlow. A theoretical comparison of texture algorithms. *IEEE Trans. Pattern Analysis and Machine Intelligence*, PAMI-2(3):204–222, 1980.

[Conners 82]   R. W. Conners, C. A. Harlow, and J. Dwayer. Radiographic Image Analysis: Past and Present. In *Proc. 6th. Int. Conf. on Pattern Recognition*, pages 1152–1169, Munich, Germany, 1982.

[Conners 83]   R. W. Conners et al. Identifying and Locating Surface Defects in Wood: Part of an Automated Lumber Processing System. *IEEE Trans. Pattern Analysis and Machine Intelligence*, PAMI-5(6):573–583, 1983.

[Conway 63]   M. E. Conway. Design of a Separable Transition-Diagram Compiler. *Comm. ACM*, 6(7):396–408, 1963.

[Cooley 67a]   J. W. Cooley, P. A. W. Lewis, and P. D. Welch. Application of the Fast Fourier Transform to Computation of Fourier Integrals. *IEEE Trans. Audio and Electroacoustics*, AU-15(2):79–84, 1967.

[Cooley 67b]   J. W. Cooley, P. A. W. Lewis, and P. D. Welch. Historical Notes on the Fast Fourier Transform. *IEEE Trans. Audio and Electroacoustics*, AU-15(2):76–79, 1967.

[Cornsweet 70]   T. N. Cornsweet. *Visual Perception*. Academic Press, 1970.

[Cortes 89]   C. Cortes and J. A. Hertz. A network system for image segmentation. In *IJCNN Int. Joint Conf. Neural Networks*, pages 121–125, IEEE, Piscataway, NJ, 1989.

[Crowley 90]   J. L. Crowley and K. Sarachik. Dynamic World Modeling Using Vertical Line Stereo. In *Computer Vision — ECCV90*, pages 241–246, Antibes, France, 1990.

[Cypher 87]   R. E. Cypher, J. L. C. Sanz, and L. Snyder. The Hough Transform Has O(N) Complexity on SIMD $N \times N$ Mesh Array Architecture. In *1987 Workshop on Computer Architecture for Pattern Analysis and Machine Intelligence (CAPAMI'87)*, pages 115–121, IEEE Computer Society Press, Seattle, Washington, 1987.

[Dailey 89]   M. J. Daily. Color image segmentation using Markov random fields. In *Proc. Image Understanding Workshop 1989*, pages 552–562, Morgan Kaufmann, San Mateo, CA, 1989.

[Danielsson 81]   P-E. Danielsson and S. Levialdi. Computer Architectures for Pictorial Information Systems. *IEEE Computer*, 14(11):53–67, 1981.

[Dasgupta 84]   S. Dasgupta. *The Design and Description of Computer Architectures.* John Wiley & Sons, New York, 1984.

[Dasgupta 90]   S. Dasgupta. A Hierarchical Taxonomic System for Computer Architectures. *IEEE Computer*, 23(3):64–74, 1990.

[Davie 86]   E. B. Davie, D. G. Higgins, and C. D. Cawthorn. An Advanced Adaptive Antenna Test-Bed Based on a Wavefront Array Processor System. In *Proc. Int'l Workshop Systolic Arrays*, 1986.

[Davis 76]   L. S. Davis. A survey of edge detection techniques. *Computer Graphics and Image Processing*, 4:248–270, 1976.

[Davis 79]   L. S. Davis, S. A. Johns, and J. K. Aggarwal. Texture analysis using generalised co-occurance matrices. *IEEE Trans. pattern Analysis and Machine Intelligence*, PAMI-1:251–259, 1979.

[Davis 81]   A. L. Davis and S. A. Lowder. A Simple Management Application Program in a Graphical Data-Driven Programming Language. In *Digest of Papers Compcon Spring 81*, pages 162–167, 1981.

[Davies 81]   E. R. Davies and A. P. N. Plummer. Thinning Algorithms: A Critique and A New Methodology. *Pattern Recognition*, 14:53–63, 1981.

[Davies 84a]   E. R. Davies. Circularity – a new principle underlying the design of accurate edge orientation operators. *Image and Vision Computing*, 2:134–142, 1984.

[Davies 84b]   E. R. Davies. Design of cost-effective systems for the inspection of certain foodproducts during manufacture. In *Proc. 4th Conf. Robot Vision and Sensory Controls*, pages 437–446, London, 1984.

[Davies 85]   E. R. Davies. Radial histogram as an aid in the inspection of circular objects. In *IEE Proceedings, Pt. D*, pages 158–163, 1985.

[Davies 86a]   E. R. Davies. Image space transforms for detecting straight edges in industrial images. *Pattern Recognition Letters*, 4:185–192, 1986.

[Davies 86b]   E. R. Davies and A. I. C. Johnstone. Engineering trade-offs in the design of a real-time system for the visual inspection of small products. In *UK RESEARCH IN ADVANCED MANUFACTURE: Proc. Inst. Mechanical Engineers*, pages 15–22, Mechanical Engineering Pub. Ltd, London, London, 1986.

[Deans 81]   S. R. Deans. Hough Transform from the Radon Transform. *IEEE Trans. Pattern Analysis and Machine Intelligence*, PAMI–3:185, 1981.

[Deeker 72]   G. F. P. Deeker and J. P. Penny. On interactive map storage and retrieval. *Information*, 10:62–74, 1972.

[Dennis 80]   J. B. Dennis. Data Flow Supercomputers. *IEEE Computer*, 48–56, November 1980.

[Dolecek 84]   Q. E. Dolocek. *Parallel Processing Systems for VHSIC*. Technical Report, Applied Physics Lab., Johns Hopkins University, Laurel, MD, 1984.

[Doo 78]   D. Doo and M. Sabin. Analysis of the behaviour of recursive division surfaces. *Computer Aided Design*, 10(6):356–360, 1978.

[Doyle 62]   W. Doyle. Operation useful for similarity-invariant pattern recognition. *J. Assoc. Comput. Mach.*, 9:259–267, 1962.

[Dreschler 81] L. Dreschler and H.-H. Nagel. Volumetric Model and 3D-Trajectory of a Moving Car Derived from Monocular TV-Frame Sequence of a Street Scene. In *Int. J. Conf. Artificial Intelligence*, pages 692–699, 1981.

[Duda 72]   R. O. Duda and P. E. Hart. Use of hough transformation to detect lines and curves in pictures. *Comm. ACM.*, 15:11–15, 1972.

[Duda 73]   R. O. Duda and P. E. Hart. *Pattern Classification and Scene Analysis*. Wiley, New York, 1973.

[Dudani 77]   S. A. Dudani, K. J. Breeding, and R. B. McGhee. Aircraft Identification by Moment Invariants. *IEEE Trans. Computers*, C-26:39–46, 1977.

[Duff 73]   M. J. B. Duff, D. M. Watson, T. J. Fountain, and G. K. Shaw. A cellular logic array for image processing. *Patt. Recog.*, 5:229–247, 1973.

[Duff 76]   M. J. B. Duff. Clip 4: a large scale integrated circuit array parallel processor. In *Proc. 3rd Int. Joint Conference Pattern Recognition*, pages 728–733, Coronado, California, 1976.

[Duff 78]   M. J. B. Duff. Review of the CLIP image processing system. *Proc. National Computer Conference*, 1055–1060, 1978.

[Duff 80]   M. J. B. Duff. Propagation in cellular logic array. In *Proc. IEEE Workshop on Picture Data Description and Management*, pages 259–262, 1980.

[Duff 89]   M. J. B. Duff. The Limits of Computing in Arrays of Pro-
                 cessors. In J. C. Simon, editor, *From Pixels to Features*,
                 pages 403–413, Elsevier Science Publishers B. V., North Hol-
                 land, 1989.

                 bibitem[Dyer 77]Dyer77 C. R. Dyer and A. Rosenfeld. *Cellu-
                 lar pyramids for image analysis.* Technical Report No. 544,
                 Computer Science Center, University of Maryland, College
                 Park, MD, 1977.

[Dyer 82]   C. R. Dyer. The space efficiency of quadtrees. *Computer
                 Graphics and Image Processing*, 19:335–348, 1982.

[Edmonds 88]   J. M. Edmonds. *Studies of Inspection Algorithms and Asso-
                 ciated Microprogrammable Hardware Implementations.* PhD
                 thesis, University of London, 1988.

[Eichmann 88]   G. Eichmann and T. Kasparis. Topologically Invariant Tex-
                 ture Descriptors. *Computer Vision, Graphics, and Image
                 Processing*, 41:267–281, 1988.

[Ein-Dor 85]   P. Ein-Dor. Grosch's law re-visited: CPU power and cost of
                 computation. *Commun. ACM*, 28(2):142–151, 1985.

[Ekhlund 79]   J. O. Ekhlund. On the use of fourier phase features for texture
                 descrimination. *Computer Graphics Image Process.*, 9:199–
                 201, 1979.

[Erman 80]   L. D. Erman et al. The Hearsay-II Speech-Understanding Sys-
                 tem: Integrating Knowledge to Resolve Uncertainty. *Com-
                 puting Survey*, 12:213–253, 1980.

[Faber 78]   R. D. Faber, C. J. Burstone, and D. J. Solonche. Comput-
                 erized interactive orthodontic treatment planning. *American
                 Journal of Orthodontics*, 73(1):36–46, 1978.

[Farhat 85]   N. Farhat, D. Psaltis, A. Prata, and E. G. Paek. Optical Im-
                 plementation of the Hopfield Model. *Applied Optics*, 24:1469,
                 1985.

[Farhat 87]   N. Farhat and D. Psaltis. Optical Implementation of Asso-
                 ciative Memory. In J. L. Horner, editor, *Optical Signal Pro-
                 cessing*, pages 129–162, Academic Press, 1987.

[Fau 87]   P. Fau and T. Hanson. Using generic geometric models for
                 intelligent shape extraction. In *Proc. DARPA Image Under-
                 standing Workshop*, pages 227–233, 1987.

[Feder 66]   J. Feder and H. Freeman. Digital curve matching using a
                 contour correlation scheme. *IEEE Int. Conv. Record*, 3:69–
                 85, 1966.

[Feldman 74]   J. A. Feldman and Y. Yakimovsky. Decision theory and arti-
                ficial intelligence: I. A semantics-based region analyzer. *Ar-
                tificial Intelligence*, 5(4):349–371, 1974.

[Feldman 85]   J. A. Feldman. Connectionist Models and Parallelism in High
                Level Vision. *Computer Vision, Graphics, and Image Pro-
                cessing*, 31:178–200, 1985.

[Fennema 79]   C. L. Fennema and W. B. Thompson. Velocity Determina-
                tion in Scenes Containing Several Moving Objects. *Computer
                Graphics and Image Processing*, 9:301–315, 1979.

[Feng 74]   T. Y. Feng. Data manipulation Functions in Parallel Pro-
                cessors and their Implementations. *IEEE Trans. Computer*,
                C-23(3):309–318, 1974.

[Fischler 81]   M. A. Fischler and R. C. Bolles. Random sample consen-
                sus: A paradigm for model fitting with applications to image
                analysis and automated cartography. *Comm. ACM*, 381–395,
                June 1981.

[Fisher 71]   C. Fisher. The New Quantimet 720. *Microscope*, 1, 1971.

[Fisher 81]   A. Fisher. Systolic Algorithms for Running Order Statistics
                in Signal and Image Processing. In H. T. Kung, R. F. Sproull,
                and G. L. Steele, Jr., editors, *VLSI Systems and Computa-
                tions*, pages 265–272, Carnegie-Mellon University, Computer
                Science Press., 1981.

[Fisher 84]   J. A. Fisher. Very Long Instruction Word Architectures and
                Eli-512. In *Proc. of the 10th Computer Architecture Conf.*,
                pages 140–150, 1984.

[Fisher 87]   A. L. Fisher and P. T. Highnam. Computing the Hough
                Transform on a Scan Line Array Processor. In *1987 Workshop
                on Computer Architecture for Pattern Analysis and Machine
                Intelligence (CAPAMI'87)*, pages 83–87, IEEE Computer So-
                ciety Press, Seattle, Washington, 1987.

[Flynn 66]   M. J. Flynn. Very high-speed computing systems. *Proc.
                IEEE*, 54:1901–1909, 1966.

[Flynn 72]   M.J. Flynn. Some computer organisations and their effective-
                ness. *IEEE Trans. on Computers*, C-21:948, 1972.

[Foley 90]   J. D. Foley, A. Van Dam, S. K. Feiner, and J. F. Hughes.
                Computer Graphics: Principles and Practice. Addison Wes-
                ley, Reading, Massachusetts, 1990.

[Forrest 72]   A. R. Forrest. Interactive interpolation and approximation
                  by bezier polynomials. *Computer Journal*, 15:71–79, 1972.

[Forshaw 88]   M. R. B. Forshaw. Speeding Up the Marr–Hildreth Edge
                  Operator. *Computer Vision, Graphics, and Image Processing*,
                  41:172–185, 1988.

[Forsythe 60]   G. Forsythe and W. Wasow. *Finite Difference Methods for
                  Partial Differential Equations*. Wiley, New York, 1960.

[Fountain 81]   T. J. Fountain. Clip4: a progress report. In M. J. B. Duff and
                  S. Levialdi, editors, *Languages and Architectures for Image
                  Processing*, pages 283–291, Academic Press, London, 1981.

[Fountain 83a]   T. J. Fountain. The development of the clip7 image processing
                  system. *Pattern Recognition Letters*, 1:331–339, 1983.

[Fountain 83b]   T. J. Fountain. *A Study of Advanced Processor Architectures
                  and Their Optimisation for Image Processing Tasks*. PhD
                  thesis, University of London, 1983.

[Fountain 86]   T. J. Fountain. Array architecture for iconic and symbolic im-
                  age processing. In *Proceedings of the 8th Int. Pattern Recog-
                  nition Conference*, pages 24–33, Paris, France, 1986.

[Fountain 87a]   T. J. Fountain. *Processor Arrays: Architecture and Applica-
                  tions*. Academic Press, New York, 1987.

[Fountain 87b]   T. J. Fountain, M. Postranecky, and G. K. Shaw. The CLIP4S
                  system. *Pattern Recognition Letters*, 5:71–79, 1987.

[Fountain 88]   T. J. Fountain, K. N. Matthews, and M. J. B. Duff. The
                  CLIP7A Image Processor. *IEEE Trans. Pattern Analysis and
                  Machine Intelligence*, PAMI-10:310–319, 1988.

[Freeman 61]   H. Freeman. On the encoding of arbitrary geometric configu-
                  rations. *IRE Trans. Electron Comput.*, EC-10:260–268, 1961.

[Freeman 74]   H. Freeman. Computer processing of line-drawing images.
                  ACM Comp. Surv., 6:57–97, 1974.

[Freeman 78]   H. Freeman. Shape description via the use of critical points.
                  Pattern Recognition, 10:159–166, 1978.

[Frei 77]   W. Frei. Image enhancement by histogram hyperbolization.
                  Computer Graphics Image Processing, 6:286, 1977.

[Fu 87]   K. S. Fu, R. C. Gonzalez, and C. S. G. Lee. *ROBOTICS
                  – Control, Sensing, Vision, and Intelligence*. McGraw-Hill,
                  New York, 1987.

[Fukushima 75]   K. Fukushima. Cognitron: A Self-Organizing Multilayered Neural Network. *Biological Cybernetics*, 20(3/4):121–136, 1975.

[Fukushima 88]   K. Fukushima. A Neural Network for Visual Pattern Recognition. *IEEE Computer*, 65–75, March 1988.

[Gajski 84]   D. Gajski et al. Ceder. In *Proc. Camcon*, pages 306–309, Spring, 1984.

[Gajski 85]   D. Gajski and J-K. Peir. Essential Issues in Multiprocessor Systems. *Computer*, 9–27, 1985.

[Galloway 75]   M. M. Galloway. Texture analysis using grey-level run lengths. *Computer Graphics Image Process.*, 4:172–179, 1975.

[Garibotto 79]   G. Garibotto and L. Lambarelli. Fast on-line implementation of two dimensional median filtering. *Electron. Lett.*, 15:24–25, January 1979.

[Gennery 79]   D. Gennery. Stereo-camera calibration. In *Proc. Image Understanding Workshop*, pages 101–107, 1979.

[Gennery 80]   D. Gennery. Object detection and measurement using stereo vision. In *Proc. Image Understanding Workshop*, pages 161–167, 1980.

[Gerritsen 83]   F. A. Gerritsen. A Comparison of the CLIP4, DAP and MPP Processor-Array Implementations. In M. J. Buff, editor, *Computer Structures for Image Processing*, pages 15–29, Academic Press, London, 1983.

[Gibson 50]   J. J. Gibson. *The perception of the visual world*. Riverside Press, Cambridge, 1950.

[Gibson 66]   J. J. Gibson. *The senses considered as perceptual systems*. Houghton Mifflin, Boston, 1966.

[Gibson 79]   J. J. Gibson. *The ecological approach to visual perception*. Houghton Mifflin, Boston, 1979.

[Gmitro 87]   A. F. Gmitro and G. R. Gindi. Optical neurocomputer for implementation of the Marr–Poggio stereo algorithm. In M. Audill and C. Butler, editors, *IEEE First Int. Conf. Neural Networks*, pages 599–606, IEEE, Picataway, NJ, 1987.

[Goetcherian 80] Vartkes Goetcherian. From binary to grey tone image processing using fuzzy logic concepts. *Pattern Recognition*, 12:7–15, 1980.

[Golay 69]   M. J. E. Golay. Hexagonal parallel pattern transformations. *IEEE Trans. Computer*, C-18:733–740, 1969.

[Gong 89]   S. G. Gong. *Parallel Computation of Visual Motion*. PhD thesis, Dept. Engineering Science, Oxford Uni., 1989.

[Gong 90]   S. Gong and M. Brady. Parallel Computation of Optic Flow. In O. Faugeras, editor, *Computer Vision — ECCV90*, pages 124–133, Springer Verlag, 1990.

[Gonzalez 84]   R. C. Gonzalez and P. Wintz. *Digital Image Processing*. Addison-Wesley, Reading, MA, 1984.

[Goodman 68]   J. W. Goodman. *Introduction to Fourier Optics*. McGraw-Hill, New York, 1968.

[Goodman 78]   J. W. Goodman, A. R. Dias, and L. M. Woody. Fully Parallel, High-Speed Incoherent Optical Method for Performing Discrete Fourier Transforms. *Opt., Lett.*, 2:1, 1978.

[Goodman 86]   T. N. T. Goodman and K. Unsworth. Manipulating shape and producing geometric continuity in beta-spline curves. *IEEE Computer Graphics and Application*, 50–56, 1986.

[Gordon 69]   W. J. Gordon. Spline-blended surface interpolation through curve networks. *J. Math. Mech.*, 18(10):931–952, 1969.

[Gordon 74]   W. J. Gordon and R. F. Riesenfeld. B-spline curves and surfaces. In R. E. Barnhill and R. F. Riesenfeld, editors, *Computer Aided Geometric Design*, pages 95–126, Academic Press, New York, 1974.

[Goshtasby 85]   A. Goshtasby. Template Matching in Rotated Images. *IEEE Trans. Pattern Analysis and Machine Intelligence*, PAMI-7:338–344, 1985.

[Gottlieb 83]   A. Gottlieb et al. The NYU Ultracomputer — Designing an MIMD Shared Memory Parallel Computer. *IEEE Trans. Computer*, C-32(2):175–189, 1983.

[Graf 88]   H. P. Graf, L. D. Jackel, and W. E. Hubbard. VLSI Implementation of a Neural Network Model. *IEEE Computer*, 41–49, March 1988.

[Graf 89]   H. P. Graf and L. D. Jackel. Analog Electronic Neural Network Circuits. *IEEE Circuits and Devices Magazine*, 44–49, July 1989.

[Graham 66]   R. E. Graham. Snow removal — A noise-stripping process for picture signal. *IRE Trans. Inf. Theor.*, 8:129–144, 1966.

[Graham 80]   D. Graham and P. E. Norgren. The diff3 analyzer: a parallel/serial golay image processor. In M. Onoe, K. Preston Jr., and A. Rosenfeld, editors, *Real Time Medical Image Processing*, pages 163–182, Plenum, London, 1980.

[Graham 83]   J. Graham. The Magiscan interactive chromosome karyotyper — MICKY. In *Proc. Vth Euro. Chromosome Analysis Workshop*, pages 6.5.1–6.5.4, Heidelberg, 1983.

[Graham 87]   J. Graham. Automation of routine clinical chromosome analysis I – karyotyping by machine. *Anal. Quant. Cytol. Histol.*, 9:383–390, 1987.

[Granum 80]   E. Granum. *Recognition Aspects of Chromosome Analysis*. PhD thesis, Technical Univ. of Denmark, 1980.

[Grasmuller 84]   H. Grasmuller et al. Local feature extraction for model-based workpiece recognition. In *Proceedings of the 7th Int. Pattern Recognition Conference*, pages 886–889, Montreal,Canada, 1984.

[Gravley 74]   J. F. Gravely and P. M. Benzies. The clinical significance of tracing error in cephalometry. *British Journal of Orhtodentics*, 1(3):95–101, 1974.

[Grimson 79]   W. E. L. Grimson and D. Marr. A computer implementation of a theory of human stereo vision. In *Proc. Image Understanding Workshop*, pages 41–47, Palo Alto, CA, 1979.

[Grimson 80]   W. E. L. Grimson. Aspects of a computational theory of human stereo vision. In *Proc. Image Understanding Workshop*, pages 128–149, 1980.

[Groen 85]   F. C. A. Groen *Prophase banding classifier*. In *Preliminary Report on EEC Work Group Meeting on Automated Chromosome Analysis*, pages 82–93, Leiden, 1985.

[Grosch 85]   H. R. J. Grosch. High speed arithmetic: The digital computer as a research tool. *J. Opt. Soc. Am.*, 43:306–310, 1985.

[Guerra 86]   C. Guerra. A VLSI algorithm for the optimal detection of a curve. *J. Parallel Distributed Computing*, 3, 1986.

[Guerra 87]   C. Guerra and S. Hambrusch. Parallel Algorithm for Line Detection on a Mesh. In *1987 Workshop on Computer Architecture for Pattern Analysis and Machine Intelligence (CA-PAMI'87)*, pages 99–106, IEEE Computer Society Press, Seattle, Washington, 1987.

[Gupta 88]   M. M. Gupta. Fuzzy Logic, Neural Networks and Computer
                  Vision. In *SPIE Conf. on Intelligent Robots and Computer
                  Vision*, pages 366–368, Cambridge, 6-11 Nov., 1988.

[Gupta 89]   M. M. Gupta and G. K. Knopf. The Percept: A Neural Model
                  for Machine Vision. In *Proc. of the Int. Conf. on New Gen-
                  eration of Computers*, pages 147–154, Beijing, 17-19 April,
                  1989.

[Gurd 85]   J. R. Gurd, C. C. Kirkham, and I. Watson. The Manch-
                  ester prototype dataflow computer. *Comm. ACM*, 28(1):34–
                  52, 1985.

[Gustafson 88a]   J. L. Gustafson. Reevaluating Amdahl's Law. *Comm. of the
                  ACM*, 31(5):532–533, 1988.

[Gustafson 88b]   J. L. Gustafson, G. R. Montry, and R. E. Benner. Devel-
                  opment of Parallel Methods for a 1024-Processor Hypercube.
                  SIAM Journal on Scientific and Statistical Computing, 9:609–
                  638, 1988.

[Guzman 69]   A. Guzman. Automatic Interpretation and Classification of
                  Images. In A. Grasseli, editor, *Automatic Interpretation and
                  Classification of Images*, Academic Press, New York, 1971.

[Hall 74]   E. Hall. Almost uniform distribution for computer image en-
                  hancement. *IEEE Trans. Computer*, C-23:207, 1974.

[Hall 79]   E. L. Hall. *Computer Image Processing and Recognition*. Aca-
                  demic Press, New York, 1979.

[Hanahara 88]   K. Hanahara, T. Maruyama, and T. Uchiyama. A Real-Time
                  Processor for the Hough Transform. *IEEE Trans. Pattern
                  Analysis and Machine Intelligence*, PAMI-10:121–125, 1988.

[Hannah 80]   M. J. Hannah. Bootstrap stereo. In *Proc. Image Understand-
                  ing Workshop*, pages 201–208, !980.

[Hanson 78]   A. R. Hanson and E. M. Riseman. VISIONS : a computer
                  system for interpreting scenes. In *Computer Vision Systems*,
                  Academic Press, 1978.

[Haralick 73]   R. M. Haralick, K. Shanmugam, and I. Dinstein. Textural
                  features for image classification. *IEEE Trans. Systems, Man,
                  Cybernet.*, SMC-3(6), 1973.

[Haralick 80]   R. M. Haralick. Edge and region analysis for digital image
                  data. *Computer Graphics and Image Processing*, 12:60–73,
                  1980.

[Haralick 84]   R. M. Haralick. Digital step edges from zero crossings of second directional derivatives. *IEEE Trans. on Pattern Recognition and Machine Intelligence*, PAMI-6:58–68, 1984.

[Haralick 85]   Robert M. Haralick and Linda G. Shapiro. Image Segmentation Techniques. *Computer Vision, Graphics, and Image Processing*, 29:100–132, 1985.

[Haralick 87]   R. M. Haralick, S. R. Sternberg, and X. Zhuang. Image Analysis Using Mathematical Morphology. *IEEE Trans. Pattern Analysis and Machine Intelligence*, PAMI-9:532–550, 1987.

[Harlow 73]   C. A. Harlow and S. A. Eisenbeis. The analysis of radiographic images. *IEEE Trans. Computer*, C-22:678–688, 1973.

[Harlow 76]   C. A. Harlow, S. J. Dwyer, and G. Lodwick. On radiographic image analysis. In A. Rosenfeld, editor, *Digital Picture Analysis*, pages 67–150, Springer-Verlag, New York, 1976.

[Harris 87]   C. G. Harris. Determination of ego-motion from matched points. In *Proc. Alvey Vision Conf.*, Cambridge (U.K), 1987.

[Hartley 42]   R. V. L. Hartley. A more symmetrical Fourier analysis applied to transmission problems. *Proc. IRE*, 30:144–150, 1942.

[Hausler 77]   G. Hausler. Optical software survey. *Opt. Acta.*, 24:965–977, 1977.

[Hayes 86]   J. P. Hayes, T. Mudge, Q. F. Stout, S. Colley, and J. Palmer. A Microprocessor-based Hypercube Supercomputer. *IEEE Micro*, 6–17, Oct. 1986.

[Hayes-Roth 85a] B. Hayes-Roth. A Blackboard Architecture for Control. *Artificial Intelligence*, 26:251-321, 1985.

[Hayes-Roth 85b] F. Hayes-Roth. Rule–based Systems. *Comm. ACM*, 28:921-932, 1985.

[Hecht-Nielson 88] R. Hecht-Nielson. Neurocomputing: picking the human brain. *IEEE Spectrum*, 25(3):36–41, 1988.

[Hennessy 84]   J. L. Hennessy. VLSI Processor Architecture. *IEEE Trans. Computer*, C-33(12):1221–1246, 1984.

[Hilditch 69]   C. J. Hilditch. Linear skeletons from square cupboards. In *Machine Intelligence IV*, pages 403–420, Edinburgh Univ. Press, Edinburgh, 1969.

[Hilditch 83]   C. J. Hilditch. Comparison of thinning algorithms on a parallel processor. *Image and Vision Computing*, 1:115–132, 1983.

[Hilditch 84]   E. C. Hildreth. *Measurement of Visual Motion*. MIT Press, Cambridge, MA, 1984.

[Hillis 85]   W. D. Hillis. *The Connection Machine*. MIT Press, Cambridge, MA, 1985.

[Hillis 87a]   W. D. Hillis. The Connection Machine. *Scientific American*, 256:108–115, June 1987.

[Hillis 87b]   W. D. Hillis and G. L. Steele. Data parallel algorithms. *Commun. ACM*, 29:1170–1183, 1987.

[Hinton 84]   G. E. Hinton *et al*. *Boltzmann machines: Constraint satisfaction networks that learn*. Technical Report No. CMU-CS-84-119, Carnegie-Mellon University, Pittsburgh, PA, 1984.

[Hinton 89]   G. E. Hinton. Connectionist Learning Procedures. *Arificial Intelligence*, 40:185–234, 1989.

[Hoare 78]   C. A. R. Hoare. Communicating sequential processes. *Communication of the ACM*, 21(8):666–677, 1978.

[Hoare 85]   C. A. R. Hoare. *Communicating Sequential Processes*. Prentice Hall, Englewood Cliffs, NJ, 1985.

[Hodgson 85]   R. M. Hodgson, D. G. Bailey, M. J. Naylor, A. L. M. Ng, and S. J. McNeill. Properties, implementations and applications of rank filters. *Image and Vision Computing*, 3:3–14, 1985.

[Homewood 87]   M. Homewood et al. The IMS T800 Transputer. *IEEE Micro*, 7(5):10–26, 1987.

[Hopfield 82]   J. J. Hopfield. Neural networks and physical systems with emergent collective computational abilities. In *Proc. Nat'l Acad. Sci., USA*, pages 2554–2558, 1982.

[Hopfield 84]   J. J. Hopfield. Neurons with graded response have collective computational properties like those of two-state neurons. In *Proc. Nat'l Acad. Sci., USA*, pages 3088–3092, 1984.

[Horn75]   B. K. P. Horn. Obtaining shape from shading information. In P. H. Winston, editor, *The Psychology of Computer Vision*, McGraw-Hill, New York, 1975.

[Horn 77a]   B. K. P. Horn. Understanding image intensities. *Artificial Intelligence*, 8(2):201–231, 1977.

[Horn 77b]   B. K. P. Horn and R. W. Sjoberg. Calculating the reflectance map. *Applied Optics*, 18(11):1770–1779, 1977.

[Horn 81]   B. K. P. Horn and B. G. Schunck. Determining Optical Flow. *Artificial Intelligence*, 17:185–203, 1981.

[Horn 86a]   B. K. P. Horn. *Robot Vision*. The MIT Press, Cambridge, Massachusetts, 1986.

[Horn 86b]   B. K. P. Horn and M. J. Brooks. The Variational Approach to Shape from Shading. *Computer Vision, Graphics, and Image Processing*, 33:174–208, 1986.

[Horn 90]   B. K. P. Horn. Height and Gradient from Shading. *Int. Journal of Computer Vision*, 5(1):37–75, 1990.

[Horowitz 74]   S. L. Horowitz and T. Pavlidis. Picture segmentation by a directed split-and-merge procedure. In *Proc. 2nd IJCPR*, pages 424–433, 1974.

[Horowitz 78]   E. Horowitz and S. Sahni. *Fundamentals of Computer Algorithms*. Pitman, London, 1978.

[Hough 62]   P. V. C. Hough. Method and means for recognizing complex patterns. U. S. Patent 3,069,654;, 1962.

[Hu 62]   M. K. Hu. Visual pattern recognition by moment invariants. *IEEE Trans. Inform. Theory*, 8:179–187, 1962.

[Huang 79]   T. S. Huang, G. Y. Gang, and G. Y. Tang. A fast two dimensional median filtering algorithm. *IEEE Trans. Acoust., Speech, Signal Process.*, ASSP–27:13–18, 1979.

[Huang 82]   T. S. Huang, editor. *Two-Dimensional Digital Signal Processing II: Transformation and Median Filters*. Springer-Verlag, New York / Heidelberg, 1982.

[Huang 87]   K. S. Huang, B. K. Jenkins, and A. A. Sawchuk. Binary Image Algebra and Digital Optical Cellular Image Processors. In *Topical Meetings on Optical Computing*, pages 20–23, Incline Village, Nevada, 1987.

[Huang 89]   K. S. Huang, B. K. Jenkins, and A. A. Sawchuk. Binary Image Algebra and Optical Cellular Logic Image Processors. *Computer Vision, Graphics, and Image Processing*, 45:295–345, 1989.

[Hubel 62]   D. H. Hubel and T. N. Wiesel. Receptive fields, binocular interaction and functional architecture in the cat's visual cortex. *J. Pysiol*, 160:106–154, 1962.

[Hubel 77]   D. H. Hubel and T. N. Wiesel. Functional architecture of macaque monkey visual cortex. *Proceedings of the Royal Society of London*, B-198:1–59, 1977.

[Hueckel 71]   M. Hueckel. An operator which locates edges in digitised pictures. *J. ACM*, 18:113–125, 1971.

[Huertas 86]   A. Huertas and G. Medioni. Edge detection with subpixel precision. *IEEE Trans. Pattern Analysis and Machine Intelligence*, PAMI-8(5):651–664, 1986.

[Huertas 88]   A. Huertas and R. Nevatia. Detecting building in aerial images. *Computer Vision, Graphics and Image Processing*, 1988.

[Huffman 71]   D. A. Huffman. Impossible objects as nonsense sentences. In R. Meltzer and D. Michie, editors, *Machine Intelligence 6*, Elsevier, 1971.

[Hummel 75]   R. Hummel. Histogram modification techniques. *Computer Graphics Image Processing*, 4:209, 1975.

[Hummel 77]   R. Hummel. Image enhancement by histogram transformation. *Computer Graphics Image Processing*, 6:184, 1977.

[Hunt 81]   D. J. Hunt. The ICL DAP and its application to image processing. In M. J. B. Duff and S. Levialdi, editors, *Languages and Architectures for Image Processing*, pages 275–282, Academic Press, London, 1981.

[Hussain 84]   Zahid Hussain. *Automatic Identification and Measurement of Cephalometric Landmarks*. Technical Report Report: 84/4, Image Processing Group, Dept. of Physics and Astronomy, University College London, 1984.

[Hussain 85]   Z. Hussain and H. H-S. Ip. Automatic identification of cephalometric features on skull radiographs. *ACTA POLYTECHNICA SCANDINAVIA (Image Science'85)*, 194–197 (Vol. 2), 1985.

[Hussain 88a]   Zahid Hussain. *Automatic Segmentation and Measurement of Cephalometric Landmarks on Lateral Skull Radiographs Using Cellular Logic Image Processors*. PhD thesis, University of London, 1988.

[Hussain 88b]   Zahid Hussain. A fast approximation to a convex hull. *Pattern Recognition Letters*, 8:289–294, 1988.

[Hussain 88c]   Zahid Hussain. Boundary Descriptors — A Critical Survey. Technical Report 88/2, Machine Vision Group, Royal Holloway & Bedford New College (University of London), Egham Hill, Egham, Surrey TW20 0EX, England, U.K., 1988.

[Hutchinson 88]   J. Hutchinsons, C. Koch, J. Luo, and C. Mead. Computing Motion Using Analog and Binary Resistive Networks. *IEEE Computer*, 52–63, March 1988.

[Hwang 78]   J. J. Hwang and E. L. Hall. Scene representation using adjacency matrices and sampled shapes of regions. In *Proc. IEEE Conf. on Pattern Recognition and Image Processing*, pages 250–261, Chicago, 1978.

[Hwang 85]   K. Hwang and F. A. Briggs. *Computer Architecture and Parallel Processing*. McGraw–Hill, New York, 1985.

[Hwang 87]   K. Hwang and J. Ghosh. Hypernet: A Communication-Efficient Architecture for Constructing Massively Parallel Computers. *IEEE Trans. Pattern Analysis and Machine Intelligence*, PAMI-36(12):1450–1466, 1987.

[Ibrahim 84]   H. A. H. Ibrahim. *Image understanding algorithms on fine-grained tree-structured SIMD machine*. PhD thesis, Dept. of Computer Science, Columbia University, 1984.

[Ibrahim 87]   H. A. H. Ibrahim, J. R. Kender, and D. E. Shaw. Low-Level Image Analysis Tasks on Fine–Grained Tree–Structured SIMD Machines. *Journal of Parallel and Distributed Computing*, 4:546–574, 1987.

[Ikeuchi 81]   K. Ikeuchi and B. K. P. Horn. Numerical shape from shading and occluding boundaries. *Artificial Intelligence*, 17:141–185, 1981.

[Inmos 84]   INMOS LTD. *Occam Programming Manual*. Prentice–Hall, Englewood Cliffs, NJ, 1984.

[Ip 80]   H. H-S. Ip. *An analysis of the two–dimensional Fourier Transform for a typical parallel processor*. Technical Report, Image Processing Group, Dept. of Physics and Astronomy, University College London, 1980.

[Ip 83a]   H. H-S. Ip. *Automatic Detection and Reconstruction of Three-Dimensional Objects Using a Cellular Array Processor*. PhD thesis, University College London, 1983.

[Ip 83b]   H. H-S. Ip. Detection and three-dimensional reconstruction of a vascular network from serial sections. *Pattern Recognition Letters*, 1(5,6):497–505, 1983.

[Ip 84]   H. H-S. Ip. *Preliminary Report on the Algorithms for Computer Recognition of Cephalometric Landmarks on Lateral Skull Radiographs*. Technical Report Report: 84/3, Image Processing Group, Dept. of Physics and Astronomy, University College London, 1984.

[Jacobus 80]   C. J. Jacobus, R. T. Chin, and J. M. Selander. Motion Detec-
                  tion and Analysis of Matching Graphs of Intermediate–Level
                  Primitives. *IEEE Trans. Pattern Analysis and Machine In-
                  telligence*, PAMI-2:465–510, 1980.

[Jahns 80]   J. Jahns. Concepts of optical digital computing — A survey.
                  Optik, 57:429–449, 1980.

[Jain 79a]   R. Jain and H.-H. Nagel. On the Analysis of Accumula-
                  tive Difference Picture from Image Sequence of Real World
                  Scenes. *IEEE Trans. Pattern Analysis and Machine Intelli-
                  gence*, PAMI-1:206–214, 1979.

[Jain 79b]   R. Jain, W. N. Martin, and J. K. Aggarwal. Extraction of
                  Moving Objects Images Through Change Detection. In *Proc.
                  6th Int. Conf. Artificial Intelligence*, pages 425–428, 1979.

[Jelinek 78]   J. Jelinek. An algebraic theory for parallel processor design.
                  Computer Journal, 22:363–375, 1978.

[Jamieson 86]   L. H. Jamieson, P. T. Muller, and H. J. Siegel. FFT Al-
                  gorithms for SIMD Parallel Processing Systems. *Journal of
                  Parallel and Distributed Computing*, 3:48–71, 1986.

[Jenkins 85]   B. K. Jenkins and A. A. Sawchuk. Optical Cellular Logic
                  Architectures for Image Processing. In *IEEE Computer Soc.
                  Workshop on Computer Architecture for Pattern Analysis and
                  Image Database management*, pages 61–65, IEEE, Miami, FL,
                  1985.

[Jenny 77]   C. J. Jenny. Process Partitioning in Distributed Systems.
                  IEEE NTC Conf. Record, Vol. 2:31:1–10, 1977.

[Jensen 74]   K. Jensen and N. Wirth. *PASCAL User Manual and Report*.
                  Springer-Verlag, New York, 1974.

[Johannsen 82]   G. Johannsen and J. Bille. A threshold selection method us-
                  ing information measures. In *Proc. 6th Int. Conf. Pattern
                  Recognition*, pages 140–143, Munich, Germany, 1982.

[Johnson 90]   R. P. Johnson. Contrast based edge detection. *Pattern Recog-
                  nition*, 23(3/4):311–318, 1990.

[Judd 79]   K. J. Judd. Compression of binary images by stroke encoding.
                  Comput. Digital Techniques, 2:41–48, 1979.

[Julesz 71]   B. Julesz. *Foundations of Cyclopean Perception*. Univ.
                  Chicago Press, Chicago, 1971.

[Julesz 75]   B. Julesz. Experiments in the visual perception of texture.
                  Scientific American, 232(4):2–11, 1975.

[Justusson 81]   B. I. Justusson. Median filtering: statistical properties. In
                  T. S. Huang, editor, *Two dimensional digital signal processing
                  II, Top. Appl. Phys.*, pages 161–196, Springer Verlag, Berlin,
                  1981.

[Kaizer 55]   H. Kaizer. *A Qualification of Textures on Aerial Photographs.*
                  Technical Report 121, A. 69484, Boston Univ. Research Lab.,
                  Boston University, 1955.

[Kapur 85a]   J. N. Kapur, P. K. Sahoo, and A. K. C. Wong. A New Method
                  for Grey-Level Picture Threshold Using the Entropy of the
                  Histogram. *Computer Vision, Graphics, and Image Process-
                  ing*, 29:273–285, 1985.

[Karp 66]   R. M. Karp and R. E. Miller. Properties of a Model for
                  Parallel Computation: Determinacy, Termination, Queueing.
                  SIAM J. Applied Math, 14:1390–1411, 1966.

[Katz 65]   Y. H. Katz. Pattern recognition of meteorological satellite
                  cloud photography. In *Proc. Third Symp. on Remote Sensing
                  of Environment*, pages 173–214, 1965.

[Keller 81]   R. M. Keller and W-C. J. Yen. A Graphical Approach to
                  Software Development Using Function Graphs. In *Digest of
                  Papers Compcon Spring 81*, pages 156–161, 1981.

[Kent 85]   E. W. Kent, M. O. Shneier, and R. Lumia. PIPE (Pipelined
                  Image–Processing Engine). *Journal of Parallel and Dis-
                  tributed Computing*, 2:50–78, 1985.

[Kernighan 78]   B.W. Kernighan and D.M. Ritchie. *The C Programming Lan-
                  guage.* Prentice-Hall, 1978.

[Kimme 75]   C. Kimme, D. Ballard, and J. Sklansky. Finding circles by an
                  array of accumulators. *Comm. ACM.*, 18:120–122, 1975.

[King 78]   E. A. King. Laser scanning. In *Proc. 4th Nondestructive
                  Testing of Wood Symp.*, pages 15–22, 1978.

[Kirkpatrick 83] S. Kirkpatrick, C. D. Gelatt Jr., and M. P. Vecci. Optimiza-
                  tion by simulated annealing. *Science*, 220:671–680, 1983.

[Kirchof 80]   W. Kirchof, P. Haberacker, E. Krauth, G. Kritikos, and R.
                  Winter. Evaluation of LANDSAT Image Data for land-Use
                  Mapping. *Acta Astronautica*, 7:243–253, 1980.

[Kirsh 57]   R. A. Kirsh, L. Cahn, C. Ray, and G. H. Urban. Experiments
                  in processing pictorial information with a digital computer. In
                  Proc. Eastern Joint Computer Conf., pages 221–229, 1957.

[Kirsh 71]   R. A. Kirsh. Computer determination of the constituent structure of biological images. *Comput. Biomed. Res.*, 4:315–528, 1971.

[Kitchen 82]   L. Kitchen and A. Rosenfeld. Grey–level corner detection. *Pattern Recognition Letters*, 1:95–102, 1982.

[Kitchen 83a]   P. W. Kitchen and A. Pugh. Processing of binary images. In *Robot Vision*, pages 21–42, Springer-Verlag, Berlin, 1983.

[Kitchen 83b]   P. W. Kitchen and A. Pugh. Processing of binary images. In A. Pugh, editor, *International Trends in Manufacturing Technology. Robot Vision*, Springer-Verlag, 1983.

[Klette 84]   R. Klette. On the approximation of convex hulls of finite grid points sets. *Pattern Recognition Letters*, 2:19–22, 1984.

[Knuth 73]   D. E. Knuth. *The Art of Computer Programming - Volume 3, Sorting and Searching*. Addison-Wesley, Reading, MA, 1973.

[Koch 86]   C. Koch, J. Marroquin, and A. Yuille. Analog neuronal networks in early vision. In *Proc. nat. Acad. Sci.*, pages 4263–4267, 1986.

[Kohler 81]   R. Kohler. A segmentation system based on threshold. *Computer Graphics Image Processing*, 15:319–338, 1981.

[Konecny 81]   C. Konecny and D. Pape. Correlation techniques and devices. *Photogramm. Eng. Remote Sensing*, 47(3):323–333, 1981.

[Kramer 75]   Henry P. Kramer and Judith B. Bruckner. Iterations of a non-linear transformation for enhancement of digital images. *Pattern Recognition*, 7:53–58, 1975.

[Krogman 57]   W. M. Krogman and Sassouni. *Syllabus in Roentgenographic Cephalometry*. Technical Report, P.A. Centre for Research in Child Growth. P.A., U.S.A., 1957.

[Kronsjo 79]   L. I. Kronsjo. *Algorithms: their complexity and efficiency*. Wiley, New York, 1979.

[Kruse 73]   B. Kruse. A Parallel Picture Processing Machine. *IEEE Trans. Computers*, C-22(12):1075–1087, 1973.

[Kruse 77]   B. Kruse. *Design and Implementation of a Picture Processor*. PhD thesis, Univ. of Linkoping, Sweden, 1977.

[Kumar 86]   D. V. K. V. Kumar and C. A. Rahenkamp. Calculation of geometric moments using Fourier plane intensities. *Applied Optics*, 25(6):997–1007, 1986.

[Kung 80]   H. T. Kung. Special–Purpose Devices for Signal and Image Processing: An Opportunity in VLSI. In *Proc. SPIE, Vol. 241, Real–Time Signal Processing III*, pages 76–84, Society of Photo–Optical Instrumentation Engineers, July, 1980.

[Kung 81a]   H. T. Kung and S. W. Song. *A Systolic 2–D Convolution Chip.* Technical Report CMU–CS–81–110, Carnegie–Mellon University Computer Science Dept., 1981.

[Kung 81b]   H. T. Kung, L. M. Ruane, and D. W. L. Yen. A Two–Level Pipelined Systolic Array for Convolution. In H. T. Kung, R. F. Sproull, and G. L. Steele, Jr., editors, *VLSI Systems and Computations*, pages 255–264, Carnegie–Mellon University, Computer Science Press., 1981.

[Kung 82]   H. T. Kung. Why Systolic Architectures? *IEEE Computer*, 15(1):37–46, 1982.

[Kung 84]   H. T. Kung. Systolic Algorithms for the CMU WARP Processor. In *Proceedings of the 7th Int. Pattern Recognition Conference*, pages 570–577, Montreal,Canada, 1984.

[Kung 87]   S. Y. Kung, S. C. Lo, S. N. Jean, and J. N. Hwang. Wavefront Array Processors — Concept to Implementation. *IEEE Computer*, 20(7):18–33, 1987.

[Kushner 82]   T. Kushner, A. Y. Wu, and A. Rosenfeld. Image Processing on ZMOB. *IEEE Trans. Computer*, C-31(10):943–951, 1982.

[Lakshmivarahan 84] S. Lakshmivarahan, S. K. Dhall, and L. L. Miller. Parallel Sorting Algorithms. In *Advances in Computers*, Academic Press, 1984.

[Land 71]   E. H. Land and J. J. McCann. Lightness and retinex theory. *J. Opt. Soc. Am.*, 61:1–11, 1971.

[Lane 80]   J. Lane and R. Riesenfeld. A theoretical development for the computer generation of piecewise polynomial surfaces. *IEEE Trans. Pattern Analysis and Machine Intelligence*, PAMI-2:35–46, 1980.

[Laws 79]   K. I. Laws. Texture energy measures. In *Proc. Image Understanding Workshop*, pages 47–51, 1979.

[Laws 80]   K. I. Laws. *Textured image segmentation.* PhD thesis, Dept. of Engineering, Univ. Southern Calif., 1980.

[Lee 81]   S. H. Lee, editor. *Optical Information Processing: Fundamentals (Topics in Applied Physics, Vol. 23)*. Springer-Verlag, Berlin, 1981.

[Lee 83]   Johg-Sen Lee. Digital Image Smoothing and the Sigma Filter. *Computer Vision, Graphics, and Image Processing*, 24:255–269, 1983.

[Lee 86]   J. S. J. Lee, R. M. Haralick, and L. G. Shapiro. Morphologic edge detection. In *Proc. 8th ICPR*, pages 369–373, Paris, 1986.

[Leese 70]   J. A. Leese, C. S. Novak, and V. R. Taylor. The determination of cloud pattern motion from geosynchronous satellite image data. *Pattern Recognition*, 2:279–292, 1970.

[Leipnik 60]   R. Leipnik. The extended entropy uncertainty principle. *Inf. Control*, 3:18–25, 1960.

[Lettvin 59]   J. Y. Lettvin et al. What the frog's eye tells the frog's brain. *Proc. IRE*, 47:1940–1951, 1959.

[Leu 85]   J-G. Leu and W. G. Wee. Detecting the Spatial Structure of Natural Textures Based on Shape Analysis. *Computer Vision, Graphics, and Image Processing*, 31:67–88, 1985.

[Levy-Mandel 86] A. D. Levy-Mandel, A. N. Venetsanopoulos, and J. K. Tsotsos. Knowledge-based landmarking of cephalograms. *Computers and Biomedical Research*, 19:282–309, 1986.

[Li 87]   H. Li and M. Maresca. Polymorphic–torus network. In *Proc. Int. Conf. on Parallel Processing*, pages 411–414, 1987.

[Li 89]   H. F. Li, D. Pao, and R. Jayakumar. Improvements and systolic implementation of the Hough transformation for straight line detection. *Pattern Recognition*, 22(6):697–706, 1989.

[Lillestrand 72] R. L. Lillestand. Techniques for Change Detection. *IEEE Trans. Computer*, C-21:654–659, 1972.

[Liow 90]   Y-T. Liow and T. Pavlidis. Use of Shadow for Extracting Buildings in Aerial Images. *Computer Vision, Graphics, and Image Processing*, 49:242–277, 1990.

[Lipkin 70]   B. Lipkin and A. Rosenfeld, editors. *Picture Processing and Psychopictorics*. Academic Press, New York, 1970.

[Lloyd 87a]   S. A. Lloyd, E. R. Haddow, and J. F. Boyce. A Parallel Binocular Stereo Algorithm Utilizing Dynamic Programming and Relaxation Labelling. *Computer Vision, Graphics, and Image Processing*, 39:202–225, 1987.

[Lloyd 87b]   D. Lloyd, J. Piper, D. Rutovitz, and G. Shippey. Multiprocessing interval processor for automated cytogenetics. *Applied Optics*, 26:3356–3366, 1987.

[Longuet-Higgins 80] H. C. Longuet-Higgins and K. Prazdny. The interpretation of moving retinal image. *Proc. R. Soc.*, B208:385–397, 1980.

[Lougheed 80] R. M. Lougheed, D. L. McCubbrey, and S. R. Sternberg. Cytocomputers: Architectures for Parallel Image Processing. In *Proc. Workshop Picture Data Descr. and Management*, pages 281–286, Pacific Grive, CA, Aug. 27-28, 1980.

[Lowe 87] D. G. Lowe. Three–Dimensional Object Recognition from Single Two–Dimensional Images. *Artificial Intelligence*, 31:355–395, 1987.

[Lu 78] S. Y. Lu and K. S. Fu. A syntactic approach to texture analysis. *Computer Graphics Image Processing*, 7:303–330, 1978.

[Luetjen 80] K. Luetjen, P. Gemmar, and H. Ischen. FLIP: A flexible multi–processor system for image processing. In *Proc. Fifth Int. Conf. on Pattern Recognition*, Miami, Florida, 1980.

[Lundsteen 86] C. Lundsteen, T. Gerdes, and J. Maahr. Automatic classification of chromosomes as part of a routine system for clinical analysis. *Cytometry*, 7:1–7, 1986.

[Lundstrom 85] S. F. Lundstrom and R. L. Larsen. Computer and Information Technology in the Year 2000 — a Projection. *IEEE Computer*, 68–79, Sept. 1985.

[Mandelbrot 82] B. B. Mandelbrot. *The Fractal Geometry of Nature*. Freeman, San Francisco, CA, 1982.

[Manning 74] J. R. Manning. Continuity conditions for spline curves. *Computer J.*, 17(2):181–186, 1974.

[Maresca 88] M. Maresca, M. A. Lavin, and H. Li. Parallel Architectures for Vision. *Proc. IEEE*, 76(8):970–981, 1988.

[Marr 76] D. Marr. Early processing of visual information. *Philosophical Trans. of the Royal Society of London*, B-275:483–524, 1976.

[Marr 77a] D. Marr and T. Poggio. *A theory of human stereo vision*. Technical Report Memo 451, AI Lab., MIT, Cambridge, MA, 1977.

[Marr 77b] D. Marr and T. Poggio. Cooperative computation of stereo disparity. *Science*, 194:283–287, 1977.

[Marr 80] D. Marr and E. Hildreth. Theory of edge detection. *Proc. R. Soc. Lond.*, D 207.187–217, 1980.

[Marr 81]   D. Marr and S. Ullman. Directional selectivity and its use in
                 early visual processing. *Proc. R. Soc. Lond.*, B 211:151–180,
                 1981.

[Marr 82]   D. Marr. *Vision: A computational investigation into the hu-
                 man representation and processing of visual information.* W.
                 H. Freeman & Co., San Francisco, 1982.

[Martin 79]   W. N. Martin and J. K. Aggarwal. Computer analysis of
                 dynamic scenes containing curvilinear figures. *Pattern Recog-
                 nition*, 11:169–178, 1979.

[Mason 75]   D. Mason, I. J. Lauder, D. Rutoritz, and G. Spowart. Mea-
                 surement of C-Bands in human chromosomes. *Comput. Biol.
                 Med.*, 5:179–201, 1975.

[Matheron 65]   G. Matheron. *Elements pour une Tiorre des Mulieux Poreux.*
                 Masson, Paris, 1965.

[Matheron 75]   G. Matheron. *Random Sets and Integral in Geometry.*
                 Springer–Verlag, New York, 1975.

[Mathews 76]   P. C. Matthews and B. H. Beech. Method and apparatus for
                 detecting timber defects. US Patent 3,976,384, 1976.

[May 83]   D. May. Occam. *SIGPLAN Not. (ACM)*, 18(4):69–79, 1983.

[McCreary 89]   C. McCreary and H. Gill. Automatic Determination of
                 Grain Size for Efficient Parallel Processing. *Comm. ACM*,
                 32(9):1073–1078, 1989.

[McCulloch 43]   W. S. McCullochs and W. Pitts. A logical calculus of the
                 ideas immanent in the nervous activity. *Bull. Math. Biophys.*,
                 5(115), 1943.

[McKee 75]   J. W. McKee and J. K. Aggarwal. Finding the edges on
                 the surfaces of three-dimensional curved objects by computer.
                 Pattern Recognition, 7:25–52, 1975.

[Meagher 82]   D. J. Meagher. Efficient synthetic image generation of arbi-
                 trary 3-D objects. In *Proc. Pattern Recognition and Image
                 Processing Conf.*, pages 473–478, 1982.

[Medioni 85]   G. Medioni and R. Nevatia. Segment–Based Stereo Matching.
                 Computer Vision, Graphics, and Image Processing, 31:2–18,
                 1985.

[Mendelson 87]   H. Mendelson. Economics of Scale in Computing: Grosch's
                 Law Re-visited. *Commun. ACM*, 30(12):1066–1072, 1987.

[Meriam 51]   J. L. Meriam. *Mechanics , Part 2 : Dynamics.* John Wiley & Sons, New York, 1951.

[Merigot 86]   A. Merigot, P. Garda, B. Zavidovique, and F. Devos. Interlayer communication in MIMD pyramidal computer. In *Proceedings of the 8th Int. Pattern Recognition Conference,* pages 954–957, Paris, France, 1986.

[Merlin 75]   P. M. Merlin and D. J. Farber. A Parallel Mechanism for Detecting Curves in Pictures. *IEEE Trans. Computers,* C-24:96–98, 1975.

[Mero 75]   L. Mero and Z. Vassy. A simplified and fast version of the hueckel operator for finding optimal edges in pictures. *Proc. 4th International Joint Conference on Artificial Intelligence,* 650–655, 1975.

[Michaels 81]   C. F. Michaels and C. Carello. *Direct perception.* Prentice Hall, Englewood Cliffs, NJ, 1981.

[Minkowski 03]   H. Minkowski. Volumen und oberfläche. *Math. Ann.,* 57:447–495, 1903.

[Minsky 69]   M. Minsky and S. Papert. *Perceptrons: An Introduction to Computational Geometry.* MIT Press, Cambridge, MA, 1969.

[Miranker 71]   W. L. Miranker. A survey of parallelism in numerical analysis. *SIAM Rev.,* 13:524–545, 1871.

[Mitchell 77]   A. Mitchell and R. Wait. *The Finite Element Method in Partial Differential Equations.* Wiley, New York, 1977.

[Montanari 71]   U. Montanari. On the optimal detection of curves in noisy image. *Comm. of ACM,* 14:335–345, 1971.

[Moravec 79]   H. P. Moravec. Automatic Visual Obstacle Avoidance. In *Int. J. Conf. Artificial Intelligence,* pages 598–600, 1979.

[Moravec 81]   H. P. Moravec. Rover Visual Obstacle Avoidance. In *Proc. 7th Int. Jt. Conf. Artificial Intelligence,* pages 785–790, 1981.

[Mundy 80]   J. L. Mundy and J. F. Jarvis. Automatic visual inspection. In K. S. Fu, editor, *Applications of Pattern Recognition,* CRS Press, New York, 1980.

[Myers 91]   W. Myers. The Drive to the Year 2000. *IEEE Micro,* 10–12, 68–74, Feb. 1991.

[Nagao 76]   M. Nagao, H. Tanabe, and K. Ito. Agricultural land use classification of aerial photographs by histogram similarity method. In *IEEE Proc. 3rd Joint Conf. on Pattern recognition,* pages 669–672, Coronado, CA, 1976.

[Nagao 79]   M. Nagao and T. Matsuyama. Edge preserving smoothing. *Computer Graphics Image Processing*, 9:394–407, 1979.

[Nagao 80]   M. Nagao and T. Matsuyama. *A Structural Analysis of Complex Aerial Photographs*. Plenum, New York, 1980.

[Nagao 82]   M. Nagao. Control Strategies in Pattern Analysis. In *Proc. 6th. Int. Conf. on Pattern Recognition*, pages 996–1006, Munich, Germany, 1982.

[Nagel 78]   H.-H. Nagel. Analysis techniques for image sequences. In *Proc. 4th IJCPR*, pages 186-211, 1978.

[Nagel 81]   H-H. Nagel. Overview on image sequence analysis. In T. S. Huang, editor, *Image Sequence Analysis*, pages 2–39, Springer-Verlag, Berlin, 1981.

[Nakagawa 78]   Yasuo Nakagawa and Azriel Rosenfeld. A Note on the Use of Local min and max Operations in Digital Picture Processing. *IEEE Trans. Systems, Man, and Cybernetics*, SMC-8:632–635, 1978.

[Nalwa 86]   V. S. Nalwa and T. O. Binford. On Detecting Edges. *IEEE Trans. Pattern Analysis and Machine Intelligence*, PAMI-8:699–714, 1986.

[Nazif 84]   A. M. Nazif and M. D. Levine. Low Level Image Segmentation: An Expert System *IEEE Trans. Pattern Analysis and Machine Intelligence*, PAMI-6:555–577, 1986.

[Neumann 66]   J. von Neumann. A system of 29 states with a general transition rule. In A. W. Burks, editor, *Theory of Self-reproducing Automata*, Univ. of Illinois, 1966.

[Nevatia 80]   R. Nevatia and K. R. Babu. Linear feature extraction and description. *Computer Graphics and Image Processing*, 13:257–269, 1980.

[Nevatia 82]   R. Nevatia. *Machine Perception*. Prentice-Hall, Englewood Cliffs, NJ, USA, 1982.

[Nishihara 84]   H. K. Nishihara. Practical real–time imaging stereo matcher. *Optical Engineering*, 23(5):536–545, 1984.

[Noble 88]   J. A. Noble. Finding corners. *Image and Vision Computing*, 6:121–128, 1988.

[Nodes 82]   Thomas A. Nodes and Neal C. Gallagher, Jr. Median Filters: Some Modifications and Their Properties. *IEEE Trans. Acoustics, Speech, and Signal Processing*, ASSP-30:739–746, 1982.

[Norton-Wayne 84] L. Norton-Wayne and P. Saraga. A set of Shapes for the Benchmark Testing of Silhouette Recognition. In *Proceedings of the 4th Int. Conf. Robot Vision and Sensory Controls*, pages 65–74, London, 1984.

[Noyce77] R. N. Noyce. Microelectronics. *Scientific Amer.*, 237:62–69, Sept. 1977.

[Nudd 88] G. R. Nudd and N. D. Francis. Architectures For Image Analysis. In *Third Int. Conf. on Image Processing and its applications*, pages 445–451, IEE, Warwick, England, 1988.

[Ohlander 75] R. Ohlander. *Analysis of Natural Scenes*. PhD thesis, Carnegie–Mellon University, 1975.

[Ohlander 78] R. Ohlander, K. Price, and D. R. Reddy. Picture segmentation using a recursive region splitting method. *Computer Graphics Image Processing*, 8(3):313–333, 1978.

[Olson 87] T. J. Olson, L. Bukys, and M. Brown. Low level image analysis on a MIMD architecture. In *Int. Conf. Computer Vision*, pages 468–475, 1987.

[Ortega 81] J. M. Ortega and W. G. Poole Jr. *An Introduction to Numerical Methods for Differential Equations*. Pitman Publishing Inc., Marshfield, MA, 1981.

[Otto 81] G. P. Otto and D. E. Reynolds. Counting Hardware for Parallel Processors. *Computer Graphics and Image Processing*, 17:185–186, 1981.

[Otto 82a] G. P. Otto and D. E. Reynolds. *IPC Subroutine Library*. Technical Report Report: 82/10, Image Processing Group, Dept. of Physics and Astronomy, University College London, 1982.

[Otto 82b] G. P. Otto. *Counting and Labelling Objects on CLIP4*. Technical Report Report: 82/11, Image Processing Group, Dept. of Physics and Astronomy, University College London, 1982.

[Otto 84] G.P. Otto. *Algorithms For Image Processing On The CLIP4 Cellular Array Processor*. PhD thesis, University of London, 1984.

[Oyster 87] J. M. Oyster, F. Vicuna, and W. Broadwell. Associative network applications to low–level machine vision. *Applied Optics*, 26(10):1919–1926, 1987.

[Pass 81] S. D. Pass. *Parallel Techniques for High Level Image Segmentation Using the CLIP4 Computer*. PhD thesis, University of London, 1981.

[Patterson 80]   D. A. Patterson and D. R. Ditzel. The Case for the RISC. *Computer Architecture News*, 8(6):25–33, 1980.

[Patterson 82]   D. A. Patterson and C. H. Sequin. A VLSI RISC. *IEEE Computer*, 15(9):8–21, 1982.

[Pau 83]   L. F. Pau. Integrated testing and algorithms for visual inspection of integrated circuits. *IEEE Trans. Pattern Analysis and Machine Intelligence*, PAMI-5:602–608, 1983.

[Pavlidis 82]   T. Pavlidis. *Algorithm for Graphics and Image Processing*. Springer-Verlag, Berlin, 1982.

[Pavlidis 83]   Theo Pavlidis. Curve Fitting with Conic Splines. *ACM Trans. Graphics*, 2:1–31, 1983.

[Pavlidis 85]   T. Pavlidis. Scan conversion of regions bounded by parabolic splines. *IEEE Computer Graphics and Application*, 47–53, 1985.

[Pease 77]   M. C. Pease. The indirect binary n–cube microprocessor array. *IEEE Trans. Computer*, C-26:458–473, 1977.

[Peleg 81]   S. Peleg and A. Rosenfeld. A min max medial axis transformation. *IEEE Trans. Pattern Recognition and Machine Intelligence*, PAMI-3:206–210, 1981.

[Peleg 84]   S. Peleg, J. Naor, R. Hartley, and D. Avnir. Multiple resolution texture analysis and classification. *IEEE Trans. Pattern Recognition and Machine Intelligence*, PAMI-6:518–523, 1984.

[Pentland 84]   A. P. Pentland. Fractal based description of natural scenes. *IEEE Trans. Pattern Analysis and Machine Intelligence*, PAMI-6:661–674, 1984.

[Perkins 78]   W. A. Perkins. A Model-Based Vision System for Industrial Parts. *IEEE Trans. Computers*, C-27:126–143, 1978.

[Persoon 81]   E. Persoon and K-S. Fu. Shape determination using fourier descriptors. *IEEE Trans. System, Man, and Cybernetics*, SMC-7, 1981.

[Piper 80]   J. Piper, E. Granum, D. Rutovitz, and H. Ruttledge. Automation of chromosome analysis. *Signal Processing*, 2:203–221, 1980.

[Piper 85]   J. Piper. Efficient implementation of skeletonisation using interval coding. *Pattern Recognition Letts.*, 3:389–397, 1985.

[Piper 86]   J. Piper and D. Rutovitz. A Parallel Processor Implementation of a Chromosome Analysis System. *Pattern Recognition Letts.*, 4:397–404, 1986.

[Piper 87]   J. Piper. The effects of zero feature correlation assumption on maximum likelihood based classification of chromosomes. *Signal Processing*, 12:49–57, 1987.

[Poggio 85]   T. Poggio. Early Vision: From Computational Structure to Algorithms and Parallel Hardware. *Computer Vision, Graphics, and Image Processing*, 31:139–155, 1985.

[Poggio 87]   T. Poggio. MIT progress in image understanding. In *Proc. DARPA IU Workshop*, pages 41–54, Los Angeles, CA, 1987.

[Potter 75]   J. L. Potter. Velocity as a Cue for Segmentation. *IEEE Trans. Systems Man, and Cybernetics*, SMC-5:390–394, 1975.

[Potter 77]   J. L. Potter. Scene Segmentation Using Motion Information. *Computer Graphics and Image Processing*, 6:558–581, 1977.

[Potter 83]   J. L. Potter. Image processing on the massively parallel processors. *Computer*, 16:62–67, 1983.

[Potter 84]   D. J. Potter. *Analysis of Images Containing Blob-like Structures Using an Array Processor*. PhD thesis, University of London, 1984.

[Potter 85]   D. J. Potter. Computer-Assisted Analysis of Two-Dimensional Electrophoresis Images Using an Array Processor. *Computers and Biomedical Research*, 18:347–362, 1985.

[Pratt 78]   W. K. Pratt. *Digital image processing*. John Wiley & Sons Inc., New York, 1978.

[Prazdny 79]   K. Prazdny. Motion and structure from optical flow. In *Proc. Int. J. Conf. on Artificial Intelligence*, pages 702–704, 1979.

[Preston 61]   K. Preston Jr. Machine techniques for automatic identification of binucleate lymphocyte. In *Proc. Fourth Int. Conf. Medical Electronics*, Washington, DC, 1961.

[Preston 73]   K. Preston Jr. Applications of Cellular Logic to Biomedical Image Processing. In *Computer Techniques in Biomedical Engineeering*, page 295, Auerbach, New York, 1973.

[Preston 79]   K. Preston Jr., M. J. B. Duff, S. Levialdi, P. E. Norgren, and J.-I. Toriwaki. Basics of cellular logic with some applications in medical image processing. *Proc. IEEE*, 67(5):826–855, 1979.

[Preston 83]   K. Preston Jr. Xi filters. *IEEE Trans. Acoust., Speech, Signal Processing*, 31:861–876, 1983.

[Prewitt 66]   J. S. M. Prewitt and M. L. Mendelsohn. The analysis of cell images. In *Ann. New York Acad. Sci.*, pages 1035–1053, New York Acad. Sci., New York, 1966.

[Prewitt 70]   J. M. S. Prewitt. Object enhancement and extraction. In B. S. Lipkin and A. Rosenfeld, editors, *Picture Processing and Psychopictorics*, Academic Press, New York, 1970.

[Psaltis 85]   D. Psaltis and N. H. Farhat. Optical information processing based on an associative memory model of neural nets with thresholding and feedback. *Opt. Lett.*, 10(2):98, 1985.

[Pun 80]   T. Pun. A new method for gray–level picture thresholding using the entropy of the histogram. *Signal Processing*, 2:223–237, 1980.

[Pun 81]   T. Pun. Entropic thresholding: a new approach. *Computer Vision, Graphics and Image Processing*, 16:210–239, 1981.

[Radin 83]   G. Radin. The 801 Minicomputer. *IBM J. R&D*, 27(3):237–246, 1983.

[Raffel 85a]   J. I. Raffel. The RVLSI approach to wafer–scale integration. In *Proc. Workshop Wafer–scale Integration*, Southampton Univ., England, 1985.

[Raffel 85b]   J. I. Raffel et al. A wafer–scale digital integrator using restructurable VLSI. *IEEE Trans. Electron. Devices*, ED-32:479–486, 1985.

[Reddaway 79]   S. F. Reddaway. The DAP approach. *Infotech State of the Art Report on Supercomputers*, 2:309–329, 1979.

[Reddy 90]   A. L. N. Reddy and P. Banerjee. Design, Analysis, and Simulation of I/O Architectures for Hypercube Multiprocessors. *IEEE Trans. Parallel and Distributed Systems*, 1(2):140–151, 1990.

[Reeves 80]   A. P. Reeves. On global feature extraction methods for parallel processing. *Computer Graphics and Image Processing*, 14:159–169, 1980.

[Reeves 82]   A. P. Reeves and A. Rostampour. Computational Cost of Image Processing with a Parallel Binary Array Processor. *IEEE Trans. Pattern Analysis and Machine Intelligence*, PAMI-4:449–455, 1982.

[Reeves 84]   A. P. Reeves. Parallel Pascal: An Extended Pascal for Parallel Computers. *Journal of Parallel and Distributed Computing*, 1:64–80, 1984.

[Reynolds 82a] D. E. Reynolds and G. P. Otto. *Software Tools for CLIP4*. Technical Report Report: 82/1, Image Processing Group, Dept. of Physics and Astronomy, University College London, 1982.

[Reynolds 82b] D. E. Reynolds and G. P. Otto. *IPC User Manual*. Technical Report Report No. 82/4, Image Processing Group Department of Physics and Astronomy, University College London, 1982.

[Reynolds 83]   D. E. Reynolds. *Automatic Generation of Image Segmentation Procedures for a Cellular Array*. PhD thesis, University of London, 1983.

[Rhodes 88]   F. M. Rhodes et al. A Monolithic Hough Transform Processor Based on Resturcturable VLSI. *IEEE Trans. Pattern Analysis and Machine Intelligence*, PAMI-10:106–110, 1988.

[Rich 83]   E. Rich. *Artificial Intelligence*. McGraw-Hill, Singapore, 1983.

[Richardson 66] A. Richardson. An investigation into the reproducibility of some points, planes and lines used in cephalometric analysis. *Amer. J. Orthodont.*, 52:637, 1966.

[Ridler 78]   T. W. Ridler and S. Calvard. Picture thresholding using an iterative selection method. *IEEE Trans. on Systems, Man and Cybernetics*, SMC-8(8):630–632, 1978.

[Riesenfeld 73] R. F. Riesenfeld. *Applications of B-spline approximation to geometric problems of computer aided design*. PhD thesis, Syracuse University, New York, 1973.

[Rives 85]   G. Rives, M. Dhome, J-T. Lapreste, and M. Richetin. Detection of patterns in images from piecewise linear contours. *Pattern Recognition Letters*, 3:99–104, 1985.

[Rives 86]   G. Rives, J-T. Lapreste, M. Dhome, and M. Richetin. Planar partially occluded objects scene analysis. In *Proceedings of the 8th Int. Pattern Recognition Conference*, pages 1076–1079, Paris, France, 1986.

[Roberts 65]   L. G. Roberts. Machine perception of three-dimensional solids. In J. P. Tippett et al., editors, *Optical and Electro-optical Information Processing*, MIT Press, Cambridge, MA, 1965.

[Robertson 73] T. V. Robertson, P. W. Swain, and K. S. Fu. *Multispectral image partitioning*. Technical Report, School of Electrical Engineering, Purdue Univ., 1973.

[Robinson 77] G. S. Robinson. Edge detection by compass gradient masks. *Computer Graphics and Image Processing*, 6:429–501, 1977.

[Rodriguez 69] J. E. Rodriguez. *A Graph Model for Parallel Computation*. Technical Report TR-64, Laboratory for Computer Science, MIT, 1969.

[Rogers 76] D. F. Rogers and J. A. Adams. *Mathematical Elements for Computer Graphics*. McGraw Hill, New York, 1976.

[Rogers 85] D. F. Rogers. *Procedural Elements for Computer Graphics*. McGraw-Hill, Singapore, 1985.

[Rose 87] J. Rose and G. Steele. *C*: an extended C language for data parallel programming*. Technical Report PL87-5, Thinking Machines Corp., 1987.

[Rosenberg 89] J. B. Rosenberg and J. D. Becher. Mapping Massive SIMD Parallelism onto Vector Architectures for Simulation. *Software — Practice and Experience*, 19(8):739–756, 1989.

[Rosenblatt 58] F. Rosenblatt. The perceptron: a probabilistic model for information storage and organization in the brain. *Psych. Rev.*, 65, 1958.

[Rosenfeld 75] A. Rosenfeld. A characterisation of parallel thinning algorithms. *Inf. Control*, 29:286–291, 1975.

[Rosenfeld 82] A. Rosenfeld and A. C. Kak. *Digital Picture Processing*. Volume 1 & 2, Academic Press, New York, 2 edition, 1982.

[Rosenfeld 88] A. Rosenfeld, J. Ornelas, and Y. Hung. Hough Transform Algorithms for Mesh-Connected SIMD Parallel Processors. *Computer Vision, Graphics and Image Processing*, 41:293–305, 1988.

[Roth 89] M. W. Roth. Neural networks for extraction of weak targets in high clutter environments. *IEEE Trans. Syst., Man., Cyber.*, SMC–19(5):1210–1217, 1989.

[Rumelhart 86] D. E. Rumelhart, G. E. Hinton, and R. J. Williams. *Learning internal representations by error propagation*, pages 318–362. Volume 1, MIT Press, Cambridge, MA, 1986.

[Rummel 82]   P. Rummel and W. Beutel. A Model-Based Image Analysis
               System for Workpiece Recognition. In *Proc. 6th. Int. Conf.
               on Pattern Recognition*, pages 1014–1017, Munich, Germany,
               1982.

[Rummel 84]   P. Rummel and W. Beutel. Workpiece recognition and in-
               spection by a model–based scene analysis system. *Pattern
               Recognition*, 17:141–148, 1984.

[Rummel 86]   P. Rummel. GSS – a fast, model-based grey-scale sensor sys-
               tem for workpiece recognition. In *Proceedings of the 8th Int.
               Pattern Recognition Conference*, pages 18–21, Paris, France,
               1986.

[Sadjadi 78]   F. A. Sadjadi and E. L. Hall. Invariant moments for scene
               analysis. In *Proc. IEEE Cont. Pattern Recognition Image
               Process*, pages 181–187, Chicago, 1978.

[Sahoo 88]   P. K. Sahoo, S. Soltani, A. K. C. Wong, and Y. C. Chen.
               A Survey of Thresholding Techniques. *Computer Vision,
               Graphics, and Image Processing*, 41:233–260, 1988.

[Samet 80]   H. Samet. Region representation: quadtrees from binary ar-
               rays. *Computer Graphics and Image Processing*, 1980.

[Samet 84]   Hanan Samet. The Quadtree and Related Hierarchical Data
               Structures. *Computing Surveys*, 16:187–260, 1984.

[Sanz 87]   J. L. C. Sanz and I. Dinstein. Projection–Based Geometri-
               cal Feature Extraction for Computer Vision: Algorithms in
               Pipeline Architectures. *IEEE Trans. Pattern Analysis and
               Machine Intelligence*, PAMI-9:160–168, 1987.

[Sardis 79]   G. N. Sardis and D. M. Brandin. An automatic surface in-
               spection system for flat rolled steel. *Automatica*, 15:505-520,
               1979.

[Sawchuk 84]   A. A. Sawchuk and T. C. Strand. Digital Optical Computing.
               IEEE Proc., 72(7):758–779, 1984.

[Sawchuk 86]   A. A. Sawchuk and B. K. Jenkins. Dynamical Optical In-
               terconnections for Parallel Processors. *Soc. Photo-Opt. Inst.
               Eng.*, 623:143, 1986.

[Schoenberg 46] I. J. Schoenberg. Contributions to the problem of approxi-
               mation of equidistant data by analytic functions. *Quarterly
               Applied Math.*, 4(1):45–99 and 112–141, 1946.

[Schwartz 80] E. L. Schwartz. Computational Anatomy and Functional Ar-
               chitecture of Striate Cortex: A Spatial Mapping Approach to
               Perceptual Coding. *Vision Research*, 20:645–669, 1980.

[Scoller 84]   I. Schollar, B. Weidner, and T. S. Huang. Image Enhancement Using the Median and the Interquartile Distance. *Computer Vision, Graphics, and Image Processing*, 25:236–251, 1984.

[Scott 88]   G. L. Scott, S. Turner, and A. Zisserman. Using a Mixed Wave/Diffusion Process to Elicit the Symmetry Set. In *Alvey Vision Conference*, 1988.

[Seit 88]   A. Seit, K. Tsui, P. Nickolls, and S. Hunyor. Use of the $\nabla^2 G$ Operator in Automated Border Extraction in Echocardiographic Images. In *SPIE Medical Imaging II*, pages 751–759, 1988.

[Seitz 85]   C. L. Seitz. The Cosmic Cube. *Commun. ACM*, 28(1):22–33, 1985.

[Seraphim 82]   D. P. Seraphim. A New Set of Printed-Circuit Technologies for the IBM 3081 Processor Unit. *IBM J. Res. Develop.*, 26:37–44, 1982.

[Serra 82]   J. Serra. *Image Analysis and Mathematical Morphology*. Academic Press, London, 1982.

[Shafer 83]   S. A. Shafer and T. Kanade. *The theory of straight homogeneous generalised cylinders and taxonomy of generalised cylinders*. Technical Report CMU-CS-83-105, Carnegie-Mellon Univ., Pittsburgh, PA, 1983.

[Shah 84]   M. A. Shah and R. Jain. Detecting time–varying corners. In *Seventh International Conf. on Pattern Recognition*, pages 2–5, IEEE Computer Society Press, Montreal, Canada, 1984.

[Shapiro 87]   E. Shapiro. *Concurrent Prolog: Collected Papers*. MIT Press, Cambridge, MA, 1987.

[Shaw 84]   D. E. Shaw. SIMD and MSIMD variants of the NON-VON supercomputer. In *Proc. of the COMCON, Spring'84*, 1984.

[Sherman 59]   *A quasi–topological method for the recognition of line patterns*, Butterworths, London, 1959.

[Shore 73]   J. E. Shore. Second thoughts on parallel processing. *Comput. Elect. Eng.*, 1:95–109, 1973.

[Siegel 79]   H. J. Siegel. Interconnection networks for SIMD machines. *Computer*, 12(6):57–65, 1979.

[Siegel 80]   H. J. Siegel. The Theory Underlying the Partitioning of Permutation Networks. *IEEE Trans. Computer*, C-29(9):791–800, 1980.

[Siegel 81]   H. J. Siegel et al. PASM: A Partitionable SIMD/MIMD System for Image Processing and Pattern recognition. *IEEE Trans. Computer*, C-30(12):934–947, 1981.

[Silberberg 85]   T. M. Silberberg. The Hough Transform on the Geometric Arithmetic Parallel Processor. In *IEEE Computer Soc. Workshop on Computer Architecture for Pattern Analysis & Image Database Management*, pages 387–393, 1985.

[Skillicorn 88]   D. B. Skillicorn. A Taxonomy for Computer Architectures. *IEEE Computer*, 21(11):46–57, 1988.

[Sklansky 72]   J. Sklansky. Measuring concavity on a rectangular mosaic. *IEEE Trans. Computer*, C-21:1355–1364, 1972.

[Sklansky 78]   J. Sklansky. On the hough technique for curve detection. *IEEE Trans. Computer*, C-20:923–926, 1978.

[Sklansky 80]   J. Sklansky et al. Toward computed detection of nodules in chest radiographs. In M. Onoe, K. Preston, Jr., and A. Rosenfeld, editors, *Real-Time Medical Image Processing*, pages 53–59, Plenum Press, New York, 1980.

[Sklansky 82]   J. Sklansky. Finding the convex hull of a simple polygon. *Pattern Recognition Letters*, 1:79–83, 1982.

[Smith 72]   E. A. Smith and D. R. Phillips. Automated Cloud Tracking Using Precisely Aligned Digital ATS Pictures. *IEEE Trans. Computer*, C-21:715–729, !972.

[Smith 82]   *The recovery of surface orientation from image irradiance*, Palo Alto, 1982.

[Solow 72]   B. Solow. Computers in cephalometric research. *Comput. Biol. Med.*, 1:41–49, 1972.

[Solow 86]   B. Solow and S. Sierbaek-Nielsen. Growth changes in head posture related to craniofacial development. *American Journal of Orthodontics*, 89(2):132–140, 1986.

[Soroka 83]   B. I. Soroka and R. K. Bajcsy. A program for describing complex three-dimensional objects using generalised cylinders as primitives. In *Proc. Pattern Recognition and Image Processing Conf.*, 1983.

[Squire 62]   J. S. Square and S. M. Palais. *Physical and Logical Design of a Highly Parallel Computer*. Technical Report, Department of Electrical Engineering, University of Michigan, Oct. 1962.

[Squire 63]   J. S. Square and S. M. Palais. Programming and Design Considerations for a Highly Parallel Computer. In *AFIPS Conf. Proc.*, pages 395–400, 1963.

[Stark 82]   H. Stark, editor. *Applications of Optical Fourier Transforms*. Academic Press, New York, 1982.

[Steier 86]   W. H. Steier and R. K. Shori. Optical Hough Transform. *Applied Optics*, 25(16):2734–2738, 1986.

[Stockman 72]   T. G. Stockman. Image processing in the context of a visual model. *Proceedings of the IEEE*, 60(7):828–842, 1972.

[Stoer 80]   J. Stoer and R. Bulirsch. *Introduction to Numerical Analysis*. Springer-Verlag, New York, 1980.

[Stone 71]   H. S. Stone. Parallel processing with the perfect shuffle. *IEEE Trans. Computer*, C-20:153–161, 1971.

[Stone 79]   J. Stone, B. Dreher, and A. Leventhal. Hierarchical and parallel mechanisms in the organisation of visual cortex. *Brain Res. Revs.*, 180:345–394, 1979.

[Stout 83]   Q. F. Stout. Sorting, merging, selecting, and filtering on tree and pyramid machines. In *Proc. Int. Conf. on Parallel Processing*, pages 214–221, 1983.

[Strang 73]   G. Strang and G. Fix. *An Analysis of the Finite Element Method*. Prentice-Hall, Englewood Cliffs, NJ, 1973.

[Strassen 69]   V. Strassen. Gaussian elimination is not optimal. *Num. Math.*, 13:354, 1969.

[Strat 79]   T. M. Strat. *A Numerical Method for Shape from Shading from a Single Image*. Master's thesis, Dept. Electrical Eng. and Computer Sci., 1979.

[Strong 91]   J. P. Strong. Computation on the Massively Parallel Processor at the Goddard Space Flight Center. *Proc. IEEE*, 79(4):548–558, 1991.

[Subbarao 87]   M. Subbarao. Solution and uniqueness of image flow equations for rigid curved surfaces in motion. In *Proc. 1st Int. Conf. Computer Vision*, pages 687–692, London, England, 1987.

[Sugawara 80]   Junji Sugawara, Yoshinari Kanamori, and Toshihiko Sakamoto. Analysis of mandibular form in orthodontics. In Lindberg and Kaihara, editors, *MEDINFO 80*, pages 1168–1172, 1980.

[Suresh 83]   B. R. Suresh, R. A. Fundakowski, T. S. Levitt, and J. E. Over-
land. A Real-Time Automated Visual Inspection System for
Hot Steel Slabs. *IEEE Trans. Pattern Analysis and Machine
Intelligence*, PAMI-5(6):563–572, 1983.

[Tang 83]   Gregory Y. Tang. A Discrete Version of Greens's Theorem. *IEEE Trans. Pattern Analysis and Machine Intelligence*,
PAMI-4:242–249, 1983.

[Tanida 85]   J. Tanida and Y. Ichioka. Optical–logic–array processor Using
Shadograms. II. Optical Digital Parallel Processing. *J. Opt.
Soc. Am.*, 2(8):1237–1244, 1985.

[Tanimoto 80] S. L. Tanimoto and A. Klinger, editors. *Structured Computer
Vision: Machine Perception through Hierarchical Computa-
tion Structures*. Academic Press, New York, 1980.

[Tanimoto 82] S. L. Tanimoto. *Sorting, histogramming, and other statistical
operations on a pyramid machine*. Technical Report 82-08-
02, Dept. Comp. Sci., Univ. Washington, 1982.

[Tanimoto 83a] S. L. Tanimoto. A boolean matching operator for hierarchical
image processing. In *CAPAIDM Workshop 83*, pages 253–
256, 1983.

[Tanimoto 83b] S. L. Tanimoto. A pyramidal approach to parallel processing. In *10th Annual Int. Symp. on Computer Architecture*,
pages 372–378, 1983.

[Thompson 77] C. D. Thompson and H. T. Kung. Sorting on a mesh connected parallel computer. *Commun. ACM*, 20:263–271, 1977.

[Thompson 80] W. B. Thompson. Combining Motion and Contrast for Segmentation. *IEEE Trans. Pattern Analysis and Machine Intelligence*, PAMI-2:543–549, 1980.

[Thorpe 84]   C. E. Thorpe. *FIDO: Vision and navigation for a robot rover*.
PhD thesis, Carnegie–Mellon Univ., Pittsburgh, PA, 1984.

[Toussaint 82] G. T. Toussaint and D. Avis. On a convex hull algorithm
for polygons and its application to triangulation problems.
Pattern Recognition, 15(1):23–29, 1982.

[Tropf 80]   H. Tropf. Analysis–by–Synthesis Search for Segmentation. In
Proc. of the 5th ICPR, Miami, USA, 1980.

[Troxel 78]   D. E. Troxel and C. Lynn. Enhancement of news photos.
Computer Graphics Image Processing, 7:266, 1978.

[Tsuji 78]   S Tsuji and F Matsumoto. Detection of Ellipses by a Modified
                  Hough Transform. *IEEE Transactions on Computers*, C-27,
                  No. 8:777–781, 1978.

[Tsuji 79]   S. Tsuji, M. Osada, and M. Yachida. Three–Dimensional
                  Movement Analysis of Dynamic Line Images. In *Proc. 6th
                  Int. Conf. Artificial Intelligence*, pages 876–901, 1979.

[Turney 85]   J. L. Turney, T. N. Mudge, and R. A. Volz. Recognizing
                  Partially Occluded Parts. *IEEE Trans. Pattern Analysis and
                  Machine Intelligence*, PAMI-7:410–421, 1985.

[Udupa 75]   K. J. Udupa and I. S. N. Murthy. Some new concepts for
                  encoding line patterns. *Pattern Recognition*, 7:225–233, 1975.

[Uhr 72]   L. Uhr. Layered "recognition cone" networks that process,
                  classify and describe. *IEEE Trans. Computer*, C-21:758–768,
                  1972.

[Uhr 79]   L. Uhr and R. Douglass. A parallel–serial recognition cone
                  system for perception. *Pattern Recognition*, 11:29–40, !979.

[Unger 58]   S. H. Unger. A computer oriented to spatial problems. *Proc.
                  IRE*, 46:1744–1750, 1958.

[Varga 62]   R. Varga. *Matrix Iterative Analysis*. Prentice-Hall, Engle-
                  wood Cliffs, NJ, 1962.

[Vasudevan 88]   S. Vasudevan, R. L. Cannon, and J. C. Bezdek. Heuristics
                  for Intermediate Level Road Finding Algorithms. *Computer
                  Vision, Graphics, and Image Processing*, 44:175–190, 1988.

[Veillon 84]   F. Veillon. Towards a Systematic Study of Shape Measures.
                  In *Proceedings of the 7th Int. Pattern Recognition Conference*,
                  pages 607–610, Montreal,Canada, 1984.

[Verbeek 88]   P. W. Verbeek, H. A. Vrooman, and L. J. van Vliet. Low-
                  level image processing by max–min filters. *Signal Processing*,
                  15:249–258, 1988.

[Wagner 87]   K. Wagner and D. Psaltis. Multilayer Optical Learning Net-
                  works. In *SPIE Vol. 752 Digital Optical Computing*, pages 86–
                  97, 1987.

[Walker 72]   G. Walker and C. J. Kowalski. Computer morphometrics in
                  craniofacial biology. *Comput. Biol. Med.*, 2:235–249, 1972.

[Wallace 88]   A. M. Wallace. Industrial applications of computer vision
                  since 1982. *IEE Proc. Part E: Computers and Digital Tech-
                  niques*, 135:117–136, 1988.

[Wallach 53]   H. Wallach and D. N. O'Connell. The kinetic depth effect. *J. Expt. Psychology*, 45(4):205–217, 1953.

[Wang 83]   D. C. C. Wang, A. H. Vagnucci, and C. C. Li. Digital image enhancement: a survey. *Computer Vision, Graphics, and Image Processing*, 24:363–381, 1983.

[Wang 90]   H. Wang, M. Brady, and I. Page. A fast algorithm for computing optic flow and its implementation on a Transputer array. In *British Machine Vision Conf.*, pages 175–180, Oxford, 1990.

[Watson 79]   I. Watson and J. Gurd. A Prototype Data Flow Computer With Token Labelling. In *AFIPS Conf. Proc., 1979*, pages 623–628, NCC, New York, 1979.

[Watt 89]   A. Watt. *Fundamentals of Three-Dimensional Computer Graphics.* Addison–Wesley Pub. Co., Wokingham, England, 1989.

[Waxman 87]   A. M. Waxman and K. Wohn. Image flow theory: A framework for 3-D inference from time–varying imagery. In C. Brown, editor, *Advances in Computer Vision*, Erlbaum Publishers, Hillside, NJ, 1987.

[Wechsler 77]   H. Wechsler and J. Sklansky. Finding the rib cage in chest radiographs. *Pattern Recognition*, 9:21–30, 1977.

[Weems 89]   C. C. Weems et al. The image understanding architecture. *Int. J. Computer Vision*, 2:251–282, 1989.

[Werman 84]   M. Werman and S. Peleg. Multiresolution texture signatures using min-max operators. In *Proceedings of the 7th Int. Pattern Recognition Conference*, pages 97–99, Montreal,Canada, 1984.

[Weszka 74]   J. S. Weszka. *Threshold Selection 4.* Technical Report TR-336, Computer Science Center, University of Maryland, 1974.

[Weszka 75]   J. S. Weszka. *Threshold Selection Techniques 5.* Technical Report TR-349, Computer Science Center, University of Maryland, 1975.

[Weszka 78]   J. S. Weszka. Survey: a survey of threshold selection techniques. *Computer Graphics and Image Processing*, 7:259–265, 1978.

[Widrow 60]   B. Widrow and M. E. Hoff. Adaptive switching circuits. In *IRE WESCON conv. Record 4*, pages 96–104, 1960.

[Widrow 90]   B. Widrow and M. A. Lehr. 30 Years of Adaptive Networks:
                  Perceptron, Madaline, and Backpropagation. *Proc. IEEE*,
                  78(9):1415–1442, 1990.

[Wiejak 85]   J. S. Wiejak, H. Buxton, and B. F. Buxton. Convolution
                  with Separable Masks for Early Image Processing. *Computer
                  Vision, Graphics, and Image Processing*, 32:279–290, 1985.

[Willis 86]   N. Willis. Architectural Considerations. In *Computer Archi-
                  tecture and Communications, Paradigm*, pages 126–133, 1986.

[Wilson 77]   H. R. Wilson and S. C. Giese. Threshold visibility of fre-
                  quency gradient patterns. *Vision Research*, 17:1177–1190,
                  1977.

[Witkin 81]   A. P. Witkin. Recovering Surface Shape and Orientation from
                  Texture. *Artificial Intelligence*, 17:17–45, 1981.

[Wong 76a]   R. Y. Wong. *Sequential Pattern Recognition as Applied to
                  Scene Matching*. PhD thesis, University of Southern Califor-
                  nia, Los Angeles, 1976.

[Wong 76b]   R. Y. Wong, E. L. Hall, and J. Rouge. Hierarchical search
                  for image matching. In *Proc. IEEE Conf. Decision Control*,
                  Clearwater, FL, 1976.

[Wong 78]   R. Y. Wong and E. L. Hall. Scene matching with invari-
                  ant moments. *Comput. Graphics Image Processing*, 8:16–24,
                  1978.

[Wood 79]   A. Wood. *CAP4 User's Manual*. Technical Report, Image
                  Processing Group, Dept. of Physics and Astronomy, Univer-
                  sity College London, 1979.

[Wood 83]   A. Wood. *Parallel Processing Techniques for Image Sequence
                  Analysis*. PhD thesis, University of London, 1983.

[Yachida 77]   M. Yachida and S. Tsuji. A Versatile Vision System for Com-
                  plex Industrial Parts. *IEEE Trans. Computers*, C-26:882–894,
                  1977.

[Yakimovsky 73]   Y. Y. Yakimovsky and J. Feldman. A semantics–based de-
                  cision theoretic region analyzer. In *Proc. 3rd Int. J. Conf.
                  Artificial Intelligence*, pages 580–588, 1973.

[Yatagai 86]   T. Yatagai. Cellular Logic Architectures for Optical Comput-
                  ers. *Applied Optics*, 25:1571–1577, 1986.

[Young 71]   D. Young. *Iterative Solution of Large Linear System*. Aca-
                  demic Press, New York, 1971.

[Yu88]   S-S. Yu, W-C. Cheng, and C. S. C. Chiang. Printed Circuit
              Board Inspection System PI/1. In *Automated Inspection and
              High Speed Vision·Architecture II*, pages 126–134, 1988.

[Zadeh 73]   L. A. Zadeh. Outline of a new approach to the analysis of
              complex systems and decision processes. *IEEE Trans. Sys-
              tem, Man and Cybernatics*, SMC-3:28–44, 1973.

[Zadeh 83]   L. A. Zadeh. Commonsense Knowledge Representation Based
              on Fuzzy Logic. *Computer*, 16(10):61–65, 1983.

[Zahn 72]   C. T. Zahn and R. Z. Roskies. Fourier Descriptors for Plane
              Closed Curves. *IEEE Trans. Computer*, C-21:269–281, 1972.

[Zakhor 87]   A. Zakhor and A. V. Oppenheim. Quantization errors in the
              computation of the discrete Hartley Transform. *IEEE Trans.
              Acoust. Speech, Signal Processing*, ASSP-35:1592–1601, 1987.

[Zhang 84]   T. Y. Zhang and C. Y. Suen. A Fast Parallel Algorithm for
              Thinning Digital Patterns. *Comm. ACM.*, 27:236–239, 1984.

[Zhou 88]   Y. T. Zhou, R. Chellappa, A. Vaid, and B. K. Jenkins. Image
              restoration using a neural network. *IEEE Trans. Acoust.,
              Speech, Signal Processing*, ASSP–36(7):1141–1151, 1988.

[Zhu 86]   M. L. Zhu and P. S. Yeh. Automatic road network detec-
              tion on aerial photographs. In *IEEE Conf. Computer Vision,
              Pattern Recognition*, 1986.

[Zuniga 83]   O. A. Zuniga and R. Haralick. Corner Detection Using the
              Facet Model. In *IEEE Proc. Conf. Pattern Recognition Image
              Processing*, pages 30–37, 1983.

Index